Germ Cell Protocols

METHODS IN MOLECULAR BIOLOGY™

John M. Walker, Series Editor

289. **Epidermal Cells**, *Methods and Applications*, edited by *Kursad Turksen, 2004*
288. **Oligonucleotide Synthesis**, *Methods and Applications*, edited by *Piet Herdewijn, 2004*
287. **Epigenetics Protocols**, edited by *Trygve O. Tollefsbol, 2004*
286. **Transgenic Plants:** *Methods and Protocols*, edited by *Leandro Peña, 2004*
285. **Cell Cycle Control and Dysregulation Protocols:** *Cyclins, Cyclin-Dependent Kinases, and Other Factors*, edited by *Antonio Giordano and Gaetano Romano, 2004*
284. **Signal Transduction Protocols,** *Second Edition*, edited by *Robert C. Dickson and Michael D. Mendenhall, 2004*
283. **Biconjugation Protocols,** edited by *Christof M. Niemeyer, 2004*
282. **Apoptosis Methods and Protocols,** edited by *Hugh J. M. Brady, 2004*
281. **Checkpoint Controls and Cancer, Volume 2:** *Activation and Regulation Protocols*, edited by *Axel H. Schönthal, 2004*
280. **Checkpoint Controls and Cancer, Volume 1:** *Reviews and Model Systems*, edited by *Axel H. Schönthal, 2004*
279. **Nitric Oxide Protocols,** *Second Edition*, edited by *Aviv Hassid, 2004*
278. **Protein NMR Techniques,** *Second Edition*, edited by *A. Kristina Downing, 2004*
277. **Trinucleotide Repeat Protocols,** edited by *Yoshinori Kohwi, 2004*
276. **Capillary Electrophoresis of Proteins and Peptides,** edited by *Mark A. Strege and Avinash L. Lagu, 2004*
275. **Chemoinformatics,** edited by *Jürgen Bajorath, 2004*
274. **Photosynthesis Research Protocols,** edited by *Robert Carpentier, 2004*
273. **Platelets and Megakaryocytes, Volume 2:** *Perspectives and Techniques*, edited by *Jonathan M. Gibbins and Martyn P. Mahaut-Smith, 2004*
272. **Platelets and Megakaryocytes, Volume 1:** *Functional Assays*, edited by *Jonathan M. Gibbins and Martyn P. Mahaut-Smith, 2004*
271. **B Cell Protocols,** edited by *Hua Gu and Klaus Rajewsky, 2004*
270. **Parasite Genomics Protocols,** edited by *Sara E. Melville, 2004*
269. **Vaccina Virus and Poxvirology:** *Methods and Protocols*,edited by *Stuart N. Isaacs, 2004*
268. **Public Health Microbiology:** *Methods and Protocols*, edited by *John F. T. Spencer and Alicia L. Ragout de Spencer, 2004*
267. **Recombinant Gene Expression:** *Reviews and Protocols, Second Edition*, edited by *Paulina Balbas and Argelia Johnson, 2004*
266. **Genomics, Proteomics, and Clinical Bacteriology:** *Methods and Reviews*, edited by *Neil Woodford and Alan Johnson, 2004*
265. **RNA Interference, Editing, and Modification:** *Methods and Protocols*, edited by *Jonatha M. Gott, 2004*
264. **Protein Arrays:** *Methods and Protocols*, edited by *Eric Fung, 2004*
263. **Flow Cytometry,** *Second Edition*, edited by *Teresa S. Hawley and Robert G. Hawley, 2004*
262. **Genetic Recombination Protocols,** edited by *Alan S. Waldman, 2004*
261. **Protein–Protein Interactions:** *Methods and Applications*, edited by *Haian Fu, 2004*
260. **Mobile Genetic Elements:** *Protocols and Genomic Applications*, edited by *Wolfgang J. Miller and Pierre Capy, 2004*
259. **Receptor Signal Transduction Protocols,** *Second Edition*, edited by *Gary B. Willars and R. A. John Challiss, 2004*
258. **Gene Expression Profiling:** *Methods and Protocols*, edited by *Richard A. Shimkets, 2004*
257. **mRNA Processing and Metabolism:** *Methods and Protocols*, edited by *Daniel R. Schoenberg, 2004*
256. **Bacterial Artifical Chromosomes, Volume 2:** *Functional Studies*, edited by *Shaying Zhao and Marvin Stodolsky, 2004*
255. **Bacterial Artifical Chromosomes, Volume 1:** *Library Construction, Physical Mapping, and Sequencing*, edited by *Shaying Zhao and Marvin Stodolsky, 2004*
254. **Germ Cell Protocols, Volume 2:** *Molecular Embryo Analysis, Live Imaging, Transgenesis, and Cloning*, edited by *Heide Schatten, 2004*

METHODS IN MOLECULAR BIOLOGY™

Germ Cell Protocols

Volume 1

Sperm and Oocyte Analysis

Edited by

Heide Schatten

Department of Veterinary Pathobiology
University of Missouri–Columbia, Columbia, MO

Humana Press Totowa, New Jersey

999 Riverview Drive, Suite 208
Totowa, New Jersey 07512

www.humanapress.com

This publication is printed on acid-free paper. ∞
ANSI Z39.48-1984 (American Standards Institute)

Permanence of Paper for Printed Library Materials.

Cover illustration: Background: Figure 2, Chapter 18, volume 1. Inset: Figure 12, Chapter 13, volume 1.

Cover design by Patricia F. Cleary.

For additional copies, pricing for bulk purchases, and/or information about other Humana titles, contact Humana at the above address or at any of the following numbers: Tel.: 973-256-1699; Fax: 973-256-8341; E-mail: humana@humanapr.com; or visit our Website: www.humanapress.com

Printed in the United States of America. 10 9 8 7 6 5 4 3 2 1

E-ISBN 1-59259-744-0
Library of Congress Cataloging in Publication Data

Germ cell protocols / edited by Heide Schatten.
p. ; cm. — (Methods in molecular biology ; v. 253-254)
Includes bibliographical references and index.
ISBN 1-58829-121-9 (v. 1 : alk. paper) — ISBN 1-58829-257-6 (v. 2 : alk. paper)
1. Veterinary embryology—Research—Technique. 2. Germ cells—Research—Technique. [DNLM: 1. Cytological Techniques. 2. Germ Cells. 3. Genetic Techniques.] I. Schatten, Heide. II. Series: Methods in molecular biology (Clifton, N.J.) ; 253-254.
SF767.5.G47 2004
636.089'264—dc22 2003023511

Preface

All sexually reproducing organisms produce primordial germ cells, a small population of cells that differentiate into gametes of either sex and carry totipotency, an ability to develop into an entire new organism. The study of germ cells has undergone enormous advances in recent years and has entered into an explosive phase of new discoveries with the introduction of transgenic technologies and nuclear cloning. Basic knowledge and techniques developed for lower vertebrate and invertebrate systems have facilitated the study of higher vertebrates, including humans. Many experiments that have first been performed on lower vertebrates provided the tools and strategies that could later be applied to other less readily available mammalian systems. The discovery of centrosomes in ascidians and sea urchin eggs now benefits studies of fertility and infertility in mammals. External in vitro fertilization, now a common technique in assisted fertilization, has only been possible as a result of numerous studies in lower systems in which external fertilization is natural. Egg activation, first explored in sea urchin and ascidian eggs, now benefits techniques designed to increase cloning efficiency in farm and domestic animals. Gene manipulations and molecular methods have added to the possibilities of producing live offspring with enormous biomedical, ecological, and economic implications.

The two volumes on germ cells combine techniques in a variety of different systems; we have selected those systems that have provided landmarks in advancing our knowledge on germ cells. Topics include sperm and egg activation, motility, fertilization, nuclear development, nuclear cloning, molecular characterization of specific events, imaging of cell structures, cryopreservation of germ cells, and other specific techniques that have advanced our knowledge of germ cells. Each chapter provides a short overview of the topic, followed by detailed techniques that are not normally described in regular research papers. Volume 1 is focused on sperm cells, oocyte analysis, oocyte maturation, fertilization, and preparation techniques, whereas volume 2 is mainly focused on molecular egg analysis, live egg imaging, nuclear cloning, oocyte techniques, and other specific methods for egg manipulation. Molecular preparations and characterizations and molecular imaging are methods treated in both volumes that will be of interest to cell and molecular biologists, reproductive biologists, physiologists, clinicians in human and veterinary medicine, and other researchers who want to become familiar with, and apply, techniques that are commonly not described in sufficient detail in research papers. This book will also be of interest to students for the study of a variety of aspects regarding research on germ cells.

A number of chapters in *Germ Cell Protocols: Volume 1: Sperm and Oocyte Analysis* are devoted to the newly emerging field of nuclear transfer that will be of interest to researchers in agriculture. The possibility of producing offspring from cultured adult cell lines enables the generation of offspring from donors of high genetic value with proven performance. In addition, it might provide a powerful tool in saving endangered species or populations and in protecting biodiversity. The diversity of chapters in this volume will allow cross-species transfer of knowledge which will be useful to study animals whose germ cells are low in number or hard to obtain. Because no recent comprehensive literature using our format is available on germ cells, this book will be of value to a wide variety of researchers.

It is a great and timely pleasure to edit two volumes on germ cells depicting specific methods that have impacted germ cells research. The methods presented here are currently in the highest demand and are expected to continue to be of value to investigators in the future. I am indebted to Dr. John Walker for inviting these two volumes on germ cells, to the publisher who recognized the value of publishing detailed methods on specific topics, with special thanks to Tom Lanigan and Craig Adams. I would like to thank the contributors who shared their intellectual and practical laboratory expertise and experiences with the present germ cell community and with those who will enter the field in the future. My sincere thanks to everybody who contributed so well.

Heide Schatten

Contents

Contributors

BADER AL-ANZI • *Division of Biology, California Institute of Technology, Pasadena, CA*

ANA PAULA ALVES-VIEIRA • *Department of Animal Sciences, University of Illinois at Urbana-Champaign, Urbana, IL*

CATHY BOUTIN • *Cambridge Research and Instrument Inc., Woburn, MA*

HAIM BREITBART • *Faculty of Life Sciences, Bar-Ilan University, Israel*

MARY M. BUHR • *Department of Animal & Poultry Science, University of Guelph, Guelph, ON, Canada*

HEATHER R. BURKIN • *Department of Animal Sciences, University of Illinois at Urbana-Champaign, Urbana, IL*

ALLISON M. CASE • *Department of Obstetrics, Gynecology and Reproductive Sciences, College of Medicine, University of Saskatchewan, Saskatoon, Saskatchewan, Canada*

AMITABHA CHAKRABARTI • *Department of Biomedical Engineering, Cleveland Clinic Foundation, Cleveland, OH*

DOUGLAS CHANDLER • *School of Life Sciences, Arizona State University, Tempe, AZ*

SYLVIE CHASTANT-MAILLARD • *Maitre de Conférences, Service de Reproduction, UMR 1198 INRA/ENVA, Biologie du Développement et Reproduction, Ecole Nationale Vétérinaire d'Alfort, France*

JOHN K. CRITSER • *Department of Veterinary Pathobiology, Comparative Medicine Center, University of Missouri–Columbia, Columbia, MO*

HARRY G. DENEER • *Department of Pathology, College of Medicine, University of Saskatchewan, Saskatoon, Saskatchewan, Canada*

WOLFGANG ENGEL • *Institut für Humangenetik, Zentrum für Hygiene und Humangenetik, der Universität Göttingen, Göttingen, Germany*

HENG-YU FAN • *State Key Laboratory of Reproductive Biology, Institute of Zoology, The Chinese Academy of Sciences, Beijing, China*

JIMMY GILL • *Houston Fertility Institute, Tomball, TX*

WILLIAM KINSEY • *Department of Anatomy and Cell Biology, University of Kansas Medical Center, Kansas City, KS*

FRANZ LAHNSTEINER • *Institute for Zoology, University of Salzburg, Salzburg, Austria*

YEHUDIT LAX • *Faculty of Life Sciences, Bar-Ilan University, Israel*

MANYU LI • *Institut für Humangenetik, Zentrum für Hygiene und Humangenetik, der Universität Göttingen, Göttingen, Germany*

Elinor MacGregor • *Department of Anatomy and Cell Biology, University of Kansas Medical Center, Kansas City, KS*

Robert W. McGaughey • *School of Life Sciences, Arizona State University, Tempe, AZ*

Kathryn J. Mengerink • *Department of Marine Biology, University of California-San Diego, La Jolla, CA*

Katrina Merkies • *(current address) Department of Equine Science, Lake Erie College, Painesville, OH*

David J. Miller • *Department of Animal Sciences, University of Illinois at Urbana–Champaign, Urbana, IL*

Steven F. Mullen • *Cryobiology Laboratory, Department of Veterinary Pathobiology, Comparative Medicine Center, University of Missouri at Columbia, Columbia MO*

Karim Nayernia • *Institut für Humangenetik, Zentrum für Hygiene und Humangenetik, der Universität Göttingen, Göttingen, Germany*

O. Olatunbosun • *Department of Obstetrics, Gynecology and Reproductive Sciences, College of Medicine, University of Saskatchewan, Saskatoon, Saskatchewan, Canada*

Karine Reynaud • *UMR 1198 INRA/ENVA, Biologie du Développement et Reproduction, Ecole Nationale Vétérinaire d'Alfort, France*

Sara Rubinstein • *Faculty of Life Sciences, Bar-Ilan University, Israel*

Marie Saint-Dizier • *UMR 1198 INRA/ENVA, Biologie du Développement et Reproduction, Ecole Nationale Vétérinaire d'Alfort, France*

Heide Schatten • *Department of Veterinary Pathobiology, University of Missouri–Columbia, Columbia, MO*

Hitoshi Sugiyama • *Department of Physiology, St. Marianna University School of Medicine, Kawasaki, Japan*

Qing-Yuan Sun • *State Key Laboratory of Reproductive Biology, Institute of Zoology, The Chinese Academy of Sciences, Beijing, China*

Peter Sutovsky • *Departments of Animal Science, and Obstetrics and Gynecology, University of Missouri–Columbia, Columbia, MO*

Mark W. Tengowski • *Pathology Technology Investigator, Drug Safety Evaluation, Pathology Central Research Division, Pfizer, Inc., Groton, CT*

Victor D. Vacquier • *Department of Marine Biology, University of California-San Diego, La Jolla, CA*

Wei-Hua Wang • *Houston Fertility Institute, Tomball Regional Hospital, Tomball, TX*

Wenjun Wu • *Department of Anatomy and Cell Biology, University of Kansas Medical Center, Kansas City, KS*

MANABU YOSHIDA • *ICORP, Japan Science and Technology Agency and Misaki Marine Biological Station, Graduate School of Science, University of Tokyo, Miura, Kanagawa, Japan*

Contents for Volume 2

1

Cryopreservation of Semen of the *Salmonidae* with Special Reference to Large-Scale Fertilization

Franz Lahnsteiner

1. Introduction

1.1. Importance of Semen Cryopreservation in Fish

Fish semen cryopreservation has important applications in the following fields: (1) in aquaculture, for synchronization of artificial reproduction, for efficient utilization of semen, and for maintaining the genetic variability of broodstocks; (2) in biodiversity, for gene banks of endangered species and of autochthon fish populations; and (3) in temporary unlimited supply with material for research as, for example, for toxicological tests or interspecific breeding.

Salmonidae are the traditionally cultured fish in many parts of the world. As a favorite game in sport fishing, restocking and conservation is necessary for many populations. Therefore, the semen cryopreservation is of particular importance in these species.

1.2. The Spermatozoa of Salmonid Fish

Salmonidae have simple, constructed spermatozoa *(1)*. They are acrosomeless and have a slightly ovoid head (length: 1.37 ± 0.15 μm; diameter 1.21 ± 0.13 μm), a cylindrical midpiece (length: 0.55 ± 0.08 μm; diameter: 0.74 ± 0.20 μm), and a flagellum that is about 40 μm long *(1)*. The midpiece contains three to five mitochondria that are fused with each other to a so-called chondriosome *(1)*. The sperm motility is inhibited by 20–40 m*M* potassium *(2)*. After motility activation, semen has high motility rates up to 100% and swimming velocities of 120–140 μm/min *(3,4)*. However, the sperm motility duration is very short. In 15–20 s, motility decreases for more than 50%, and it stops completely within 1 min *(3,4)*. Fertilization is external and also a very

From: *Methods in Molecular Biology, vol. 253: Germ Cell Protocols: Vol. 1 Sperm and Oocyte Analysis*
Edited by: H. Schatten © Humana Press Inc., Totowa, NJ

quick process. It occurs within 15–20 s after the gametes have been released into water *(5)*. For successful fertilization, spermatozoa have to swim into the micropyle of the egg *(5)*. Therefore, sperm motility and fertility are correlated and motility is often used for viability determination *(3,4)*, especially as fertilization assays and the subsequent hatching of eggs is very time-consuming in the *Salmonidae* *(6)*.

1.3. Specific Problems in the Cryopreservation of Salmonid Spermatozoa

The cryopreservation of spermatozoa of the *Salmonidae* and of fish in general faces the following problems:

1. Fish semen cryopreservation is not a laboratory method but has to be applied under field conditions with a minimum of technical supply. Therefore, the basic cryopreservation protocol must be adapted for easy and reliable outdoor use *(6,7)*.
2. As external fertilizing species with a limited annual reproduction period, *Salmonidae* have a high egg production. During natural spawning and in artificial insemination, several thousand eggs are fertilized simultaneously. Therefore, the cryopreservation methods also have to be adapted for large-scale fertilization by developing special techniques for freezing of large semen volumes and for insemination of numerous eggs *(6)*.
3. Fish semen reveal wide quality differences depending on fish age, spawning state, and general conditions *(8)*. To obtain consistent and good cryopreservation results, it is necessary to test semen for suitability for cryopreservation. Useful parameters have been calculated from regression models *(3)* and are (for >50% postthaw fertility) as follows: fresh semen motility rate >80%, average path swimming velocities between 80 and 100 μm/s, seminal plasma pH < 8.2, and seminal plasma osmolality >330 mosmol/kg *(3)*.

1.4. Cryopreservation

The described semen cryopreservation method was tested on *Hucho hucho*, *Oncorhychus mykiss*, *Salvelinus alpinus*, *Salvelinus fontinalis*, *Salmo trutta f. lacustris*, *Salmo trutta f. fario*, *Thymallus thymallus*, and *Coregonus lavaretus*. The method requires the following working steps: the dilution of semen in the extender, filling of semen in freezing vessels, freezing and thawing, and semen handling for fertilization *(6)*. An equilibration of semen in the extender is not required, as the sperm cells are small and permeable and the cryoprotectants penetrate the cells in less than 1 min *(6)*.

The extender composition was determined in a series of motility and fertility tests *(6)*. The extender inhibits the sperm motility as a result of high potassium concentrations, maintains the sperm viability because of its balanced ionic composition, and protects the spermatzoa during freezing and thawing by a combination of cryoprotectants and additives (methanol, hen egg yolk, bovine serum albumin).

Semen must be frozen as concentrated as possible, as high amounts are necessary for large-scale insemination. Therefore, low dilution ratios of semen in the extender should be used. However, too low dilution rates lead to cell compressions (critical concentration [2.0–3.0] × 10^9 cells/mL extender) during freezing and thawing and, subsequently, to a loss of semen viability ***(6)***. Additionally, semen density and therefore also the dilution rates are species-specific in the *Salmonidae* ***(6)***.

As salmonid semen cryopreservation is a field technique for freezing, a simple procedure is required. It is done in the vapor of liquid nitrogen in an isolated box whereby the distance of straws from the level of liquid nitrogen determines the freezing rates ***(6)***. Freezing rates are species-specific ***(7)***. In practice, semen volumes of at least 20–30 mL must be handled at once. However, only in straws with volumes of 0.5 mL and 1.2 mL were the freezing and thawing rates optimal. In larger straws and plastic bags, the results were very inconsistent. Therefore, a method was developed to build straw packages consisting of 1.2 mL straws by connecting them in flexible racks ***(7)***.

Thawing is performed in water of adequate temperature. To recover fertility, frozen salmonid semen must be warmed to 20°C, a temperature higher than the physiological optimum (4–6°C). The membranes or the metabolism are possibly better stabilized by this thawing procedure ***(6)***.

The thawed semen has several types of alteration ***(9)***. Therefore, it is nonstable after thawing ***(3)*** and must be handled very accurately, and in comparison to fresh semen, modified insemination procedures have to be applied ***(6,7)***: To compensate cell lesions originating during freezing and thawing and to reactivate sperm motility and fertility, special saline solutions are necessary ***(6,9)***. Also, the insemination itself is of importance: The wet fertilization (eggs and semen are placed in fertilization solution and mixed) is only suited for fertilization of small egg quantities up to 50 g ***(7)***. To inseminate larger egg quantities, dry fertilization (eggs and semen are mixed before fertilization solution is added) has to be applied ***(7)***, as it results in more homogeneous mixing of gametes ***(7)***.

1.5. Quality and Fertilizing Capacity of Frozen–Thawed Semen

There exist no species-specific differences in semen quality after cryopreservation. Representative changes in motility parameters and fertilizing capacity of rainbow trout semen after cryopreservation have been investigated ***(9)*** and are shown in **Table 1**. When compared to untreated semen, the percentage of immotile spermatozoa is significantly increased and the rate of motile spermatozoa is decreased. At low sperm-to-egg ratios also, the fertilization rate (evaluated in the embryo stage before hatching) is significantly decreased (*see* **Table 1**). However, the decrease in fertilizing capacity can be completely com-

Table 1
Motility and Fertility of Untreated and Cryopreserved Rainbow Trout Semen

Parameter	Fresh	Frozen/Thawed
Immotile (%)	4.8 ± 3.3[a]	71.3 ± 9.4[b]
Local motile (%)	9.5 ± 4.4[a]	5.3 ± 3.6[a]
Motile (%)	85.7 ± 12.2[a]	23.5 ± 5.5[b]
Circular motile (%)	52.8 ± 10.6[a]	20.1 ± 10.3[b]
Nonlinear motile (%)	29.1 ± 8.0[a]	17.1 ± 8.1[b]
Linear motile (%)	18.1 ± 11.1[a]	62.8 ± 14.0[b]
Average path sperm velocity (μm/sec)	90.3 ± 17.21[a]	94.5 ± 16.5[a]
Fertility at sperm-to-egg ratio (3–4) × 10^6	79.6 ± 12.2[a]	76.3 ± 7.4[a]
Fertility at sperm-to-egg ratio (1.5–2) × 10^6	79.8 ± 10.5[a]	62.4 ± 6.6[b]

Note: Data are mean ± SD, $n = 20$. Data in one row superscripted by the same letter are not significantly different.

pensated by higher sperm-to-egg ratios (*see* **Table 1**) ***(6)*** and then fertilization rates in the range of fresh semen control are obtained ***(6,7)***. The percentage of embryonic malformations is similar with cryopreserved and untreated semen ***(6)***.

2. Materials

All chemicals are of analytical grade. Distilled water is used.

1. Hen egg yolk is prepared freshly and carefully separated from the white, which causes agglutination of spermatozoa.
2. Extender: Dissolve 600 mg NaCl, 315 mg KCl, 15 mg $CaCl_2 \cdot 2H_2O$, 20 mg $MgSO_4 \cdot 7H_2O$, and 470 mg HEPES (sodium salt) in approx 80 mL water. Adjust to pH 7.8 with NaOH or HCl; fill with water to 100 mL. Add the following cryoprotectants and additives: 10% (v/v) methanol, 0.5% (w/v) sucrose, 1.5% (w/v) bovine serum albumin, 7% (v/v) hen egg yolk. Measure required egg yolk volumes in plastic syringes without needles. At < –20°C, the extender without hen egg yolk is stable for an unlimited time; the extender with egg yolk is prepared freshly.
3. Fertilization solution: Dissolve 500 mg $NaHCO_3$ and 600 mg Tris in approx 80 mL of water, adjust pH to 9.0 with NaOH or HCl, and fill with water to 100 mL.
4. Freezing vessels: Straws with volumes of 0.5 mL and 1.2 mL are commercially available. Preparation of straw packages (*see* **Figs. 1** and **2**): Use 1.2-mL straws and 0.45 to 0.55-mm-thick plastic foil from commercially available plastic bags that remains flexible at liquid-nitrogen temperature (test!). Cut two plastic ribbons to a width of 1.5 cm and a length depending on the desired number of straws that should be connected (required length per straw = 1.3 cm) (*see* **Fig. 1A**). Place the ribbons on top of each other (*see* **Fig. 1A**). Seal them together at their wide side with a commercial plastic bag sealing apparatus (sealing width of 6 mm) in a way that a 6-mm sealed portion is followed by a 7-mm unsealed

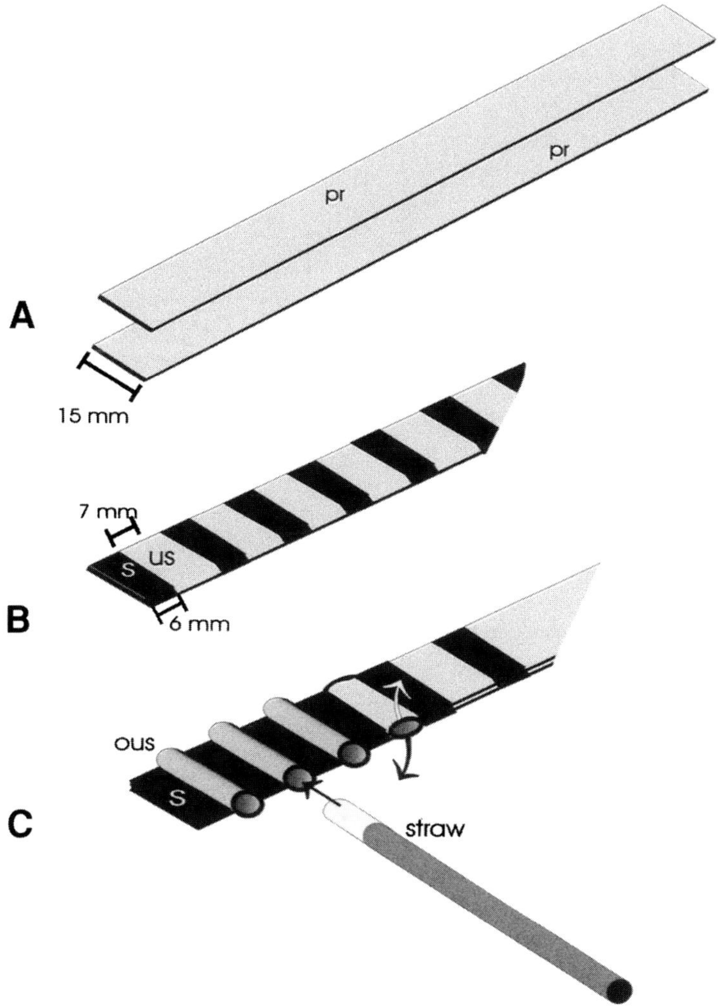

Fig. 1. Steps in the production of flexible plastic racks. ous = opened unsealed portion, pr = plastic ribbons, s = sealed portion, us = unsealed portion. **(A)** Plastic ribbons are placed one above each other; **(B)** plastic ribbons are sealed together; **(C)** the unsealed portions are opened and straws are fitted in.

portion (*see* **Fig. 1B**). Place the 1.2-mL straws in the unsealed portions with their plugged side (*see* **Fig. 1C**). Fit them tightly. In case the fit is not proper, the openings must be adjusted by additional sealing or enlargement. The distance between the single straws is 0.5 cm (*see* **Fig. 2A,B**).

5. Freezing apparatus: Freezing is done in a self-constructed insulated box (inner dimensions: base-27 × 18 cm, height = 33 cm) on a tray (*see* **Fig. 3**). This tray can

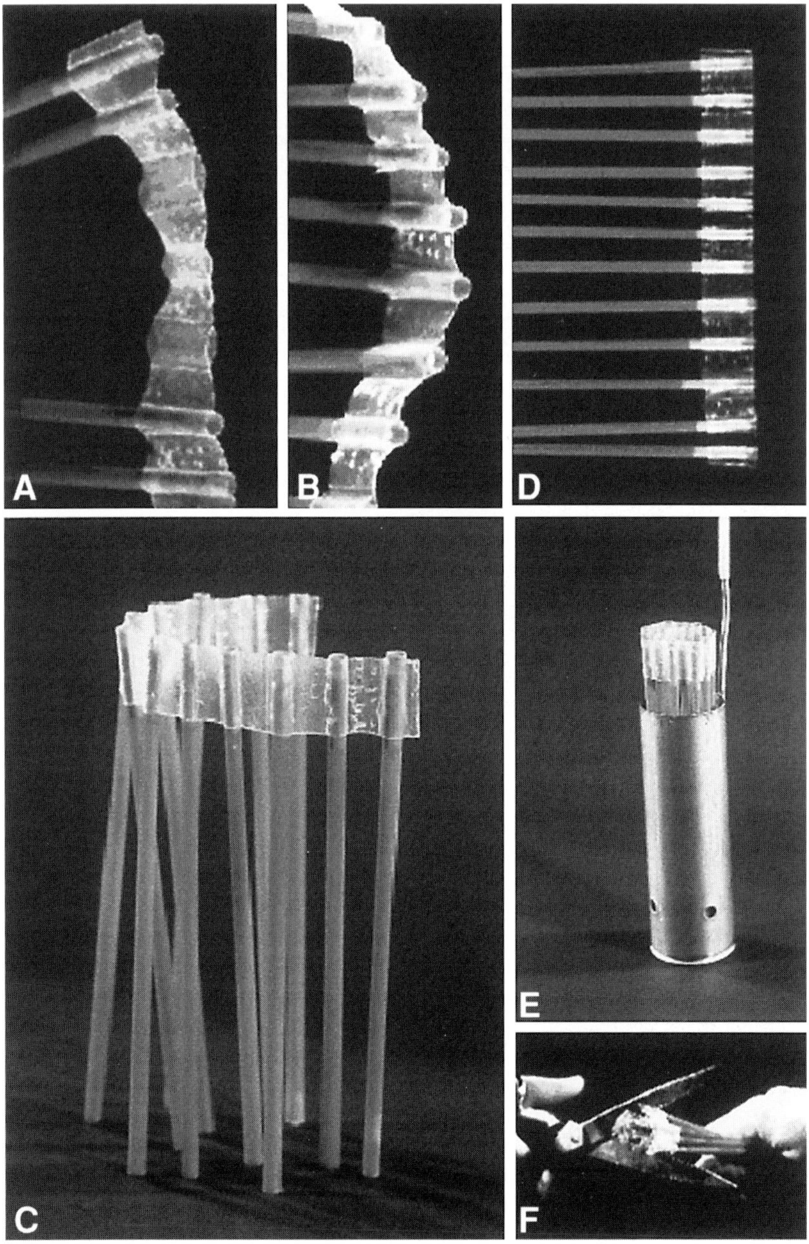

Fig. 2. Straw packages consisting of 1.2-mL straws: **(a)** plastic rack during fitting in the straws; **(b)** plastic rack loaded with straws; **(c)** straw package, placed horizontally on a plane as done for freezing; **(d)** straw package rolled together as used for storage and for cutting open; **(e)** straw package in the can of a liquid-nitrogen storage container; **(f)** straw package during cutting open.

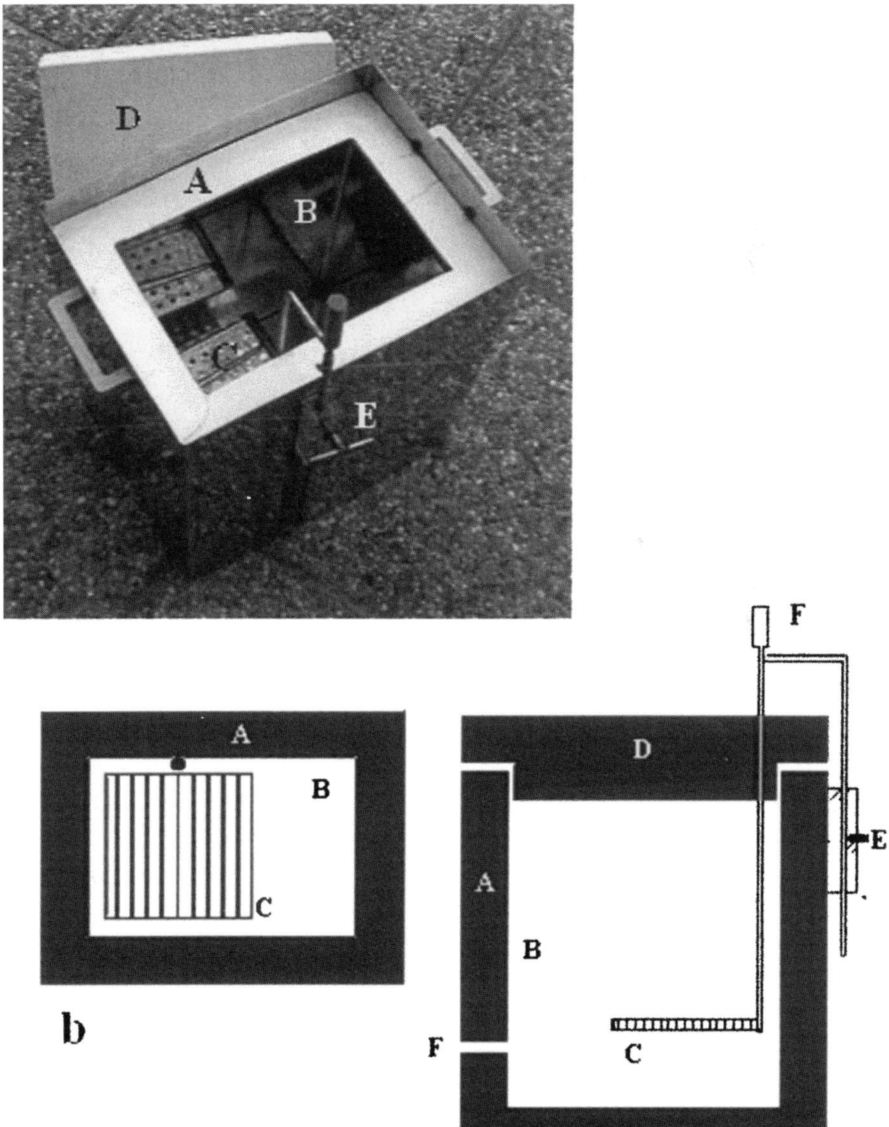

Fig. 3. Freezing box: A = insulated wall; B = freezing chamber; C = tray for straws; D = cover; E = overflow trap for liquid nitrogen. **(a)** Freezing box ready for use; **(b)** scheme of the freezing box. *Left*: cross-section, *right*: longitudinal section; E = set screw; F = tray holder.

Table 2
Optimal Dilution Ratios of Semen in the Extender and Sperm Density in the *Salmonidae*

Species	Dilution ratio (semen:extender)	Sperm density (cells/mL)
Salvelinus alpinus	1:2	8.0×10^8–2.5×10^9
Hucho hucho, Oncorhynchus mykiss, Salvelinus fontinalis, Thymallus thymallus	1:3	4.0×10^9–8×10^9
Salmo trutta f. fario	1:5	9.0×10^9–1.5×10^{10}
Salmo trutta f. lacustris	1:7	1.0×10^{10}–2×10^{10}

be adjusted to different distances (0–10 cm) above the surface of liquid nitrogen (*see* **Fig. 3A,B**), allowing the application of various freezing conditions ***(6)***. Adjustable trays are of advantage when freezing levels have to be changed frequently. Otherwise, floating trays are also useful ***(8)***.

3. Methods

3.1. Semen Collection

Dry the genital papilla of the fish from adhering water. Place a collection tube of adequate size under the genital pore and collect the semen by pressure on the abdomen (*see* **Note 1**). Store the semen on ice.

3.2. Semen Dilution and Filling of Semen into Straws

Dilute the semen in 4°C cold extender at the required ratio. Reliable dilution ratios and maximal and minimal sperm densities are listed in **Table 2** (*see* **Note 2**). A 1-min equilibration is sufficient. Do not extend equilibration to >10 min.

Cool straws to 4°C. Fill straws with micropipets. As the straws have a stopper that avoids liquid penetration, the sperm suspension can be sucked in by mouth. As this is the quickest way to fill the straws, it is of advantage in routine fieldwork.

3.3. Freezing of Straws

Cooling of the freezing box and equilibration to stable conditions requires between 15 and 30 min at room temperature. This must be considered for semen processing.

1. Cool the interior of the box with liquid nitrogen.
2. Fill up with liquid nitrogen until the overflow trap is reached. When ready for freezing, the box contains a volume of 2.67 L liquid nitrogen.

Table 3
Freezing and Thawing Conditions for Semen of the *Salmonidae*

Species	Straw type	Freezing level and temperature	Thawing
Hucho hucho,	0.5 mL	1.5 cm (–110 ± 2°C)[x]	25°C, 30 s[xxx]
Oncorhynchus mykiss,	1.2 mL	1.0 cm (–130 ± 2°C)[x]	30°C, 30 s[xxx]
Salmo trutta f. fario,			
Salmo trutta f. lacustris,			
Thymallus thymallus,			
Esox lucius			
Salvelinus fontinalis,	0.5 mL	2.5 cm (–92 ± 2°C)[xx]	25°C, 30 s[xxx]
Salvelinus alpinus	1.2 mL	2.0 cm (–100 ± 2)[xx]	30°C, 30 s[xxx]

Note: Similar superscripts indicate that freezing rates or thawing rates were similar under these conditions. Freezing temperature was measured with a thermoelectrode inserted in the straws.

3. Adjust the tray to the desired freezing level and equilibrate for 5 min to reach the appropriate temperature.
4. Place the straws or straw packages on the tray (*see* **Fig. 2C**) and freeze for 10 min. Freezing levels and freezing temperatures are species-specific and are shown in **Table 3**; freezing rates are shown in **Fig. 4**.
5. Cover to avoid extensive nitrogen evaporation (*see* **Note 3**).
6. Plunge straws into liquid nitrogen. When using an adjustable tray, the whole tray can be immersed into liquid nitrogen.

3.4. Storage of Straws

Transfer the single straws into the cans of commercial liquid-nitrogen containers. Because the straw packages remain flexible in liquid nitrogen, they can be rolled together (*see* **Fig. 2D**) and are placed in the cans also (*see* **Fig. 2E**).

3.5. Thawing of Straws

1. Thaw the 0.5-mL straws in 25°C water for 30 s and the 1.2-mL straws in 30°C water for 30 s (for thawing rates, *see* **Fig. 4**) (*see* **Note 4**). Thawing is not species-specific. Take the single straws out of the liquid-nitrogen container and transfer immediately into water. Gently agitate during thawing. After thawing, cut away the straw stopper and release the sperm suspension onto the eggs.
2. Process straw packages in a similar way. Take them out of the container, roll them out quickly, and place in water. Thereafter, roll together again and cut away the plugs of the straws with scissors (*see* **Fig. 2F**). Release sperm suspension onto the eggs.

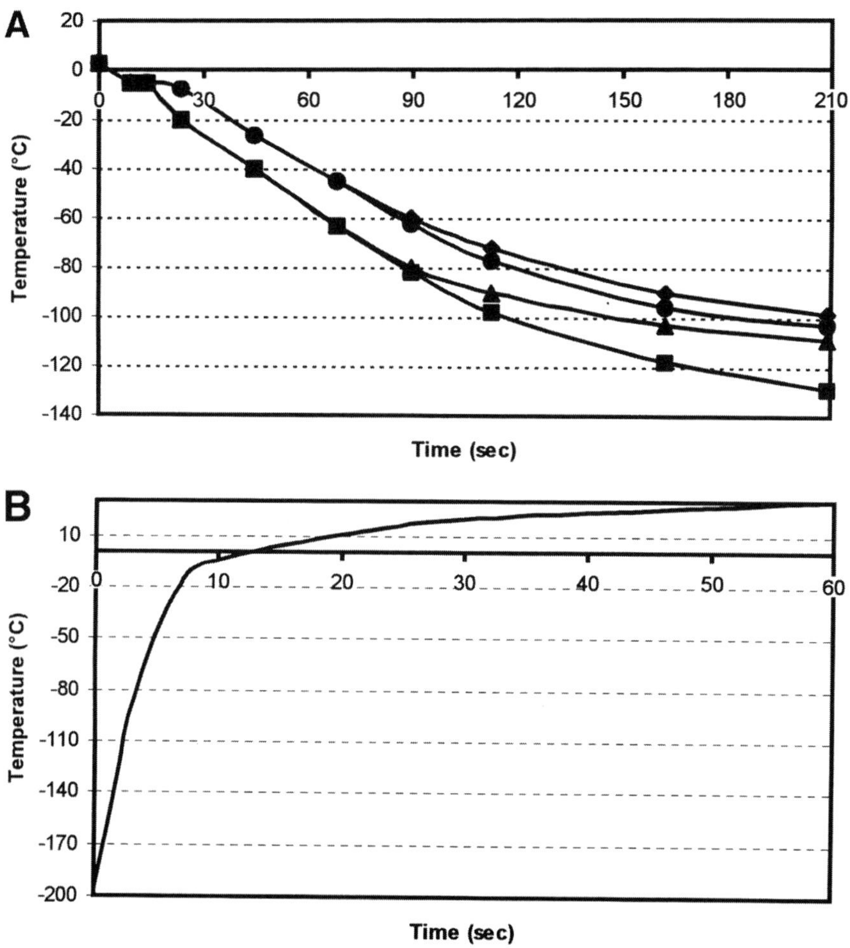

Fig. 4. (**a**) Freezing rates. ■: 1.2-mL straws, freezing level 1 cm above liquid nitrogen; ▲: 0.5-mL straws, freezing level 1.5 cm above liquid nitrogen; ●: 1.2-mL straws, 1.5 cm above liquid nitrogen; ◆: 0.5-mL straws, 2.5 cm above liquid nitrogen. (**b**) Thawing rates. Thawing rates were similar for 1.2-cm straws at 30°C for 30 s and for 0.5-mL straws at 25°C for 30 s.

3. When many straws are thawed or when the temperature gradient between the water and environment is high, a thermostat-regulated water bath has advantages for keeping the temperature constant.

3.6. Insemination with Cryopreserved Semen

1. Perform fertilization at 4–6°C. One-half milliliter of diluted cryopreserved semen is a reliable dose to fertilize 12.5 mL of eggs for all species *(6)*. Sperm-to-egg ratios differ because of species-specific differences in egg size.

2. For fertilization of egg quantities ≤50 g, wet or dry fertilization can be used; for higher egg quantities, dry fertilization is obligatory (*see* **Note 5**).
3. Wet fertilization: Place eggs in suitable beakers and add fertilization solution in a ratio of 2:1 (eggs:fertilization solution). Distribute the eggs in the fertilization solution. Thaw the semen and mix with the eggs. Eggs are stable for at least 2 min in the fertilization solution.
4. Dry fertilization: Place eggs in suitable beakers. Thaw the semen and mix with the eggs immediately. Immediately thereafter add the fertilization solution in the same ratio as for wet fertilization and mix.

4. Notes

1. Avoid contamination of semen with water and urine. They can activate sperm motility. Store semen at 4°C. At a higher temperature, viability decreases quickly. Also at 4°C, semen quality starts to decrease within 1–2 h after collection ***(10)***. If semen cannot be used within 1 h, special techniques are recommended to store unfrozen semen, such as cooling to 0°C, gassing with oxygen, or storage in thin layers ***(8,11)***.
2. According to our experience, semen density may deviate from the reported range and the dilution ratio must be changed. Semen of very high or low density is collected in the beginning and at the end of a spawning season. Also, older males have often semen of higher density. Semen density can differ between fish populations. Methods to determine the semen density have been described ***(12)***.
3. Open systems have the disadvantage that liquid nitrogen evaporates continuously. Liquid-nitrogen levels should be controlled. A deviation of 0.5 cm from the optimal freezing level resulted in a significant decrease in the postthaw fertilization rate ***(6)***.
4. Salmonid spermatozoa react very sensitively to thawing. Slight deviations from optimal conditions (changes in thawing period for 5–10 s, changes in thawing temperature for 5°C) significantly reduce the fertilization success ***(6)***. When using unsealed straws, take care that water does not enter the straw during thawing. Keep the open end of the straw over the water surface. Water inflow into the straw can lead to motility activation and osmotic damages.
5. Egg quality affects the fertilization success. Egg distribution in a single layer or in multiple layers has no influence on the fertilization success. However, quick and homogenous mixing of semen and eggs is essential.

Acknowledgments

The author is grateful to Austrian BMLF for financial support and to the Bundesanstalten in Wels and Scharfling for research cooperation.

References

1. Billard, R. (1983) Ultrastructure of trout spermatozoa: changes after dilution and deep freezing. *Cell Tissue Res.* **228,** 205–218.
2. Morisawa, M., Suzuki, K., Shimizu, H., et al. (1983) Effects of osmolality and potassium on spermatozoan motility of fresh water salmonid fishes. *J. Exp. Biol.* **107,** 105–113.

3. Lahnsteiner, F., Berger, B., Weismann, T., et al. (1996) Physiological and biochemical determination of rainbow trout, *Oncorhynchus mykiss,* semen quality for cryopreservation. *J. Appl. Aquaculture* **6,** 47–73.
4. Lahnsteiner, F., Weismann, T., and Patzner, R. A. (1998) Evaluation of the semen quality of the rainbow trout, *Oncorhynchus mykiss*, by sperm motility, seminal plasma parameters, and spermatozoal metabolism. *Aquaculture* **163,** 163–181.
5. Hart, N. H. (1990) Fertilization in teleost fishes: mechanisms of sperm-egg interactions. *Int. Rev. Cytol.* **121,** 1–66.
6. Lahnsteiner, F. (2000) Semen cryopreservation in the Salmonidae and in the Northern pike. *Aquaculture Res.* **31,** 245–258.
7. Lahnsteiner, F., Mansour, N., and Weismann, T. (2002) A new technique for insemination of large egg batches with cryopreserved semen in the rainbow trout. *Aquaculture* **209,** 359–367.
8. Jamieson, B. G. M. (1991) *Fish Evolution and Systematics: Evidence from Spermatozoa.* Cambridge University Press, Cambridge.
9. Lahnsteiner, F., Weismann T., and Patzner R. A. (1996) Changes in morphology, physiology, metabolism and fertilization capacity of semen of rainbow trout followed cryopreservation. *Prog. Fish-Cult.* **58,** 149–159.
10. Lahnsteiner, F., Weismann, T., and Patzner, R. A. (1998) Aging processes in semen of the rainbow trout, *Oncorhynchus mykiss. Prog. Fish-Cult.* **59,** 272–279.
11. Stoss, J. and Reftsie, T. (1983) Short term storage and cryopreservation of milt from Atlantic salmon and sea trout. *Aquaculture* **30,** 229–236.
12. Ciereszko, A. and Dabrowski, K. (1993) Estimation of sperm concentration of rainbow trout, whitefish and yellow perch using spectrophotometric technique. *Aquaculture* **109,** 367–373.

2

Fertilization and Sperm Chemotaxis in Ascidians

Manabu Yoshida

1. Introduction

1.1. General Introduction

Prior to fertilization, spermatozoa of many animals and plants show chemotactic behavior toward eggs. Chemotactic behavior was first described in ferns *(1)*, and their attractant was identified as the bimalate ion *(2)*. In animals, sperm chemotaxis to the egg was first observed in the hydrozoan *Spirocodon saltatrix* *(3)* and is now widely recognized in all species from cnidarians to human (for reviews, *see* **refs. *4–6***).

Spermatozoa of the ascidians *Ciona intestinalis* and *Ciona savignyi* were immotile or slightly motile when they were suspended in seawater, and if an unfertilized egg was placed in the sperm suspension, sperm near the egg were intensely activated and then showed chemotactic behavior toward the egg *(7–9)*. Egg seawater (ESW) that is a supernatant of seawater incubated with the ascidian egg has both sperm-activating and sperm-attracting activities, indicating that the ascidian egg releases some sperm-activating and sperm-attracting factors around the egg *(9,10)*. The release of the attractant from the egg seems to stop after fertilization *(9)*.

1.2. Chemical Nature and Source of the Sperm Chemoattractants

Where are the sperm chemoattractants released? Fern sperm show a chemotactic response to secretions from the female reproductive structures *(1)*. A sperm attractant of the sea urchin *Arbacia punctulata* is derived from the egg jelly *(11)*, and the source of sperm attractant of the hydrozoan, the siphonophore, is a cupule, the extracellular structure of the egg *(12)*. Therefore, the sperm attractant is released from the egg accessory organs or female gametes in these

From: *Methods in Molecular Biology, vol. 253: Germ Cell Protocols: Vol. 1 Sperm and Oocyte Analysis*
Edited by: H. Schatten © Humana Press Inc., Totowa, NJ

species. On the other hand, in the ascidians *C. intestinalis* and *C. savignyi*, sperm-attracting activity does not originate from the overall egg coats, as a layer of jelly surrounds the eggs, but originate from the vegetal pole of the egg ***(9)***. This indicates that in the ascidian the eggs themselves release the chemoattractant for the sperm.

Even though sperm chemotaxis is known in many phyla of animals and plants, the chemical nature of chemoattractants has been identified in only a few species. Sperm chemoattractants in plants were identified as organic compounds with a low molecular weight: bimalate ion in the bracken fern ***(2)*** and unsaturated cyclic or linear hydrocarbon in algae ***(13)***. On the contrary, in animals, most of the known chemoattractants and candidates for them were considered to be proteins or peptides. The chemoattractant of the sea urchin *A. punctulata* called resact is a 14-amino-acid peptide ***(11)*** and that of the amphibian *Xenopus laevis* is a 21-kDa protein ***(14)***. The chemoattractants of hydrozoa and a starfish are thought to be proteins, because the sperm-attracting activity of these species was lost by protease treatment ***(15,16)***. On the other hand, chemoattractant from the eggs of the coral *Montipora digitata* is considered to be an unsaturated fatty alcohol, dodeca-2,4-diynol ***(17)***.

In the ascidians *C. intestinalis* and *C. savignyi*, ESW has both sperm-activating and sperm-attracting activities. When the sperm attractant was purified from ESW, sperm-activating and sperm-attracting activities always comigrated, suggesting that both activities are derived from a single molecule; thus, we named the attractant of *Ciona* sperm as SAAF (sperm-activating and -attracting factor) ***(10)***. Recently, we determined the chemical nature of SAAF, a novel sulfated steroid, 3,4,7,26-tetrahydroxycholestane-3,26-disulfate ***(18)***.

1.3. Analysis of Chemotactic Behavior of Spermatozoa

Precise observations of the chemotactic behavior of sperm have been performed in several species. In the hydrozoan siphonophore, the radius of curvature of the sperm trajectory reduces as the spermatozoon closes to the cupule, the source of sperm attractant ***(19)***. On the other hand, a quick turning movement is observed in the sperm of other hydrozoans, when the sperm exhibit chemotactic behavior ***(20,21)***. During the turning movement, sperm demonstrate a temporary asymmetrical flagellar beating ***(8,21)***. The ascidian sperm also show the turning movement during chemotactic behavior ***(7–9)***. The turning movement of sperm is a typical example of the behavior seen during sperm chemotaxis; therefore, it is called the "chemotactic turn." The chemotactic turn of the hydrozoan sperm is often observed when sperm bear off the attractant ***(21)***.

As an activity assay for sperm attraction, the most commonly used method is the micropipet assay. The sample whose activity is to be examined is enclosed in the tip of a glass micropipet, the micropipet is placed in the sperm

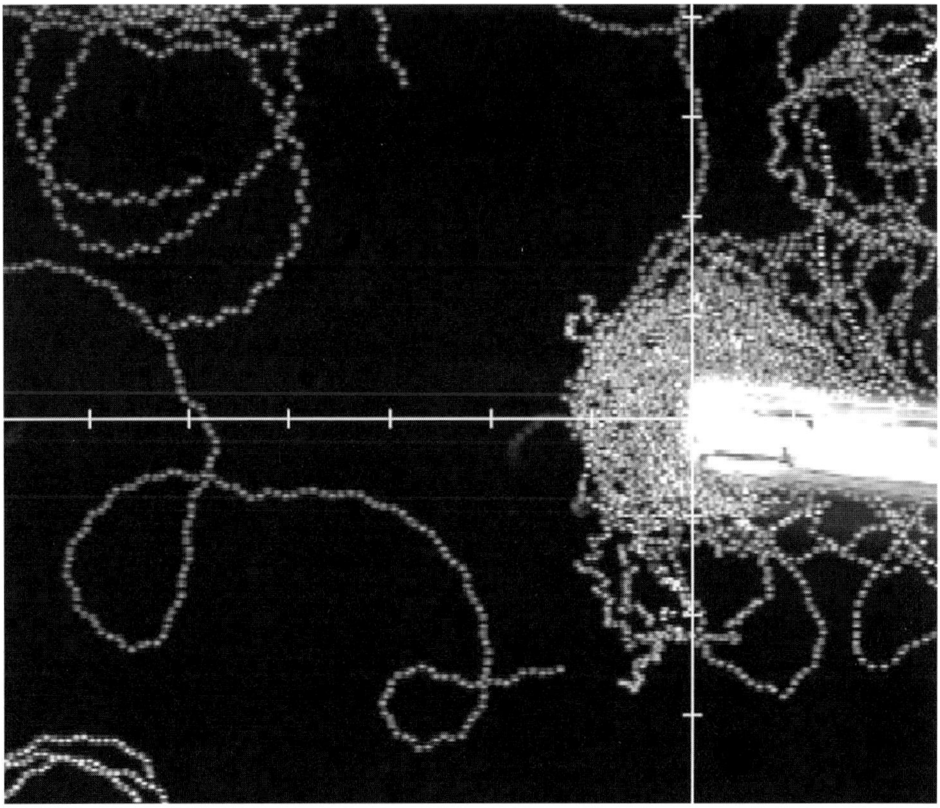

Fig. 1. Photograph of sperm trajectories near a micropipet containing ESW. The photo was integrated from 200 images taken every 20 ms. Lines show the axes for analysis. Ticks of the axes are labeled every 50 μm.

suspension, and then the sperm trajectories around the micropipet tip are observed. When a micropipet containing ESW or purified SAAF is inserted in the suspension of *Ciona* sperm, spiral trajectories of sperm toward the micropipet tip are seen, with the chemotactic turn (*see* **Fig. 1**).

Although molecular structures of the attractants and candidates have been proposed in several species, quantitative evaluation of sperm chemotaxis has not been well established. Furthermore, in the ascidian, it has been impossible to examine the activity of SAAF while eliminating side effects on sperm activation, because the ascidian sperm suspended in seawater has little motility. In order to distinguish the chemotactic behavior from the activation of motility in spermatozoa, spermatozoa are treated first with 1 m*M* theophylline for 1 min, which increases intracellular cAMP and results in sperm activation with a circular movement ***(10)***.

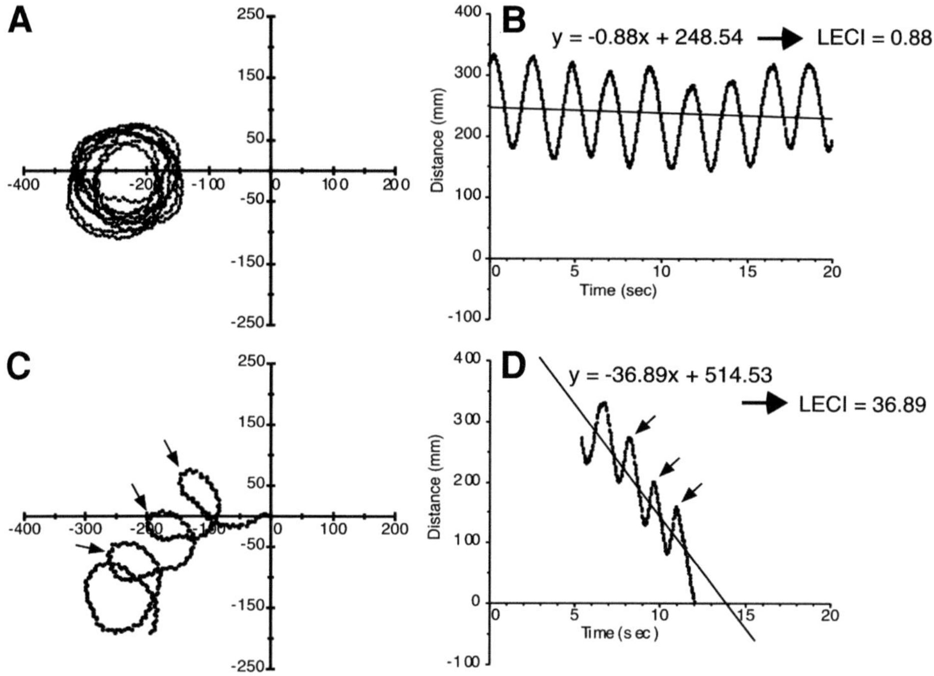

Fig. 2. Trajectories of the sperm (**A**, **C**) and plots of the distance between the sperm head and the micropipet tip against time (**B**, **D**) around the tip of micropipets containing (**A**, **B**) artificial seawater (ASW) and (**C**, **D**) ESW. The tip of the micropipet was set as the origin of coordinates (0). Arrows indicate points of the chemotactic turn. The line and formula represent the linear equation and the coefficient of time versus *D*, respectively. The negative value of the coefficients of the equation indicates the chemotaxis index (LECI).

My group has established a new method for quantitative evaluation of sperm chemotaxis using the linear equation chemotaxis index (LECI) ***(18)***. The LECI is a parameter that is derived from the negative value of the coefficient ($-a$) in a linear equation ($y = ax + b$) of time (abscissa in **Fig. 2**) versus the distance between the micropipet tip and the sperm head (D) (ordinate in **Fig. 2**). When the trajectories of sperm around the micropipet containing ESW were analyzed, D decreased with oscillation, although the parameter did not decrease with the addition of artificial seawater (*see* **Fig. 2**). The parameter LECI can represent the strength of sperm-attracting activity and will offer reliable aspects of the quantification of sperm chemotaxis.

In addition to LECI, other two parameters are useful for detailed analysis of chemotaxis in the *Ciona* sperm ***(18)***. The first is the differential quotient of D

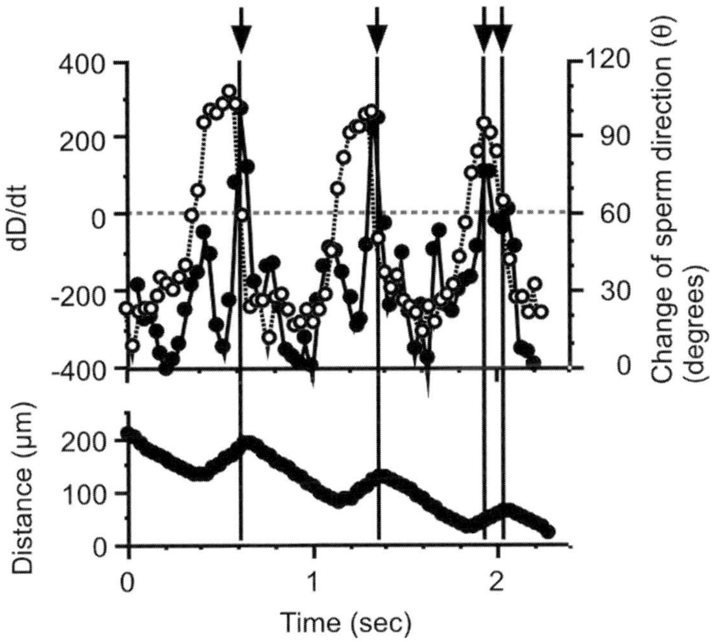

Fig. 3. Quantitative analysis of sperm chemotaxis. Changes in three parameters of a spermatozoon that showed chemotactic behavior toward purified SAAF. The points at which θ rose above 60° (i.e., the turning movement) occurred just after the peak of *dD/dt* (arrow) and coincided with the decrease of *dD/dt*.

with time (*dD/dt*), which represents the velocity of sperm approaching the micropipet tip. The second is the change of sperm direction (θ). The chemotactic turn can be quantitatively characterized as a quick increase in the value of θ. When *Ciona* sperm do not show chemotaxis, the sperm usually move with a constant curvature and θ is almost constant in the 0°–30° range (*see* **Fig. 3**). However, when the ascidian sperm show chemotaxis, the sperm show a quick change in their swimming direction (the chemotactic turn) and thus the θ periodically increased to values of over 60° (*see* **Fig. 3**). Furthermore, θ always started to increase just after the peak of *dD/dt* and decreased coincidentally with the decrease in *dD/dt* (*see* **Fig. 3**). Because the positive value of *dD/dt* indicates that a sperm is moving away from the micropipet tip, these results suggest that the chemotactic turn occurs when sperm move away from the micropipet tip. A similar strategy was considered in the hydrozoan sperm ***(21)***. The chemotactic behavior of sperm may be controlled by the chemotactic turn that is triggered when sperm detect any decrease in the concentration of the chemoattractant.

1.4. Signaling Mechanisms of Sperm Chemotaxis

How do the sperm chemoattractants act on the movement of sperm? The requirement of extracellular Ca^{2+} for chemotaxis has been known in the bracken fern ***(22)***, hydroids ***(12,23)***, sea urchin ***(11)***, and ascidians ***(7–9)***; Ca^{2+}-chelating agents (e.g., EDTA) completely suppress chemotactic behavior of sperm. Analytical studies in the hydrozoans siphonophores showed that the diameters of the trajectories decrease upon approach of the sperm to the cupule ***(19)***. No modification of the sperm trajectories was seen in the sperms' chemotactic behavior toward the cupule in the absence of Ca^{2+}, suggesting that Ca^{2+} regulates the motility pattern of the flagellum ***(19)***. In the sea urchin *A. punctulata*, resact, the sperm-activating and sperm-attracting peptide, binds to the receptor guanylyl cyclase ***(24)*** and seems to induce an increase in $[Ca^{2+}]_i$ through cGMP and cAMP ***(25)***. The $[Ca^{2+}]_i$ elevation may be controlled by a channel like the sperm-specific cyclic nucleotide-gated and voltage-dependent Ca^{2+} channels that were recently found in the mouse ***(26,27)*** and seem to induce the asymmetry of the flagellum waveform ***(28–30)***. The same role of extracellular Ca^{2+}-induced flagellar asymmetry on induction of sperm chemotaxis has been reported in hydrozoa ***(19,21)***. On the other hand, sperm chemotaxis of the ascidians *C. intestinalis* and *C. savignyi* does not seem to require intracellular cAMP changes. When SAAF acts on sperm, it induces entry of extracellular Ca^{2+} and an increase in intracellular cAMP ***(10,31)***. This results in protein-kinase A (PKA)–dependent phosphorylation on 21-kDa and 26-kDa axonemal proteins and activation of sperm motility ***(32)***. On the other hand, the chemotactic behavior of the ascidian sperm also requires extracellular Ca^{2+}, but theophylline-activated sperm, in which cAMP increases because theophylline blocks cAMP phosphodiesterase, show the same chemotactic behavior as normal sperm ***(9,10)***. Therefore, changes in cAMP are not required for sperm chemotaxis, and the mechanism of sperm chemotaxis is different from sperm activation, even though SAAF induces both phenomena. Recently, we found that the store-operated Ca^{2+} channel looks to mediate the asymmetrical flagellar waveform of the ascidian sperm, resulting in chemotactic behavior ***(33)***.

2. Materials

1. *Ciona intestinalis.*
2. Micromanipulator.
3. Phase-contrast or dark-field microscope (*see* **Note 1**).
4. High-speed video camera system (HAS-200 and HAS-PCI; Ditect, Tokyo, Japan).
5. Image-analyzing application.
6. Micropipet puller.
7. High-performance liquid chromatograph (HPLC) equipment.

8. Rotary evaporator.
9. Freeze-dryer.
10. Centrifugal vaporizer.
11. Centrifuge.
12. Artificial seawater (ASW): 462 m*M* NaCl, 9 m*M* KCl, 11 m*M* $CaCl_2$, 48 m*M* $MgCl_2$, and 10 m*M* HEPES–NaOH, pH 8.2.
13. 20 mg/mL Bovine serum albumin (BSA, Fraction V).
14. 25 m*M* theophylline.
15. 2% agar.
16. Ethanol (HPLC grade).
17. Chloroform (HPLC grade).
18. Acetonitrile (HPLC grade).
19. Methanol (HPLC grade).
20. Sep-Pak C18 resin (Waters, Milford, MA).
21. Reversed-phase HPLC ODS column.
22. Empty column (25 mm diameter × 150 mm).
23. Glass slide.
24. Glass capillaries with inner glass fiber (outer diameter = 1 mm) (Narisige GD-1).
25. Pasteur pipet.
26. Scissors.
27. Plastic dish (diameter = 3.5 cm).

3. Methods

The following methods outline (1) the collection of sperm and eggs, (2) the purification of SAAF, and (3) the assay for sperm chemotaxis.

3.1. Collection of Gametes

1. Keep collected *C. intestinalis* in aquaria with continuously flowing seawater. To prevent spontaneous spawning, light the animals continuously until use.
2. Remove the tunic and open the body with scissors. Collect eggs and semen from the oviduct and vas deferens respectively, with Pasteur pipets. Keep eggs and semen at 16–18°C and 4°C, respectively.

3.2. Purification of SAAF

The purification schema is shown in **Fig. 4**.

1. Wash the collected eggs once with ASW and suspend them in 40 vol of ASW. Incubate the egg suspension for 14–20 h at 4°C.
2. Centrifuge the egg suspension at $1.6 \times 10^3 g$ for 15 min and obtain the supernatant. Centrifuge the supernatant at $2.2 \times 10^4 g$ for 30 min at 4°C. The obtained supernatant is the ESW.
3. Lyophilize the ESW.
4. Add 1/10 vol of absolute ethanol to the residues of the ESW and vortex the mixture. Centrifuge at $2.2 \times 10^4 g$ for 15 min at 4°C and transfer the supernatant to a

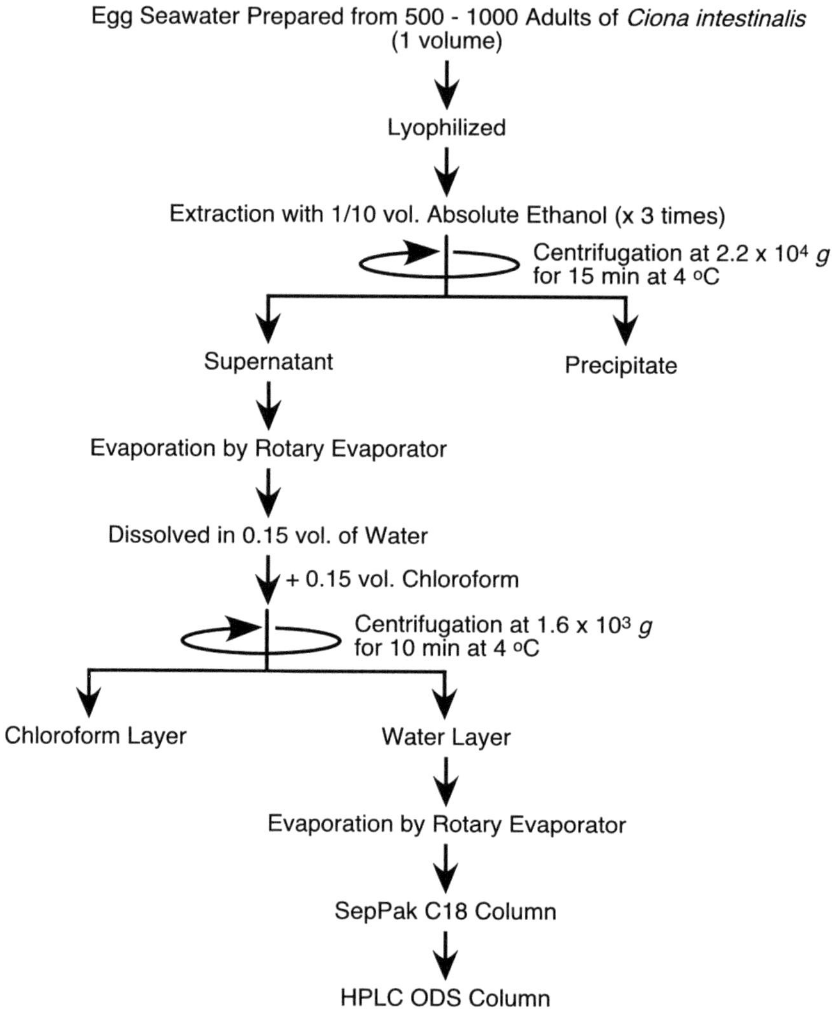

Fig. 4. Schematic drawing of purification procedures.

fresh tube. Perform ethanol extraction from the precipitate twice more and combine the obtained supernatant.

5. Evaporate the ethanol from the supernatant with a rotary evaporator.
6. Dissolve the residue in a volume of deionized water equal to the volume of ethanol used for the extraction.
7. Add an equal volume of chloroform and stir.
8. Centrifuge at $1.6 \times 10^3 g$ for 10 min. Transfer the upper water layer to a fresh tube.
9. Evaporate the layer with the rotary evaporator and dissolve the residue in the volume of deionized water equal to one-half of the volume of water used for dissolving the ethanol extract. This sample is the crude SAAF solution.

10. Column chromatography using reversed-phase resin, Sep-Pak C18.
    a. Add 100 mL of methanol to 40 mL of Sep-Pak C18 resin (Waters, Milford, MA) and gently mix. Keep it for 5 min until the resin settles.
    b. Discard the supernatants by decanting. Add 60 mL of methanol to the resin and gently mix it until the resin is uniformly suspended.
    c. Place the empty column on the stand. Fill the column with methanol.
    d. Immediately add a uniform suspension of Sep-Pak C18 resin to the column with a pipet or decantation. Wash the column with 200 mL of methanol. Then, equilibrate the column with 200 mL of deionized water.
    e. Apply the crude SAAF solution from **step 9** to the Sep-Pak C18 column by decanting or pipetting. Discard the flowthrough.
    f. Wash the column by adding 150 mL of deionized water. Discard the flowthrough.
    g. Wash the column by adding 150 mL of 20% methanol. Discard the flowthrough.
    h. To elute the adsorbed materials, add 150 mL of 60% methanol.
    i. Dry the collected elute with a freeze-dryer or a rotary evaporator. Dissolve the residue in 1 mL of deionized water.
    j. Check the sperm-activating and sperm-attracting activity.
11. Column chromatography using an ODS HPLC column.
    a. Wash the C18 HPLC column with a 10-bed volume of 80% acetonitrile, and then equilibrate the column with a 10-bed volume of 20% acetonitrile.
    b. Load the sample from **step 10** onto the HPLC column.
    c. Wash the column with a 1-bed volume of 20% acetonitrile; then elute with a 5-bed volume of a 20–30% acetonitrile linear gradient. Collect fractions, dry the fraction with the centrifugal vaporizer, and dissolve the residue with 1 mL of deionized water.
    d. Assay the sperm-activating and sperm-attracting activity of each fraction. Usually the activity is eluted at around 25–28% acetonitrile fractions. The obtained active fraction is a purified SAAF.

3.3. Observation of Sperm Chemotaxis

1. Wash glass slides thoroughly to remove oil drops on the surface. Dry the glass slides.
2. To prevent sperm sticking on the glass surface, coat glass slides with 2% BSA using a pipet (*see* **Note 2**).
3. Make the micropipets from 1-mm-outer diameter glass capillaries using a micropipet puller. From the shoulder to the tip, the length should be about 1 cm. If too long or too short, adjust the micropipet puller. Cut the tip of micropipets to 50–100 μm.
4. Mix each sample to examine the chemotactic activity with the same volume of 2% agar and keep the mixture at 50–60°C to prevent coagulation.
5. Put the micropipet into the mixture. The mixture should rise into the tip of the micropipet by capillary action. Leave the packed micropipet until the agar coagulates.

6. Set the micropipet on a micromanipulator.
7. Dilute the stocked semen to 1/10,000–1/5000 with ASW containing 1 m*M* theophylline and incubate for 1 min at room temperature for preactivation of sperm motility.
8. Set the glass slide on a phase-contrast microscope and put the sperm suspension on the BSA-coated glass slide.
9. Insert the micropipet in the sperm suspension. Set the tip of the micropipet in the center of the field of view.
10. Upload images of sperm around the micropipet tip onto a personal computer every 20 ms using a high-speed change-coupled device (CCD) camera (HAS-200, Ditect; or similar instrumentation) and a video card (HAS-PCI, Ditect; or similar instrumentation) (*see* **Note 3**). To observe flagellar formation, set the electrical shutter speed to 1/500 s or faster.

3.4. Analysis of Sperm Chemotaxis

Analyze the data for position of each sperm obtained in **step 10** of **Subheading 3.3.** as follows:

1. Digitize the position of each sperm using an image-analyzing application (Dipmotion 2D, Ditect; or similar instrumentation). Locate the micropipet tip to the origin (0).
2. Calculate the distance between the micropipet tip and sperm head (*D*) for every point of the sperm as follows (*see* **Fig. 5**):

$$D = (X_t^2 + Y_t^2)^{1/2}$$

3. Plot the value for *D* against time. Calculate the LECI as a negative value of the coefficient (–a) in the linear equation $y = ax + b$ of the time-vs-*D* plots.
4. Calculate the differential quotient of *D* with respect to time (*dD*/*dt*) as $(D_{P2} - D_P)/Dt$ (*see* **Fig. 4** and **Note 4**).
5. Calculate the change of sperm direction (θ) as follows (*see* **Fig. 5**):

$$\theta = \pi - \arccos \frac{(x_{t+\Delta t} - x_t)^2 + (y_{t+\Delta t} - y_t)^2 + (x_t - x_{t-\Delta t})^2 + (y_t - y_{t-\Delta t})^2 - (x_{t+\Delta t} - x_{t-\Delta t})^2 - (y_{t+\Delta t} - y_{t-\Delta t})^2}{2\sqrt{(x_{t+\Delta t} - x_t)^2 + (y_{t+\Delta t} - y_t)^2}\,\sqrt{(x_t - x_{t-\Delta t})^2 + (y_t - y_{t-\Delta t})^2}} \text{ (rad)}$$

(*see* **Note 4**).

6. Examine each parameter. The chemotactic turn is represented as θ > π/3 (60°).

4. Notes

1. A dark-field microscope is more appropriate to observe sperm flagella than a phase-contrast microscope. A phase-contrast microscope with a negative-phase lens (NH series lens in Olympus, or BM series lens in Nikon) is the most appropriate system for observing the sperm flagella, but it is very difficult to get these lenses now, because major microscope companies have stopped making this type of lens.

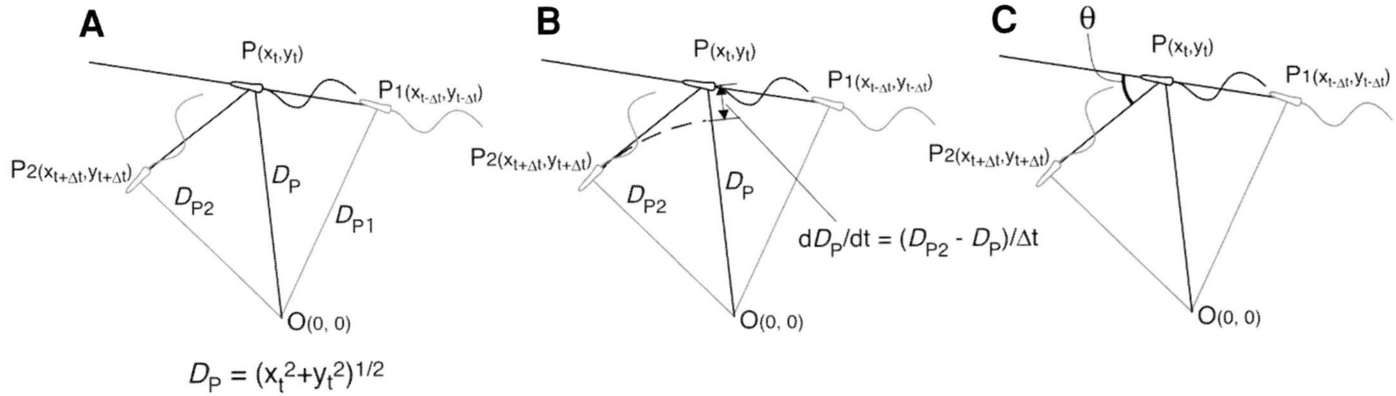

$$\theta = \pi - \arccos\frac{(x_{t+\Delta t} - x_t)^2 + (y_{t+\Delta t} - y_t)^2 + (x_t - x_{t-\Delta t})^2 + (y_t - y_{t-\Delta t})^2 - (x_{t+\Delta t} - x_{t-\Delta t})^2 - (y_{t+\Delta t} - y_{t-\Delta t})^2}{2\sqrt{(x_{t+\Delta t} - x_t)^2 + (y_{t+\Delta t} - y_t)^2}\sqrt{(x_t - x_{t-\Delta t})^2 + (y_t - y_{t-\Delta t})^2}} \text{ (rad)}$$

Fig. 5. Definition of parameters of the sperm chemotaxis: **(A)** D, **(B)** dD/dt, **(C)** θ. Formulas show the calculation method for each parameter of the sperm on the point $P(x_t, y_t)$. $P_1(x_{t-\Delta t}, y_{t-\Delta t})$ and $P_2(x_{t+\Delta t}, y_{t+\Delta t})$ represent the points of the sperm before and after Δt time, respectively.

2. To remove particles from the BSA, it is better to filtrate with a 0.45-μm syringe filter.
3. The method of capturing the sequential images varies and depends on the High-speed video system. Please follow the instruction manual of your own system.
4. To smooth the sperm trajectories, it is better to configure Δt as 40–60 ms (two or three frames).

References

1. Pfeffer, W. (1884) Locomotorische Richtungsbewegungen durch chemische Reize. *Unters. Bot. Inst. Tübingen* **1,** 363–482.
2. Brokaw, C. J. (1958) Chemotaxis of bracken spermatozoids. The role of bimalate ions. *J. Exp. Biol.* **35,** 192–196.
3. Dan, J. C. (1950) Fertilization in the medusan, *Spirocodon saltatrix. Biol. Bull. Mar. Biol. Lab. Woods Hole* **99,** 412–415.
4. Miller, R. L. (1985) Sperm chemo-orientation in metazoa, in *Biology of Fertilization* (Metz, C. B. and Monroy, A., eds.), Academic, New York, Vol. 2, pp. 275–337.
5. Cosson, M. P. (1990) Sperm chemotaxis, in *Controls of Sperm Motility: Biological and Clinical Aspects* (Gagnon, C., ed.), CRC, Boca Raton, FL, pp. 104–135.
6. Eisenbach, M. (1999) Sperm chemotaxis. *Rev. Reprod.* **4,** 56–66.
7. Miller, R. L. (1975) Chemotaxis of the spermatozoa of *Ciona intestinalis. Nature* **254,** 244–245.
8. Miller, R. L. (1982) Sperm chemotaxis in ascidians. *Am. Zool.* **22,** 827–840.
9. Yoshida, M., Inaba, K., and Morisawa, M. (1993) Sperm chemotaxis during the process of fertilization in the ascidians *Ciona savignyi* and *Ciona intestinalis. Dev. Biol.* **157,** 497–506.
10. Yoshida, M., Inaba, K., Ishida, K., et al. (1994) Calcium and cyclic AMP mediate sperm activation, but Ca^{2+} alone contributes sperm chemotaxis in the ascidian, *Ciona savignyi. Dev. Growth Differ.* **36,** 589–595.
11. Ward, G. E., Brokaw, C. J., Garbers, D. L., et al. (1985) Chemotaxis of *Arbacia punctulata* spermatozoa to resact, a peptide from the egg jelly layer. *J. Cell Biol.* **101,** 2324–2329.
12. Carré, D. and Sardet, C. (1981) Sperm chemotaxis in siphonophores. *Biol. Cell* **40,** 119–128.
13. Maier, I. and Müller, D. G. (1986) Sexual pheromones in algae. *Biol. Bull.* **170,** 145–175.
14. Olson, J. H., Xiang, X., Ziegert, T., et al. (2001) Allurin, a 21-kDa sperm chemoattractant from *Xenopus* egg jelly, is related to mammalian sperm-binding proteins. *Proc. Natl. Acad. Sci. USA* **98,** 11,205–11,210.
15. Cosson, J., Carré, D., and Cosson, M. P. (1986) Sperm chemotaxis in siphonophores: identification and biochemical properties of the attractant. *Cell Motil. Cytoskeleton* **6,** 225–228.
16. Punnett, T., Miller, R. L., and Yoo, B.-H. (1992) Partial purification and some chemical properties of the sperm chemoattractant from the forcipulate starfish *Pycnopodia helianthoides* (Brandt, 1835). *J. Exp. Zool.* **262,** 87–96.

17. Coll, J. C., Bowden, B. F., Meehan, G. V., et al. (1994) Chemical aspects of mass spawning in corals. I. Sperm-attractant molecules in the eggs of the scleractinian coral *Montipora digitata*. *Mar. Biol.* **118,** 177–182.
18. Yoshida, M., Murata, M., Inaba, K., and Morisawa, M. (2002) A chemoattractant for ascidian spermatozoa is a sulfated steroid. *Proc. Natl. Acad. Sci. USA* **99,** 14,831–14,836.
19. Cosson, M. P., Carré, D., and Cosson, J. (1984) Sperm chemotaxis in siphonophores. II. Calcium-dependent asymmetrical movement of spermatozoa induced by attractant. *J. Cell Sci.* **68,** 163–181.
20. Miller, R. L. (1966) Chemotaxis during fertilization in the hydroid *Campanularia*. *J. Exp. Zool.* **162,** 23–44.
21. Miller, R. L. and Brokaw, C. J. (1970) Chemotactic turning behaviour of *Tubularia* spermatozoa. *J. Exp. Biol.* **52,** 699–706.
22. Brokaw, C. J. (1974) Calcium and fragellar response during the chemotaxis of bracken spermatozoids. *J. Cell. Physiol.* **83,** 151–158.
23. Cosson, M. P., Carré, D., Cosson, J., et al. (1983) Calcium mediates sperm chemotaxis in siphonophores. *J. Submicrosc. Cytol.* **15,** 89–93.
24. Singh, S., Lowe, D. G., Thorpe, D. S., et al. (1988) Membrane guanylate cyclase is a cell-surface receptor with homology to protein kinases. *Nature* **334,** 708–712.
25. Cook, S. P., Brokaw, C. J., Muller, C. H., and Babcock, D. F. (1994) Sperm chemotaxis: egg peptides control cytosolic calcium to regulate flagellar responses. *Dev. Biol.* **165,** 10–19.
26. Ren, D., Navarro, B., Perez, G., et al. (2001) A sperm ion channel required for sperm motility and male fertility. *Nature* **413,** 603–609.
27. Quill, T. A., Ren, D., Clapham, D. E., and Garbers, D. L. (2001) A voltage-gated ion channel expressed specifically in spermatozoa. *Proc. Natl. Acad. Sci. USA* **98,** 12,527–12,531.
28. Brokaw, C. J., Josslin, R., and Bobrow, L. (1974) Calcium ion regulation of flagellar beat symmetry in reactivated sea urchin spermatozoa. *Biochem. Biophys. Res. Commun.* **58,** 795–800.
29. Brokaw, C. J. (1979) Calcium-induced asymmetrical beating of triton-demembranated sea urchin sperm flagella. *J. Cell Biol.* **82,** 401–411.
30. Brokaw, C. J. and Nagayama, S. (1985) Modulation of the asymmetry of sea urchin sperm flagellar bending by calmodulin. *J. Cell Biol.* **100,** 1875–1883.
31. Izumi, H., Márian, T., Inaba, K., Oka, Y., and Morisawa, M. (1999) Membrane hyperpolarization by sperm-activating and -attracting factor increases cAMP level and activates sperm motility in the ascidian *Ciona intestinalis*. *Dev. Biol.* **213,** 246–256.
32. Nomura, M., Inaba, K., and Morisawa, M. (2000) Cyclic AMP- and calmodulin-dependent phosphorylation of 21 and 26 kDa proteins in axoneme is a prerequisite for SAAF-induced motile activation in ascidian spermatozoa. *Dev. Growth Differ.* **42,** 129–138.
33. Yoshida, M., Ishikawa, M., Izumi, H., De Santis, R., and Morisawa, M. (2003) Store-operated calcium channel regulates the chemotactic behavior of ascidian sperm. *Proc. Natl. Acad. Sci. USA* **100,** 149–154.

3

Assays for Vertebrate Sperm Chemotaxis

Hitoshi Sugiyama, Bader Al-Anzi,
Robert W. McGaughey, and Douglas E. Chandler

1. Introduction

1.1. A Short History of Sperm Chemotaxis

Although sperm motility has been studied for over 300 yr, sperm chemotaxis in animals has been documented and quantified only in the last 40 yr. A clear demonstration of chemotactic guidance of sperm to eggs, thereby increasing the likelihood of fertilization, was first provided in a number of invertebrate species, most notably ascidians and sea urchins (for a review, *see* **ref. *1***). Studies of the sperm chemoattractant resact, a small peptide released from the jelly coat of sea urchin eggs, led to molecular characterization of the peptide, its receptor, and the intracellular signaling pathways used to modify sperm movement ***(2–5)***. More recently, evidence for sperm chemotaxis in vertebrates was obtained. Work from several laboratories demonstrated that human follicular fluid contains a sperm chemoattractant that acts preferentially on capacitated spermatozoa ***(6–16)***. The chorion of teleost eggs, especially that of herring, was shown to release proteins that initiate sperm motility close to the egg micropyle ***(17,18)***, thus serving a role functionally similar to that of a sperm chemoattractant. Recently, allurin, a 21-kDa protein demonstrated to be a sperm chemoattractant, was isolated from *Xenopus laevis* egg jelly and sequenced ***(19,20)***. Sequence analysis indicated that allurin is homologous to several mammalian sperm-binding proteins and studies are underway to identify an allurin ortholog in mammals.

Success of the above studies depended on the use of an in vitro assay for sperm chemotaxis. In this chapter, we discuss the major types of assays available and provide detailed methods for two of the assays we have used successfully.

From: *Methods in Molecular Biology, vol. 253: Germ Cell Protocols: Vol. 1 Sperm and Oocyte Analysis*
Edited by: H. Schatten © Humana Press Inc., Totowa, NJ

1.2. What Is Chemotaxis and How Is It Detected?

Chemotaxis, the stimulated movement of cells toward the source of an attractant, has been studied extensively in bacteria, slime molds, sperm, immune and defense cells, and migrating embryonic and cancer cells ***(6,7,21–28)***. Conceptually, this cell behavior—directed movement—is thought to arise by the action of a signal transduction system that links sensing of the chemoattractant with alterations and decision-making in the cytoskeleton. It is presumed and, in some cases, has been demonstrated that sensing of the directional source of the chemical requires the presence of a chemical gradient, the chemical being at a higher concentration near its source. Thus, it may be argued that in order to prove the presence of chemotaxis, one must demonstrate that (1) a concentration gradient of the attractant exists, (2) cells move in a directed and nonrandom manner within the gradient, and (3) cells accumulate at or near the source of the chemical. These criteria, therefore, are those used in assays that detect and quantitate chemotaxis.

Therefore, chemotaxis assays incorporate the following: (1) A concentration gradient of the chemoattractant agent must be set up, (2) motile cells must be introduced into the gradient, (3) cells must be observed moving within the gradient and their trajectories tracked, and (4) cells must be observed to accumulate at or near the high-concentration end of the gradient. Because of technical difficulties and time limitations, most assays evaluate requirement 3 or 4, but not both. The first requirement—setting up a chemical gradient—is critical. Numerous chemicals are known that stimulate cell motility in a random manner but not in a directed manner. Increase in the velocity of cell movement without regard to orientation of the cell or direction of movement is termed "chemokinesis." Chemicals that make cells move faster but in a random, undirected manner generally do not have to be present as a concentration gradient in order to exert their effect. In contrast, chemotaxis requires a chemical gradient to be used by the cell as an environmental cue during the orientation and reorientation that occurs as cell motility proceeds ***(29)***.

A chemical gradient can be set up in three basic ways. First, a micropipet or capillary can serve as source of chemoattractant when dipped into a buffer solution that contains no chemoattractant. Second, a small droplet of chemoattractant may be positioned within a chamber of agent-free buffer. This chamber may be substantially three dimensional and macroscopic, as in the two-chamber assay described below or may be microscopic, with the source of chemoattractant being either a well or depression in the slide. Third, a filter may subdivide two compartments, one of which contains a uniform concentration of chemoattractant; in this case, the gradient is set up entirely within the thin layer of solution present near and within the filter.

Observation and quantitation of cell movement is usually carried out in one of two ways. The first measures the *accumulation* of cells on the uphill side of a gradient or near the source of chemoattractant and can be either macroscopic or microscopic. For example, if a porous cellulose or polycarbonate membrane separates the low and high ends of the gradient, the number of cells entering the membrane or crossing the membrane into a separate chamber can be counted by a hemocytometer or quantitated by a radioactive or fluorescent probe ***(30–32)***; or, for example, if a capillary tube containing the chemoattractant is placed in a cell suspension, the number of cells entering or approaching the capillary tip can be determined.

The second type of chemotaxis assay observes microscopically the orientation, behavior, and trajectory of cells moving in a chemoattractant gradient. Both recent and classical observations suggest that cells moving in a chemoattractant gradient undergo periods of reorientation in which the direction of the gradient is sensed by differences in receptor occupation on opposite sides of a cell [spacial sensing ***(33)***] or by differences in receptor occupation from one moment to the next (temporal sensing). In either case, the trajectory of a cell undergoing such sensing is characterized by distinct changes in direction toward the high-concentration end of the gradient. Such changes in direction are typically monitored by video microscopy of cells moving within a defined physical plane of focus. One might refer to such microscopic assays as *observational* or *direct-view* assays. These assays usually incorporate not only a source of chemoattractant but also a control or mock source that does not contain a chemoattractant. Data from observational assays include trajectories, velocities, and direction changes of individual cells (together referred to as "tracking" assays) as well as determinations of cell orientation, cell–cell interactions, cell morphology, flagellar wave patterns, cytoskeletal reorganization, and repeated motility behavior patterns such as stopping and starting.

Table 1 lists examples of devices and protocols that have been used to detect chemotaxis. Accumulation assays are considerably easier to quantitate because the data are in the form of numbers of cells found within a predetermined space or area. Generally, data collection is faster, can be analyzed by typical statistical methods, and is well suited to automation or relatively high throughput. Thus, accumulation assays are commonly used for determining dose–response curves, chemoattractant structure–function relationships, and pharmacological interactions with the chemotactic response. Although accumulation assays are perhaps the easiest way to detect chemotaxis, they can produce false-positive results for at least two reasons. First, if a chemical at a high concentration immobilizes cells for any reason (aggregation, adhesion, inhibition of motility), randomly moving cells over time would tend to accumulate near the source regardless of whether the chemical was an attractant. This phenomenon is

Table 1
Devices for Measuring Chemotaxis

Device	Features	Assay type	Cell types	Ref.
Boyden chamber	Cells enter or cross cellulose membrane	Accumulation	Neutrophils	*34–37*
Capillary tube source	Cells move toward and enter capillary filled with attractant	Accumulation	Bacteria, human sperm, bull sperm	*8,38–41*
Capillary tube rejection	Cells avoid capillary filled with control buffer	Accumulation	Sperm	*8,42*
Capillary tube gradient	Cells migrate in a gradient set up within a capillary tube	Accumulation	Human sperm	*9–12,43*
Costar transwell or Neuro Probe chamber	Cells cross a porous polycarbonate filter	Accumulation	Human sperm; frog sperm	*13,14,19,20,44,45*
Biological target attraction	Cells move toward cellular source of chemoattractant	Observational	Ascidian sperm	*1,46–48*
Micropipet attraction	Cells reorient and move toward attractant-filled micropipet	Observational	Slime molds, Neutrophils; Human, frog, sea urchin, ascidian, and siphonophore sperm	*1,2,14,19,47,49*
Zigmond chamber	Chemotaxis occurs in a gradient formed between cover glass and slide across a linear bridge separating chemoattractant and buffer reservoirs; cells adhere to cover glass during migration	Observational	Neutrophils, mouse sperm	*15,44,50–53*
Makler chamber modified for chemotaxis	Cells migrate between cover glass and slide; cells and chemoattractant released from wells	Observational	Human and rodent sperm, bacteria	*8,16,54–56*
Dunn chamber	Cells migrate across a circular bridge separating cell and chemoattractant reservoirs	Observational	Neutrophils	*29,57*

referred to as "trapping." Second, if a chemical stimulates cells to move in random directions but at higher velocities (i.e., the agent stimulates chemokinesis), cells may arrive at the source of the chemical rapidly and this event could be mistaken as accumulation.

Therefore, many experts in the field emphasize that "observational" assays must be used in order to verify that cell accumulation is, in fact, a result of chemotaxis and not chemokinesis or trapping. Such "observational" assays, however, require a chamber for microscopic observation that provides for the formation of a chemical gradient within a flat, waferlike space that is almost two dimensional. In addition, observational methods require the ability to track cells over many video frames. Image data of this type are generally acquired one trial at a time, require large amounts of storage space, and can require substantial image processing capabilities. Thus, throughput of data is low and analysis of trajectory data somewhat subjective. Nevertheless, these types of data can be critical when chemotaxis must be distinguished from chemokinesis and when mechanisms that link chemotactic receptor occupation to cell-orientating behavior are of interest.

1.3. The Two-Chamber Sperm-Accumulation Assay for Chemotaxis

In principle, this assay is conducted by placing a sperm suspension in the top chamber and a test chemoattractant in the bottom chamber (*see* **Fig. 1A**). The two chambers are separated from each other by a polycarbonate membrane having pores through which the sperm can swim. In practice, we use a Corning–Costar Transwell plate for such chambers, the plastic base plate containing an array of 12 wells. Each well acts as a bottom chamber and into this is placed a plastic insert that acts as the top chamber. Across the bottom of the insert is stretched the porous polycarbonate barrier that separates the two chambers. The lip of the insert is cut away so as to produce three leglike supports that rest on the plate and three arclike slots that provide for micropipet tip access to the bottom chamber when the device is assembled (*see* **Fig. 1B**). Typically, we fill all of the bottom chambers of a plate with an appropriate physiological buffer; then, each assay is started by placing the top chamber insert into a well, adding a sperm suspension, and then, by micropipet, applying carefully a 100-μL drop of chemoattractant solution to the floor of the bottom chamber. Diffusion of the chemoattractant from the drop into the lower chamber sets up a concentration gradient over the next 10–15 min that extends up to and includes the polycarbonate filter and upper chamber buffer. Sperm swimming into this gradient accumulate in the *bottom* chamber buffer that, after an appropriate incubation time, is collected, the sperm pelleted, and the sperm number counted by hemocytometer. We have used this assay to detect chemotaxis in *X. laevis* sperm in response to the egg jelly protein allurin ***(19,20)***

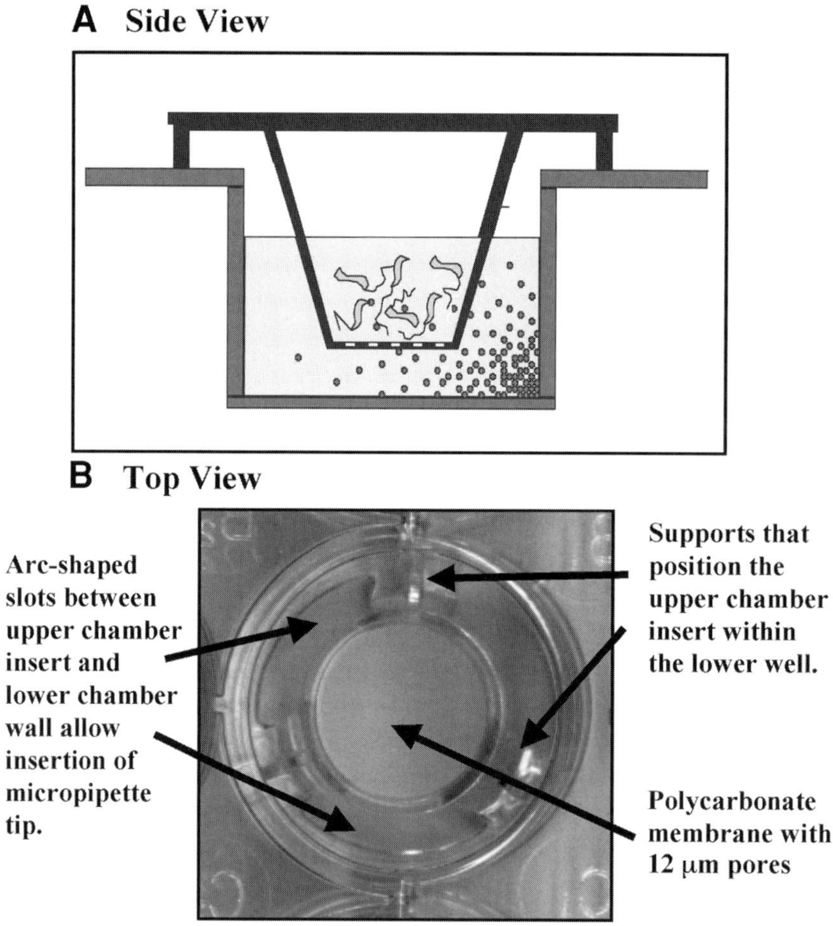

Fig. 1. **(A)** Side view: diagram of a Corning–Costar Transwell plate showing an upper chamber resting on the rim of a well that acts as the lower chamber. The upper chamber contains *Xenopus* sperm (S-shaped with tails) and its floor consists of a porous polycarbonate filter. The lower chamber contains a droplet of chemoattractant (bottom right) that is dispersing by diffusion. **(B)** Top view: photograph of a Corning–Costar Transwell plate from above.

and preliminary results suggest that this assay can also detect stimulated movement of mouse sperm.

Although the two-chamber assay is of the accumulation type, one can distinguish chemotaxis from chemokinesis and trapping using appropriate protocols. First, instead of placing a drop of chemoattractant on the floor of the bottom chamber to act as a "point" source of the gradient, one can *premix* the

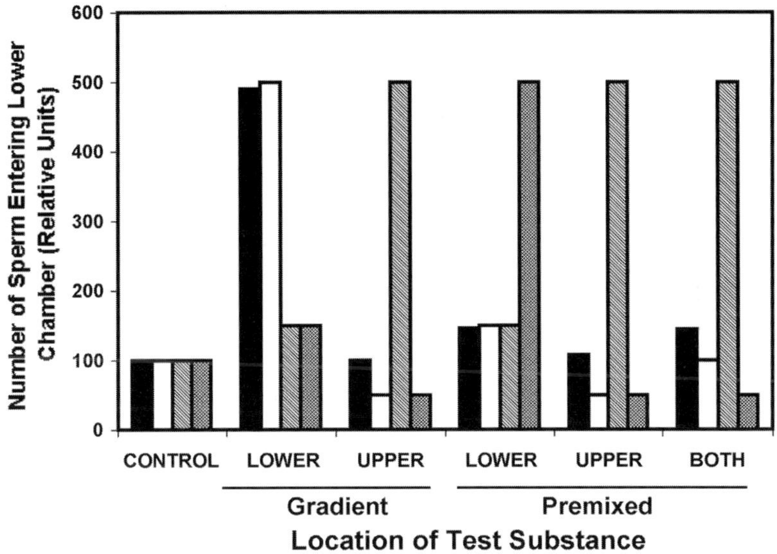

Fig. 2. Comparison of actual data for allurin (solid bars) and predicted data for a chemotactic agent (open bars), a chemokinetic agent (slashed bars), and a trapping agent (checkered bars) in a two-chamber assay. *Xenopus* sperm entering the lower chamber during a 40-min period was quantitated under six conditions (from left to right): 1, control buffer added as a droplet on the lower chamber floor; 2, active agent added as a droplet on the lower chamber floor; 3, active agent added as a droplet on the upper chamber floor; 4, active agent premixed with lower chamber buffer; 5, active agent premixed with upper chamber buffer; 6, active agent premixed with both upper and lower chamber buffers. For predicted data, responses to control buffer are set to 100, maximal responses to the active agent are set to 500, weak positive responses are set to 150, and responses lower than control are set to 50. Each type of agent is predicted to give a panel of data that is clearly distinguishable from other types. Data for allurin (from **ref.** ***19***) appear nearly identical to that predicted for a chemoattractant.

chemoattractant in the bottom chamber buffer so as to produce a nearly uniform concentration of the agent in this chamber. Second, one can add the chemoattractant to the top chamber or to both chambers to produce a reverse gradient or no gradient between chambers.

Figure 2 illustrates what these manipulations, either separately or in combination, would be expected to do if chemokinesis or trapping was occurring rather than chemotaxis. Chemotaxis (open bars, **Fig. 2**) would be characterized by maximal sperm entry when a highly localized source of attractant is present in the lower chamber (set 2); a weaker response should be seen when the attractant is premixed in the lower chamber buffer because the gradient is no longer as steep or as extended (set 4); a negative response (below control) or no response should be seen when the attractant is placed in the upper chamber or in both chambers because the gradient is reversed or absent (sets 3, 5, and 6). In contrast, chemokinesis (slashed bars, **Fig. 2**) would be characterized by a maximal response whenever the chemokinetic agent is present in the upper chamber along with the sperm at the start of the assay (sets 3, 5, and 6). The resulting increase in sperm motility would propel more sperm through the pores into the bottom chamber. This should occur even if the agent is present in both chambers because chemokinesis is dependent only on the concentration of the agent, not on the presence or direction of a gradient. In addition, *weakly* positive responses might be seen when the chemokinetic agent is premixed in the lower chamber buffer because diffusion of low concentrations of the agent across the membrane may stimulate sperm motility in the upper chamber (set 4). Finally, trapping (checkered bars, **Fig. 2**) would result in maximal sperm accumulation in the bottom chamber when the trapping agent is present in the lower chamber at high concentration (therefore premixed, set 4). The trapping agent must *not* be present in the upper chamber because this would lead to sperm immobilization before their passage through the membrane (sets 3, 5, and 6).

Thus, we believe that, by using the two-chamber assay with the described panel of conditions, one can distinguish chemotaxis from chemokinesis and trapping. Indeed, allurin-containing egg jelly extract from *X. laevis* (solid bars, **Fig. 2**) exhibits a data profile that is clearly like that of chemotaxis and is not like that of chemokinesis or trapping.

1.4. The Modified Makler Chamber for Observational Assays of Sperm Chemotaxis

The Makler chemotaxis chamber, as originally described ***(55,56)***, consists of a circular glass plate, 10 mm in diameter and 3 mm thick, encased within a hermetically sealed metal chamber. The glass plate has four wells drilled into it at the corners of a square pattern, each side of the square being 6 mm in length. Two of the wells are loaded with a sperm suspension, one well is loaded with a presumed chemoattractant, and the fourth well is loaded with a control buffer. Surrounding the square well array is a circular platform, 10 μm in height, that is machined flat on top such that the cover glass, when applied, will rest on it, leaving a 10-μm-thick layer of fluid. In this manner, sperm emerging from a source well and swimming to the chemoattractant well will be confined to swim within a thin plane of focus between the cover glass and slide. The cover glass is applied by a cap that is screwed onto the bottom assembly to maintain an even pressure on all sides.

We use a modification of the Makler chamber combined with video microscopy to observe sperm behavior and trajectories. Our chamber, constructed of Lexan, a transparent plastic, is shown in **Fig. 3**. The base is the same size as a

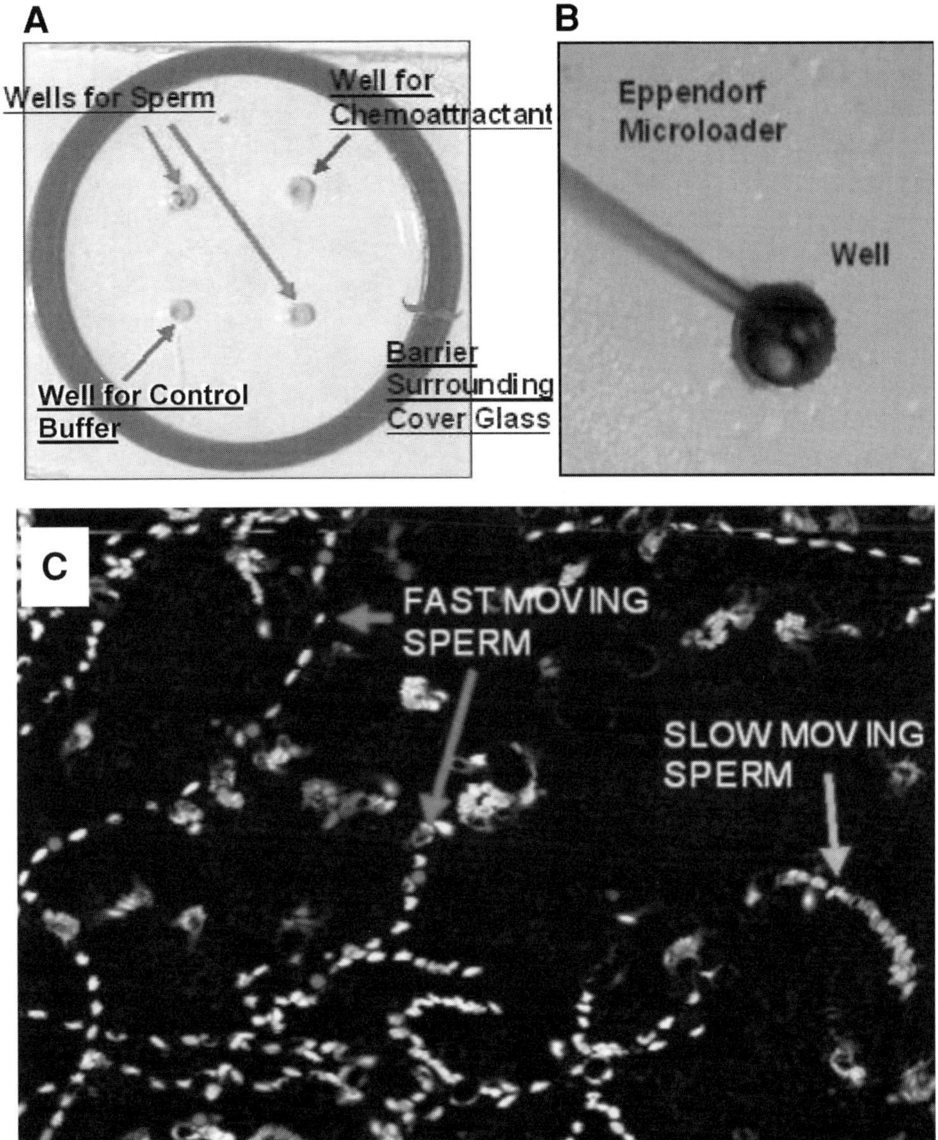

Fig. 3. **(A)** Chamber for observational assay of sperm trajectories, top view. Four wells, each 1 mm in diameter and 2 mm deep, are arranged at the corners of a square. **(B)** Wells may be filled by using Eppendorf "Microloader" pipet tips. **(C)** Video frames, 2 s apart, have been digitized by a Scion CG-7 frame grabber, imported into Photoshop 6.0, stacked as layers and then collapsed to form a composite image. Successive dark-field images of motile sperm trace out trajectories of individual sperm.

standard glass microscope slide (25 × 75 mm and 3 mm thick) and has four wells drilled into it that are arranged at the corners of a square, a pattern similar to that of the Makler chemotaxis chamber ***(54–56)***. The square is 6 mm on a side and each well is 1 mm in diameter and 2 mm deep and has a calculated volume of 1.57 μL (*see* **Fig. 3A**). The square pattern of wells is positioned with its diagonals parallel to the vertical and horizontal axes of the base to allow the microscope field to be moved along diagonals with one stage micrometer adjustment. The wells are centered within a ring of Lexan that is 19 mm in diameter inside and 2 mm high and is glued to the base.

During use, two of the wells are filled with a sperm suspension, the third well is filled with a chemoattractant, and the fourth well is filled with buffer only and serves as a negative control (*see* **Fig. 3A**). After the addition of chamber buffer and application of a round cover glass, the chamber is observed using either phase-contrast or dark-field microscopy and sperm movement is recorded on videotape. Later, the videotape is analyzed using single frames to count sperm numbers within a field and using a series of frames to trace the trajectory of individual sperm.

Trajectory tracing can be done with electronic images in one of two ways. First, for video acquired by dark-field optics, frames can be digitized using a Scion CG-7 frame grabber and Scion Image software, and then, imported into Photoshop 6.0, frames are superimposed and the background is subtracted so as to remove any immotile cells or debris. The result, as shown in **Fig. 3C**, is very similar to the classic strobe-lighting techniques ***(46,48)*** whereby the trajectory of a moving sperm is traced out by a series of images, each image recording the sperm position at a set time interval from the previous recording. Second, for video acquired by phase contrast, the frames can be digitized as above, then analyzed using custom tracking software that collects positional data with each "click" of the computer mouse and then plots the result (Burnett, unpublished data).

Evidence of chemotaxis in an observational assay is characterized not only by larger numbers of motile sperm near the chemoattractant well but also by trajectories that contain sharp changes in direction as individual sperm swim toward the well (*see* **Fig. 4C**). These abrupt changes in direction are interpreted as "reorientation events" in response to the sperm sensing the chemoattractant gradient. These trajectories are quite different from those of sperm near wells that do not contain a chemoattractant (*see* **Fig. 4B**). In this case, the paths followed are smooth curves that lead (by chance) or do not lead to the well. Whereas such trajectory data are frequently provided in image or diagrammatic form, they can also be expressed graphically, as shown in **Fig. 4D**. Both sperm numbers and the number of reorientation events are expected to be higher near the chemoattractant well (solid and open squares, **Fig. 4**) and lower near the well containing control buffer (solid and open circles, **Fig. 4**).

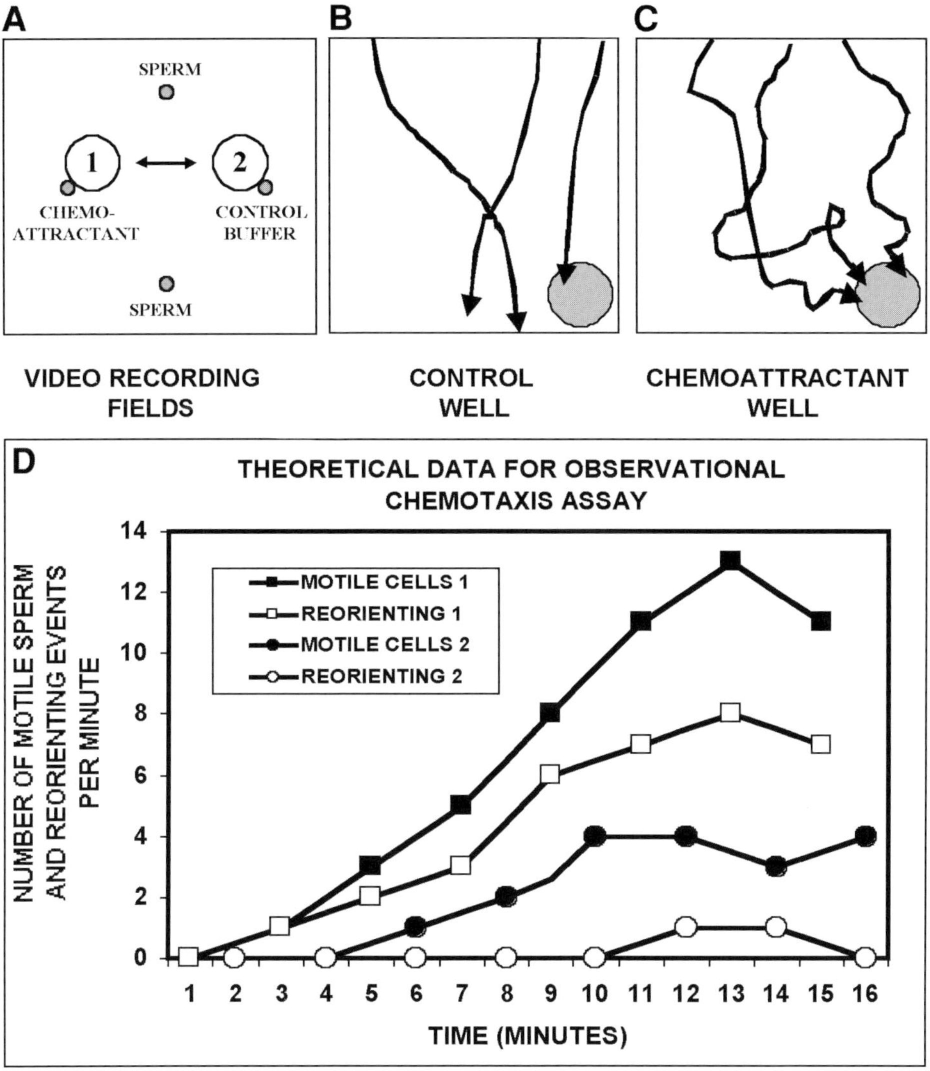

Fig. 4. Predicted data for a microscopic (observational) chemotaxis assay using a modified Makler chamber: **(A)** videotape recording alternating between field 1 (adjacent to the chemoattractant well) and field 2 (adjacent to the control buffer well) every 60 s; **(B)** typical sperm trajectories near the control buffer well are smoothly curved or straight; **(C)** typical sperm trajectories near the chemoattractant well exhibit tight turns and loops that give evidence of motile sperm reorientation; **(D)** graph showing that sperm numbers and reorientation events near the chemoattractant well (field 1) increase with time. In contrast, sperm numbers are lower and reorientation events almost absent near the control buffer well (field 2).

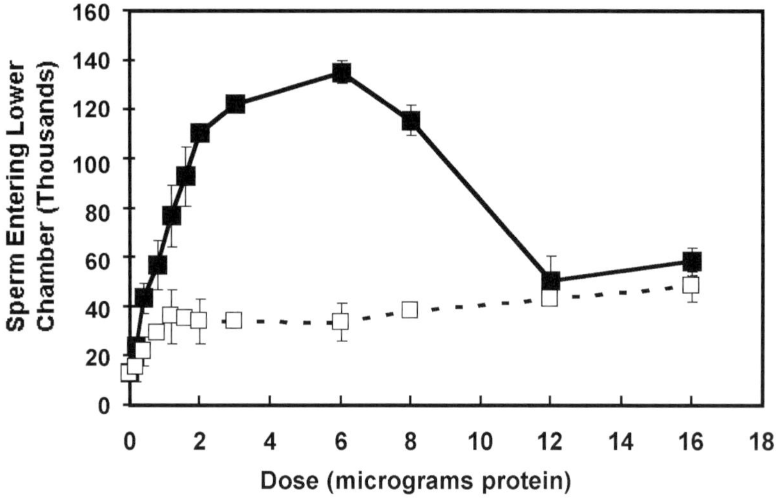

Fig. 5. Dose–response curve for chemotaxis in *Xenopus* sperm stimulated by allurin-containing egg water (solid squares) and human serum albumin (open squares). Optimal dose of egg water is 2–3 μg protein using a two-chamber assay and Corning–Costar Transwell plates. (Modified from **ref. *19*** and reprinted by permission of Academic Press.)

1.5. Detection of Chemotaxis Requires a Broad Dose–Response Curve and Replicate Measurements

Because the actual gradient set up in a chemotaxis assay is difficult to observe, it is seldom known at the outset what dose of the chemotactic agent should be used to achieve a maximal, indeed even an observable response. This problem is compounded by the fact that many chemotactic agents exhibit a biphasic dose–response curve; that is, they are inactive at concentrations either too low or too high compared to the optimal dose. An example of such a dose–response curve is seen in **Fig. 5**, where *Xenopus* egg water stimulates sperm movement in the two-chamber assay. Similar biphasic curves have been observed in human sperm responses to follicular fluid and atrial natriuretic peptide ***(14,40,42)***. In addition, the dose–response curve can be dependent on the geometry of the assay chamber, the thickness, porosity, and chemical makeup of the filter, the timing of agent and sperm addition, and the molecular nature of the chemoattractant. All of these variables can influence the steepness and distribution of the gradient or the mechanical nature of the barrier to sperm passage. Thus, the dose of chemoattractant that is used as a source of the gradient is only *empirically* related to the actual affinity of the sperm surface receptor for the chemoattractant.

A second source of variability is biological. In our experience, replicate determinations in the two-chamber assay usually fall within 10% of each other if the person conducting the assay is experienced. However, there can be greater variability between determinations using multiple batches of sperm from different frogs. Maximal response of sperm from one frog may differ as much as 100% from that of another, and negative controls may differ as much as threefold to fourfold between one batch of sperm and another. The result of this variability is that the ratio of maximal response to control response may be anywhere from 3 to 10, with a ratio of 4–7 being typical.

Recognizing the empirical nature of these assays and the biological variability encountered, we recommend that testing of an unknown agent utilize initially a broad range of doses using duplicate determinations and sperm from several frogs. Even with a familiar agent, it is best to carry out each experiment with two different doses—one "optimal" and one twofold to threefold more dilute. In some cases, one will need to demonstrate a finding by a single experiment that is "typical" of a group of experiments or to normalize data obtained from different batches of sperm.

2. Materials

2.1. General Materials/Materials for Preparation of Egg Water

1. 2500 IU human chorionic gonadotropin (hCG) in mannitol and phosphate buffer (Sigma, cat. no. 9002-61-3).
2. OR-2 buffer (1.5X): 124 m*M* NaCl, 3.77 m*M* KCl, 1.5 m*M* $CaCl_2$, 1.5 m*M* $MgCl_2$, 1.5 m*M* Na_2HPO_4, 10 m*M* HEPES, adjusted to pH 7.8 with NaOH. Use reagent-grade salts.
3. F-1 buffer: 41.25 m*M* NaCl, 1.25 m*M* KCl, 0.25 m*M* $CaCl_2$, 0.06 m*M* $MgCl_2$, 0.5 m*M* Na_2HPO_4, 2.5 m*M* HEPES, adjusted to pH 7.8 with NaOH. Use reagent-grade salts.
4. *X. laevis*, adult females (Carolina Biological or Pacific Biotech).
5. Bichinconic Acid (BCA) Protein Assay kit (Pierce, cat. no. 23227).

2.2. Two-Chamber Chemotaxis Assay for Frog Sperm

1. Corning–Costar Transwell plates with polycarbonate membranes having 12-μm-diameter pores (cat. no. 3403).
2. Formaldehyde, 37% (Fisher Scientific, cat. no. BP531).
3. *X. laevis*, sexually mature males (Carolina Biological or Pacific Biotech).
4. Benzocaine, prepared as a stock solution of 6 g/100 mL in ethanol.

2.3. Two-Chamber Assay for Mouse Sperm

1. Swiss Albino (CD-1) mice from Charles River.
2. Modified HTF buffer (M-HTF), sterile, from Irvine Scientific (cat. no. 9963).

2.4. Observational Assay for Frog or Mouse Sperm

1. Lexan microscopic chamber, custom made; based on Makler chamber described in **refs. *50*** and ***51***.
2. Cover glasses, #1, circular, 18 mm in diameter (Fischer Scientific, cat. no. 12-545-100 18CIR-1).
3. Pipet tips, "Microloader" (Eppendorf, cat. no. 930-00-100-7).
4. Purified agarose, low gelling temperature; we use Sea Kem ME (FMC Bio-Products, Rockland, ME).

3. Methods

3.1. Preparation of "Egg Water" from Xenopus laevis Egg Jelly

1. Inject *X. laevis* females at the dorsal lymph sac with 1000 IU of hCG in 500 µL of distilled water.
2. Eight to twelve hours later, strip jellied eggs from the female into a dry preweighed Petri dish. Stripping is accomplished by a repeated gentle milking pressure on the abdomen. Weigh the eggs and the dish, calculate the weight of the eggs, and add F1 buffer to the dish, a volume of 4 mL for every gram of eggs. Gently swirl the eggs in the buffer for 3 h at room temperature, taking care that the eggs do not break.
3. Remove the conditioned buffer from the eggs carefully, determine the total protein content by the BCA assay, place 300-µL aliquots in microcentrifuge tubes, and store frozen at –80 to –40°C.

3.2. The Two-Chamber Chemotaxis Assay for Frog Sperm

3.2.1. Preparation of Frog Sperm

1. Anesthetize the frog by immersion in water containing 0.03% benzocaine. This anesthetic is prepared by adding a stock solution of 6% (w/v) benzocaine in ethanol to water using a 1:200 dilution. Some precipitation of benzocaine is normal.
2. Decapitate the frog with sharp scissors and pith the brain and spinal cord.
3. Open the abdomen by a medial cut of the ventral abdominal wall and search for the testes, left and right. These organs are bean shaped, white to beige in color, 6–10 mm in length, and lie under the intestines at the level of the bottom of the stomach. Avoid bleeding, as blood is detrimental to sperm function.
4. Cut out each testis carefully with curved, sharp-pointed scissors, taking care to remove any excess fat or connective tissue. Wash the testis in ice-cold 1.5X OR-2 buffer and roll on a clean napkin to remove the last traces of blood.
5. Place each testis in 1 mL of clean ice-cold 1.5X OR-2 buffer and macerate with a razor blade. Alternatively, use a 25 gauge needle and 5-mL syringe to make a hole at one end of the testis, insert the needle into the other end, and forcefully flush the interior of the organ with 1 mL of 1.5X OR-2 buffer.
6. Transfer the cloudy sperm suspension (by micropipet using a cutoff tip) to a 15-mL conical plastic centrifuge tube and put on ice. If you used the maceration

technique (**step 5**), allow the large tissue chunks to settle by gravity for 5 min and then transfer the supernatant to a clean tube. Repeat this a second time.

7. Determine the sperm density. Dilute 20 μL of the sperm suspension 1:100 with 1.5X OR-2 and place an aliquot in a hemocytometer and count sperm in a 1-mm or 2-mm square area. Calculate the sperm density in cells/mL using the manufacturer's directions remembering to take into account the 1:100 dilution. If you used the maceration technique (**step 5**), the density should be $(1–10) \times 10^7$ cells/mL. Because the volume is about 2–3 mL total (both testes), the total yield of sperm is about $(2–20) \times 10^7$ cells. If you used the syringe blowout method (**step 5**), the yield will be lower, but the sperm preparation will be cleaner (i.e., it will have fewer testis cells).
8. Based on the sperm count, dilute the stock sperm suspension with 1.5X OR-2 buffer to obtain a working stock suspension of 2×10^7 sperm/mL.

3.2.2. The Assay

1. Prepare the chemoattractant samples in microcentrifuge tubes. Make enough of each chemoattractant to use 100 μL per assay plus a small amount more to accommodate any loss during micropipetting.
2. Set up the Costar Transwell plates containing inserts with polycarbonate membranes having 12-μm-diameter pores. Set the inserts aside and micropipet into each bottom chamber 1.5 mL of F1 buffer.
3. Prepare sperm aliquots, one for each assay, consisting of 60 μL of stock sperm suspension buffer in a microcentrifuge tube; place these tubes on ice.
4. Start each assay by placing an insert into the proper well, allowing it to float on the lower chamber buffer. Immediately prepare motile sperm by adding 540 μL of F1 buffer to a sperm aliquot tube, mix by gently pipetting up and down a few times using a cutoff tip, and then add 500 μL of the motile sperm suspension to the upper well using a cutoff micropipet tip (*see* **Note 1**). Discharge the motile sperm gently near the side of the chamber, being careful not to touch the membrane. Finally, micropipet a 100-μL aliquot of chemoattractant (or control buffer) into the bottom chamber. To do this, carefully insert the micropipet tip through the slot between the insert and the bottom chamber wall, bring the tip all the way to the bottom, and deposit the chemoattractant as a single drop at the junction between the bottom and side wall of the lower chamber (*see* **Note 2**).
5. Repeat **step 4** to start each assay on a set schedule, usually 1 min apart.
6. Terminate each assay after 40 min of incubation in the order that they were started. Remove several aliquots of fluid from the upper chamber (800–1000 μL total volume), taking care not to touch the membrane, and then carefully remove the insert from the well (*see* **Note 3**).
7. Once all inserts in a plate have been removed, transfer the bottom chamber buffer from each well (containing the sperm that passed through the membrane) to a clean, labeled, 1.5- or 2-mL microcentrifuge tube. Add 100 μL of 16 or 37% formaldehyde to each tube to fix the sperm. Fixing the sperm is optional and is done only if counting of sperm must be delayed by more than 24 h.

3.2.3. Sperm Counting

1. Centrifuge each microcentrifuge tube by a brief (3-s) spin at 12,000*g*. Remove the supernatant in each tube until a total of 100 µL of volume is left, as indicated by the 0.1-mL mark on the side of the tube. Resuspend the pellet by micropipetting up and down several times with a cutoff tip.
2. Remove a 20-µL aliquot from each tube and count on a hemocytometer slide. Count a 1- or 2-mm square area and then back-calculate to find the total number of sperm in the lower chamber.

3.3. Two-Chamber Assay for Mouse Sperm

3.3.1. Obtaining Mouse Sperm

1. Euthanize a male CD-1 mouse (10–15 wk old) by asphyxiation with CO_2 followed by cervical dislocation.
2. Dissect the mouse using a transverse cut through the posterior ventral abdominal wall. At room temperature, expose right and left caudae epididymides with attached vasa deferentia. Remove to the central well of an organ culture dish (Falcon, cat. no. 3037) containing 1.0 mL of M-HTF.
3. Isolate each vas deferens and milk sperm from it by means of a watchmaker's forceps. Make several cuts across each cauda epididymis with iris scissors to release sperm.
4. Allow sperm to disperse from the epididymis for about 15 min before removing all of the organ tissues from the dish.
5. Pipet the sperm suspension into a capped polystyrene tube (Falcon, cat. no. 2054) and allow debris to sediment for 5 min. Decant the supernatant into a second tube. Repeat three times to clarify the sperm suspension.
6. Incubate sperm suspension (should be approx 0.5 mL in M-HTF) undisturbed for 1 h at 37°C to attain capacitation. Remove the upper two-thirds of the incubate and place in a separate tube to obtain sperm of high motility.
7. Count the sperm suspension on a hemocytometer and then dilute the suspension with prewarmed M-HTF to obtain a density of 2×10^6 sperm/mL.

3.3.2. The Assay

1. Warm M-HTF buffer and a glass casserole dish filled with fine artist's sand to 37°C in an oven or incubator. This step is best done about 1 h before the assay.
2. Transfer the warm casserole dish with sand to the top of a slide warmer and burrow the bottom of a Corning–Costar Transwell plate into the sand to achieve good thermal contact. Fill each bottom well with 1.5 mL of prewarmed M-HTF. Maintain the slide warmer at a temperature that has been empirically determined to keep the chamber buffer at 37 ± 1.5°C over a 1-h period.
3. To begin each assay, place an upper chamber insert into a well, add 0.5 mL of stock sperm suspension to the top chamber, and finally place (by micropipet) 100 µL of chemoattractant onto the floor of the bottom well as described in **Subheading 3.2.2., step 4**.

4. Start assays at 1-min intervals. After 40 min of incubation, stop each assay by withdrawing about 400 μL of buffer from the top chamber twice; then, remove the insert completely.
5. Remove bottom chamber buffer from each well and micropipet into a 1.5- or 2.0-mL microcentrifuge tube and fix by adding 100 μL of 37% formaldehyde.
6. Microcentrifuge bottom chamber samples at 12,000*g* for 3 s to form a pellet and remove enough supernatant to reduce volume to 100 μL. Resuspend sperm pellet by gently pipetting up and down several times (use cutoff tip) and then count the sperm using a hemocytometer. Back-calculate to find the total number of sperm in the bottom well.

3.4. Observational Assay for Frog or Mouse Sperm Chemotaxis (see Note 4)

1. Prepare frog or mouse sperm as described in **Subheading 3.2.1.** or **3.3.1.** Dilute cell suspension to 2×10^7 cells/mL.
2. Fill chemoattractant well and control buffer well using a 0.2- to 5-μL range micropipet equipped with an Eppendorf "Microloader" pipet tip. Do not to overfill the well.
3. Fill two diagonal wells with the sperm suspension, again using a "Microloader" pipet tip. Again, be careful not to overfill the well.
4. Place a 35-μL drop of buffer directly in the center of the square pattern of wells. For *Xenopus* sperm, the buffer should be F-1; for mouse sperm, the buffer should be M-HTF. Do not let the fluid boundary touch any of the wells.
5. Apply a round, 18-mm-diameter cover glass, taking care to ensure that the cover glass contacts the central drop first so that the fluid is expressed evenly outward. The volume of the drop is critical to ensure that no extra fluid is present at the periphery of the cover glass once it is applied (*see* **Note 5**).
6. Place the slide on the microscope stage and position the slide such that the field of view is centered on a point directly between a sperm well and the chemoattractant well (*see* **Fig. 4A**). Record sperm movements using a video camera and video cassette recorder for 60 s. Reposition the slide such that the field of view is centered on a point directly between the sperm well and the control buffer well (*see* **Fig. 4A**). Record sperm movements for 60 s. Repeat this cycle for a total of 10–20 min.
7. Repeat **steps 2–6** for each chemoattractant sample to be tested.
8. Analyze videotaped data. Use a Scion CG-7 frame grabber and Scion Image software to digitize 50 s of video during each 60-s observation period in the cycle described above. Grab 1 frame every 2 s and import 25 frames into Photoshop 6.0 software. Make a composite image by adding each frame as a "layer" and then collapsing the stack into a single image. The first or last frame should be subtracted from the composite to eliminate background debris or nonmotile cells from the image.
9. Plot the number of motile cells found in each field and the number of cells showing reorientation in their trajectories versus time. The presence of chemotaxis is

indicated by a greater number of motile cells and a greater number of reorienting cells in the field near the chemoattractant well than in the field near the control buffer well. A theoretical example of data suggesting chemotaxis is shown in **Fig. 4C**.

4. Notes

1. *Xenopus* sperm are immotile in 1.5X OR-2, a high-salt buffer. Sperm motility is activated by osmotic shock, here achieved by a 1:10 dilution of the sperm stock with F-1, a low-salt buffer. *Xenopus* sperm remain motile only for a short time—about 20 min. Thus, it is critical that sperm motility is initiated immediately before each assay. Our rule of thumb is that *Xenopus* sperm used in a chemotaxis assay should have been diluted for no longer than 2 min before the start of the assay; shorter times are preferable. The procedure as written prescribes making a new sperm dilution for each assay and persons new to the assay might best follow this advice. Once proficiency in the assay has been gained, one may actually perform a set of assays more efficiently if a single sperm dilution, scaled up appropriately, is used for two to four assays. **Care should be taken, however, never to use *Xenopus* sperm that have been motile for more than 2 min.**
2. Careful deposition of the chemotactic agent on the floor of the bottom well at its side is critical for achieving a "point source" of the agent. Accidental pipet tip movements or the deposition of air bubbles can degrade the localization of the agent. Localization may be aided by making the agent solution more dense. We have tried adding 10% glycerol or sucrose to the solution to increase density. Glycerol, unfortunately, seems to increase sperm movement through the filter in the presence of nonspecific proteins like serum albumin by mechanisms that are as yet unclear. Sucrose seems to work well without nonspecific effects. Alternatively, the chemotactic agent can be mixed 1:1 with agar and pipetted into the bottom well to solidify. If one uses agar to achieve localization, one should use purified, low-temperature melting agar rather than microbiological-grade agar because the latter results in nonspecific sperm movements. When using agar, one should add bottom chamber buffer just before each assay rather than adding buffer to all wells before a set of assays. Exercising this caution prevents setting up the chemotactic gradient prematurely.
3. The initial volume of fluid in the upper chamber is 500 μL. During the course of the assay, fluid slowly leaks into the upper chamber; by the end of the assay, the upper chamber volume has increased to about 1000 μL and the lower chamber volume has decreased to about 1000 μL. This slow flow from bottom to top prevents sperm from entering the bottom chamber by gravity and thereby keeps the control levels low.
4. We are in the preliminary stages of developing this assay, and the protocol described is based on our experience to date. Although we have not yet tested this chamber in a definitive manner, we provide this protocol to aide other laboratories in developing similar assays.

5. Even when one is careful in applying the cover glass, studies with dye-filled wells suggest that well contents can be forced into the observational field on occasion. Recently, we have avoided this problem by mixing chemoattractant or control buffer in agarose (1%, purified, low-temperature gelling, Sea Kem ME), filling the wells, and, after solidification, trimming the agarose flush to the top of the well. Dye studies show that release of chemicals from agarose-filled wells is even and gradual, leading to a nicely formed symmetrical gradient.

Acknowledgments

We would like to thank Lindsay Burnett and Dr. Dennis McDaniel for helpful discussions and sharing of preliminary observations. Video microscopy and image analysis was carried out in the W.M. Keck Bioimaging Laboratory. These studies were supported by grants IBN-9807862 and IBN-0130001 from the National Science Foundation. Dr. Sugiyama was supported by a grant from the St. Mariana University School of Medicine, Kawasaki, Japan.

References

1. Miller, R. L. (1985) Sperm chemo-orientation in the metazoa, in *Biology of Fertilization* (Metz, C. B. and Monroy, A., eds.), Academic, New York, pp. 275–337.
2. Ward, G. E., Brokaw, C. J., Garbers, D. L., et al. (1985) Chemotaxis of *Arbacia punctulata* spermatozoa to resact, a peptide from the egg jelly layer. *J. Cell Biol.* **101,** 2324–2329.
3. Shimomura, H., Dangott, L. J., and Garbers, D. L. (1986) Covalent coupling of a resact analogue to guanylate cyclase. *J. Biol. Chem.* **261,** 15,778–15,782.
4. Garbers, D. L. (1989) Molecular basis of signaling in the spermatozoan. *J. Androl.* **10,** 99–107.
5. Wedel, B. and Garbers, D. (2001) The guanylyl cyclase family at Y2K. *Annu. Rev. Physiol.* **63,** 215–233.
6. Eisenbach, M. and Tur-Kaspa, I. (1999) Do human eggs attract spermatozoa? *BioEssays* **21,** 203–210.
7. Eisenbach, M. (1999) Sperm chemotaxis. *Rev. Reprod.* **4,** 56–66.
8. Ralt, D., Manor, M., Cohen-Dayag, A., et al. (1994) Chemotaxis and chemokinesis of human spermatozoa to follicular factors. *Biol. Reprod.* **50,** 774–785.
9. Vallanueva-Diaz, C., Vadillo-Ortega, F., Kably-Ambe, A., et al. (1990) Evidence that human follicular fluid contains a chemoattractant for spermatozoa. *Fertil. Steril.* **54,** 1180–1182.
10. Villanueva-Diaz, C., Arias-Martinez, J., Bustos-Lopez, H., et al. (1992) Novel model for study of human sperm chemotaxis. *Fertil. Steril.* **58,** 392–395.
11. Silwa, L. (1993) Effect of heparin on human spermatozoa migration in vitro. *Arch. Androl.* **30,** 177–181.
12. Cohen-Dayag, A., Ralt, D., Tur-Kaspa, I., et al. (1994) Sequential acquisition of chemotactic responsiveness by human spermatozoa. *Biol. Reprod.* **50,** 786–790.

13. Gnessi, L., Ruff, M. R., Fraioli, F., et al. (1985) Demonstration of receptor-mediated chemotaxis by human spermatozoa. A novel quantitative bioassay. *Exp. Cell Res.* **161,** 219–230.
14. Ralt, D., Goldenberg, M., Fetterolf, P., et al. (1991) Sperm attraction to a follicular factor(s) correlates with human egg fertilizability. *Proc. Natl. Acad. Sci. USA* **88,** 2840–2844.
15. Oliveira, R. G., Tomasi, L., Rovasio, R. A., et al. (1999) Increased velocity and induction of chemotactic response in mouse spermatozoa by follicular and oviductal fluids. *J. Reprod. Fertil.* **115,** 23–27.
16. Cohen-Dayag, A., Tur-Kaspa, I., Dor, J., et al. (1995) Sperm capacitation in humans is transient and correlates with chemotactic responsiveness to follicular factors. *Proc. Natl. Acad. Sci. USA* **92,** 11,039–11,043.
17. Vines, C. A., Yoshida, K., Griffen, F. J., et al. (2002) Motility initiation in herring sperm is regulated by reverse sodium–calcium exchange. *Proc. Natl. Acad. Sci. USA* **99,** 2026–2031.
18. Oda, S., Igarashi, Y., Manaka, K., et al. (1998) Sperm-activating proteins obtained from the herring eggs are homologous to trypsin inhibitors and synthesized in follicle cells. *Dev. Biol.* **204,** 55–63.
19. Al-Anzi, B. and Chandler, D. E. (1998) A sperm chemoattractant is released from *Xenopus* egg jelly during spawning. *Dev. Biol.* **198,** 366–375.
20. Olson, J. H., Xiang, X., Ziegert, T., et al. (2001) Allurin, a 21-kDa sperm chemoattractant from *Xenopus* is related to mammalian sperm-binding proteins. *Proc. Natl. Acad. Sci. USA* **98,** 11,205–11,210.
21. Bourret, R. B. and Stock, A. M. (2002) Molecular information processing: lessons from bacterial chemotaxis. *J. Biol. Chem.* **277,** 9625–9628.
22. Comer, F. I. and Parent, C. A. (2002) PI 3-Kinases and PTEN: how opposites chemoattract. *Cell* **109,** 541–544.
23. Iijima, M. and Devreotes, P. (2002) Tumor suppressor PTEN mediates sensing of chemoattractant gradients. *Cell* **109,** 599–610.
24. Schroder, J. M. (2000) Chemoattractant as mediators of neutrophilic tissue recruitment. *Clin. Dermatol.* **18,** 245–263.
25. Bochner, B. S. and Schleimer, R. P. (2001) Mast cells, basophils, and eosinophils: distinct but overlapping pathways for recruitment. *Immunol. Rev.* **179,** 5–15.
26. Locascio, A. and Nieto, M. A. (2001) Cell movements during vertebrate development: integrated tissue behavior versus individual cell migration. *Curr. Opin. Genet. Dev.* **11,** 464–469.
27. Comoglio, P. M. and Trusolino, L. (2002) Invasive growth: from development to metastasis. *J. Clin. Invest.* **109,** 857–862.
28. Moore, M. A. (2001) The role of chemoattraction in cancer metastases. *BioEssays* **23,** 674–676.
29. Wilkinson, P. C. (1998) Assays of leukocyte locomotion and chemotaxis. *J. Immunol. Methods* **216,** 139–153.
30. Sims, J. L., Geissler, F. T., and Page, R. C. (1985) An improved multimembrane microassay for quantitating the motility of granulocytes and monocytes labeled with chromium-51. *J. Immunol. Methods* **78,** 279–291.

31. Penno, M. B., Hart, J. C., Mousa, S. A., et al. (1997) Rapid and quantitative in vitro measurement of cellular chemotaxis and invasion. *Methods Cell Sci.* **19,** 189–195.
32. Frevert, C. W., Wong, V. A., Goodman, R. B., et al. (1998) Rapid fluorescence-based measurement of neutrophil migration in vitro. *J. Immunol. Methods* **213,** 41–52.
33. Parent, C. A. and Devroetes, P. N. (1999) A cell's sense of direction. *Science* **284,** 765–770.
34. Boyden, S. V. (1962) The chemotactic effect of mixtures of antibody and antigen on polymorphonuclear leucocytes. *J. Exp. Med.* **115,** 453.
35. Zigmond, S. H. and Hirsh, J. G. (1973) Leukocyte locomotion and chemotaxis. New methods for evaluation and demonstration of a cell-derived chemotactic factor. *J. Exp. Med.* **137,** 387–410.
36. Bignold, L. P. (1988) Measurement of chemotaxis of polymorphonuclear leukocytes in vitro. *J. Immunol. Methods* **108,** 1–18.
37. Keller, H.-U., Borel, J. F., Wilkinson, P. C., et al. (1972) Reassessment of Boyden's technique for measuring chemotaxis. *J. Immunol. Methods* **1,** 165–168.
38. Adler, J. (1973) A method for measuring chemotaxis and use of the method to determine optimum conditions for chemotaxis by *E. coli*. *J. Gen. Microbiol.* **74,** 77–91.
39. Iqbal, M., Shivaji, S., Vijayasarathy, S., et al. (1980) Synthetic peptides as chemoattractant for bull spermatozoa: structure activity correlations. *Biochem. Biophys. Res. Commun.* **96,** 235–242.
40. Anderson, R. A., Feathergill, K. A., Rawlins, R. G., et al. (1995) Atrial natriuretic peptide: a chemoattractant of human spermatozoa by a guanylate cyclase-dependent pathway. *Mol. Reprod. Dev.* **40,** 371–378.
41. Lee, S-L, Kao, C-C., and Wei, Y-H. (1994) Antithrombin III enhances the motility and chemotaxis of boar sperm. *Comp. Biochem. Physiol. A* **107,** 277–282.
42. Zamir, N., Roven-Kreitman, R., Manor, M., et al. (1993) Atrial natriuretic peptide attracts human spermatozoa in vitro. *Biochem. Biophys. Res. Commun.* **197,** 116–122.
43. Isobe, T., Minoura, H., Tanaka, K., et al. (2002) The effect of RANTES on human sperm chemotaxis. *Hum. Reprod.* **17,** 1441–1446.
44. Zatylny, C., Marvin, L., Gagnon, J., et al. (2002) Fertilization in *Sepia officinalis*: the first mollusk sperm-attracting peptide. *Biochem. Biophys. Res. Commun.* **296,** 1186–1193.
45. Tacconis, P., Revelli, A., Massobrio, M., et al. (2001) Chemotactic responsiveness of human spermatozoa to follicular fluid is enhanced by capacitation but is impaired in dyspermic semen. *J. Assist. Reprod. Genet.* **18,** 36–44.
46. Yoshida, M., Inaba, K., and Morisawa, M. (1993) Sperm chemotaxis during the process of fertilization in the ascidians *Ciona savignyi* and *Ciona intestinalis*. *Dev. Biol.* **157,** 497–506.
47. Giojalas, L. C. and Rovasio, R. A. (1998) Mouse spermatozoa modify their motility parameters and chemotactic response to factors from the oocyte microenvironment. *Int. J. Androl.* **21,** 201–206.

48. Miller, R. L. (1966) Chemotaxis during fertilization in the hydroid Campanularia. *J. Exp. Zool.* **162,** 22–44.
49. Cosson, M. P. (1990) Sperm chemotaxis, in *Controls of Sperm Motility: Biological and Clinical Aspects* (Gagnon, C., ed.), CRC, Boca Raton, FL, pp. 103–135.
50. Zigmond, S. H. (1977) Ability of polymorphonuclear leukocytes to orient in gradients of chemical factors. *J. Cell Biol.* **75,** 606–616.
51. Zigmond, S. H. (1988) Orientation chamber in chemotaxis. *Methods Enzymol.* **162,** 65–72.
52. Bultmann, B. D. and Gruler, H. (1983) Analysis of the directed and non-directed movement of human granulocytes: influence of temperature and ECHO 9 virus on *N*-formylmethionyl-leucylphenyalanine-induced chemokinesis and chemotaxis. *J. Cell Biol.* **96,** 1708–1716.
53. Fabro, G., Rovasio, R. A., Civalero, S., et al. (2002) Chemotaxis of capacitated rabbit spermatozoa to follicular fluid revealed by a novel directionality-based assay. *Biol. Reprod.* **67,** 1565–1571.
54. Makler, A. (1991) Sealed mini-chamber of variable depth for direct observation and extended evaluation of sperm motility under the influence of various gases. *Hum. Reprod.* **6,** 1275–1278.
55. Makler, A., Reichler, A., Stoller, J., et al. (1992) A new model for investigating in real time the existence of chemotaxis in human spermatozoa. *Fertil. Steril.* **57,** 1066–1074.
56. Makler, A., Makler-Shiran, E., Stoller, J., et al. (1992) Use of a sealed mini-chamber to investigate human sperm motility in real time under aerobic and anaerobic conditions. *Arch. Androl.* **29,** 255–261.
57. Zicha, D., Dunn, G. A., and Brown, A. F. (1991) A new direct-viewing chemotaxis chamber. *J. Cell Sci.* **99,** 769–775.

4

Stallion Spermatozoa Viability

Comparison of Flow Cytometry with Other Methods

Katrina Merkies and Mary M. Buhr

1. Introduction

The major obstacle in standardizing techniques to evaluate stallion fertility is the huge variation in ejaculates from different stallions and even among ejaculates collected from the same stallion *(1,2)*. Spermatozoa concentration, total volume, motility, percent live spermatozoa, and percent normal spermatozoa can range greatly *(3)*, and any or all may have an effect on the ability of the ejaculate to fertilize. Cryopreservation of semen results in partial, irreversible membrane damage to the spermatozoa, reducing fertility rates even further. Using current industry protocols for cryopreservation of stallion semen, 50–75% of stallions do not produce ejaculates suitable for freezing *(4)*. Pregnancy rates from natural or artificial insemination have shown little improvement in recent years and this has a significant impact on the economic feasibility of the horse breeding industry. Simple, accurate assessment of the viability of fresh and cryopreserved stallion spermatozoa prior to insemination could improve the reproduction rate in horses *(5)*.

Many diverse methods have been employed in an effort to repeatably and confidently assess the quality of an ejaculate *(6)*. Light microscopy has been used to visually assess both motility *(7,8)* and membrane integrity using supravital stains *(9)*. Fluorescently labeled antibodies and lectins can be used to identify membrane or intracellular components either by flow cytometry or fluorescence microscopy *(10)*. A dual-fluorescent stain that has recently become popular consists of SYBR-14, which stains living cells green, and propidium iodide (PI), which enters through disrupted membranes of presumably dead cells and stains them red *(11)*. As an alternative method of viability

From: *Methods in Molecular Biology, vol. 253: Germ Cell Protocols: Vol. 1 Sperm and Oocyte Analysis*
Edited by: H. Schatten © Humana Press Inc., Totowa, NJ

assessment, the fluorescent dye Hoechst 33258 enters through disrupted plasma membranes and binds to intracellular DNA as an indicator of nonviable cells in a manner similar to PI ***(12–14)***. Traditionally, a semen sample is assessed, using any of these tests, by a human operator using a microscope to evaluate two lots of 100 sperm; the resulting average is then assigned as the rating of the whole ejaculate. The advent of flow cytometry has allowed the objective evaluation of tens of thousands of spermatozoa from a single ejaculate to replace the rather subjective counting of a few hundred spermatozoa with light microscopy ***(15,16)***. Thus, flow cytometry estimates the quality of the entire sperm population based on 10-fold to 30-fold more sperm than light microscopy.

The use of three laboratory instruments to assess the quality of a stallion ejaculate using a variety of methods is described here. The light microscope can quickly assess the motility of a sperm sample and viability using supravital stains such as eosin–nigrosin. The fluorescence microscope and the flow cytometer can determine sperm viability using the fluorescent dyes SYBR-14, PI, and H33258.

2. Materials

2.1. Semen Collection, Processing, and Motility Evaluation

1. Missouri model artificial vagina (Nasco, Fort Atkinson, WI) equipped with an in-line milk filter to remove gel and debris.
2. EZ Mixin (Animal Reproductive Services, Chino, CA), an extender (buffer) for equine semen stored dried at –20°C and prepared immediately prior to use per manufacturer's directions in double-distilled water warmed to 37°C.
3. 0.2% Gluteraldehyde solution; 1 mL of 50% gluteraldehyde (Fisher Scientific, Unionville, ON) plus 249 mL phosphate-buffered saline (PBS). Store at 4°C.
4. PBS: 125 m*M* NaCl, 8 m*M* Na_2HPO_4, 2 m*M* NaH_2PO_4, 5 m*M* KCl, 5 m*M* dextrose, pH 7.4.
5. Neubauer hemocytometer.

2.2. Cryopreservation

1. Glycine freezing extender. Fraction A: 49 m*M* glucose, 49 m*M* fructose, 51 m*M* sodium citrate, 93 m*M* glycine, 20% (by volume; i.e., 20 mL egg yolk to 80 mL extender) egg yolk (*see* **Note 1**), 0.8 % (by volume) orvus paste (Equex; Nova Chemicals, Scituate, MA). Fraction B: Same as Fraction A but containing 12.6% glycerol (by volume; i.e., 87.4 mL Fraction A + 12.6 mL glycerol).
2. 0.5 mL Commercially available cryopreservation straws.
3. Straw-filling device (mouth pipet, vacuum pump, or automatic straw filler).

2.3. Viability Evaluation

1. Eosin–nigrosin supravital stain, buffered ***(17)***; 17 m*M* sodium citrate, 6.67% nigrosin (wt:vol; i.e., 6.67 g nigrosin in a final total volume of 100 mL sodium citrate),

1.1% (wt:vol) eosin-Y (Fisher Scientific, Unionville, ON). Dissolve eosin and nigrosin in the buffer with constant stirring over gentle heat. This can take several hours. Determine pH by litmus paper and adjust to 6.8–7.0 using 1 *N* NaOH or 1 *N* HCl if necessary. Allow to stand at room temperature for 3–4 d. Filter through Reeve Angel Grade 802 filter paper. The prepared stain can then be stored for several years at 4°C. Entellan mountant (BDH, Toronto, ON).

2. SYBR-14/PI (Live/dead Sperm Viability Kit, Molecular Probes, Eugene, OR). Prepare a working solution of SYBR-14 by diluting the dye to 0.1 mg/mL in dimethyl sulfoxide. Store both SYBR-14 (working solution) and PI (from the kit) in the dark (*see* **Note 2**) at –20°C until use.
3. Hoechst 33258 (H33258; Sigma, Oakville, ON) fluorescent dye. Prepare a working solution of H33258 by diluting the dye to 10 µg/mL in 2.9% (wt:vol) sodium citrate. Store aliquots (approx 100 µL) in the dark at –20°C (*see* **Note 2**).

3. Methods

3.1. Semen Collection, Processing, and Motility Evaluation

1. Collect stallion ejaculate using an artificial vagina (AV). The water in the water jacket of the AV should be between 37°C and 50°C when presented to the stallion (*see* **Note 3**), which means adding water at 38–55°C, depending on the environmental temperature and the length of time from preparation of the AV to actual collection. Time from preparation of AV to collection should not, under any circumstances, exceed 5 min.
2. Discard gel portion and debris captured by the in-line filter.
3. Immediately extend entire ejaculate 1:1 (vol:vol) in EZ Mixin warmed to 37°C.
4. Hold extended semen at 30–37°C for transport to the laboratory.
5. Centrifuge extended semen (500*g*, 10 min, 25°C) and discard supernatant.
6. Resuspend sperm pellet in the minimum effective volume of EZ Mixin.
7. Determine concentration by diluting a 10 µL volume into 990 µL of 0.2% gluteraldehyde solution and counting two chambers on a Neubauer hemocytometer (*see* **Note 4**).
8. Dilute spermatozoa (from **step 6**) with EZ Mixin to a concentration of 50×10^6 sperm/mL.
9. Place 10 µL of diluted semen (*see* **step 8**) on a clean glass slide warmed to 37°C and cover with a warm coverslip. Using a light microscope with a heated stage set to 37°C, observe under 400× magnification. Count at least 200 sperm and classify them as progressively motile, nonprogressively motile, or nonmotile.

3.2. Cryopreserved Semen

1. Collect stallion ejaculate and process as in **Subheading 3.1., steps 1–4**.
2. Determine concentration (**Subheading 3.1., steps 5–7**). Resuspend to 250×10^6 sperm/mL in Fraction A of glycine freezing extender (**Subheading 2.2., step 1**).

3. Dilute 1:1 (vol:vol) in Fraction B of the glycine freezing extender containing 12.6% glycerol. This will produce final concentrations of 125×10^6 sperm/mL in 6.3% glycerol. Allow to equilibrate 15 min at room temperature.
4. Load extended semen into 0.5 mL straws (*see* **Note 5**). Lay filled straws flat on a freezing rack 3 cm above the surface of liquid nitrogen for 15 min. This height and time cools the straws at a rate of –25°C/min as measured by thermocouple; rate of cool is critical to successful freezing.
5. Plunge straws into liquid nitrogen and store.
6. To thaw straws, immerse in 37°C water bath for 30 s (*see* **Note 6**). Holding straws by the tip and working as quickly as possible, dry the straws and cut off the ends; then, drain the semen into a 37°C container. Add an equal volume of EZ Mixin warmed to 37°C.

3.3. Viability Evaluation

Studies have shown that there is a strong correlation among the various methods used to determine viability of stallion semen *(18)*. Therefore, it is possible to use a single method most convenient to the researcher based on availability of equipment to confidently assess the ejaculate quality. However, results of sperm quality assessments may vary more when obtained with subjective methods such as visual estimates made with a microscope, rather than objective methods such as flow cytometry.

3.3.1. Eosin–Nigrosin

1. With all materials at 35–37°C, add 10 μL extended semen (from **Subheading 3.1., step 8** or **Subheading 3.2., step 6**) to 100 μL eosin–nigrosin in a glass spotting plate and stir gently for 3–5 s. Let stain for 30–60 s. Place a 5–10 μL drop of stained sperm on one edge of a clean glass slide. Push a second slide through the drop to smear (*see* following diagram and **Note 7**):

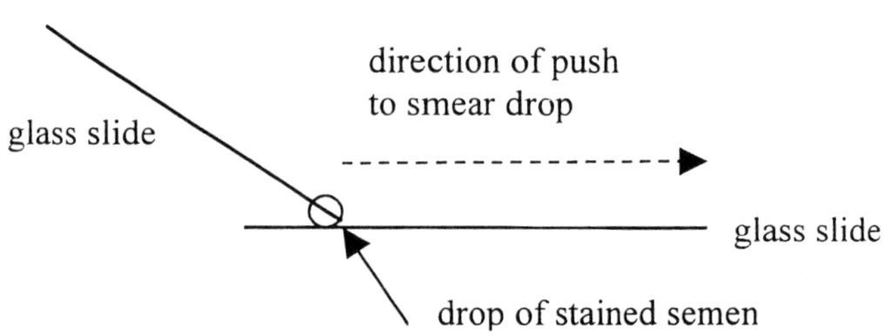

2. Allow the smear to air-dry. Apply small drop of Entellan mountant and cover with a cover slip. Allow Entellan to harden at least 1 h before scoring slides (*see* **Note 8**). View slides under light or phase-contrast microscopy using 400–1000× magnification (*see* **Fig. 1**). One operator should count at least 2 × 100 sperm per slide, classifying them as live (unstained/white) or dead (pink). (*See* **Notes 9–11**.)

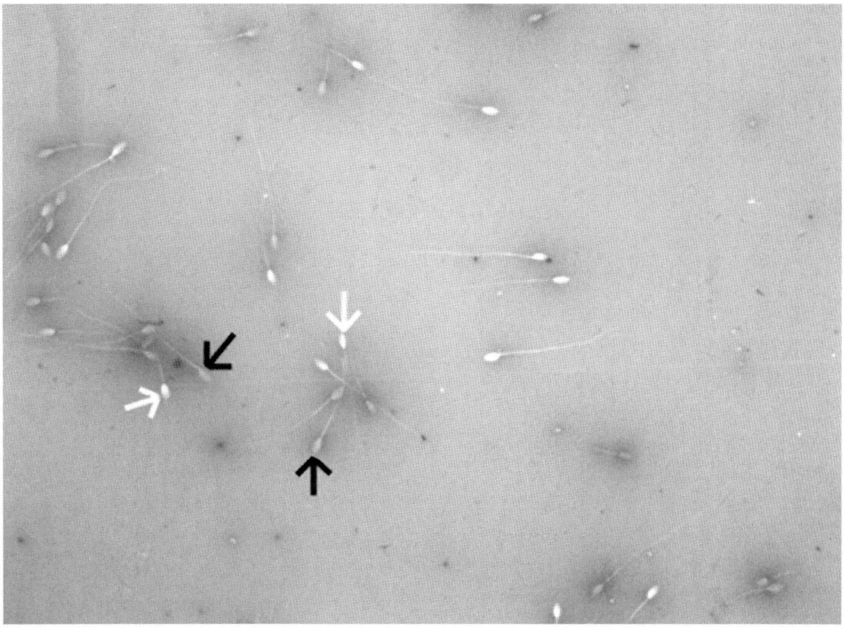

Fig. 1. Stallion spermatozoa stained with eosin–nigrosin and viewed under light microscope (magnification: 400×). Intact sperm (white arrows) do not take up the dye, have heads that are an even overall white or light shade against the background, and are considered live. Dead sperm (black arrows) take up the dye completely and are dark.

3.3.2. SYBR-14/PI

1. To 1.5 mL of extended spermatozoa (from **Subheading 3.1.**, **step 8** or **Subheading 3.2.**, **step 6**) add 5 μL SYBR-14 followed by 3 μL PI, mix by gentle swirling, and incubate for 15 min in the dark (*see* **Note 2**). Viability measurements are made immediately after incubation.
2. To count sperm with a microscope, one operator (*see* **Note 11**) counts at least 2 × 100 sperm/treatment using a fluorescence microscope with excitation at 450–490 nm and emission at 520 nm. Categorize spermatozoa as green (live), red (dead), or dual-stained (transitional) (*see* **Fig. 2**).
3. Carry out flow cytometric analysis on a flow cytometer (we used a Coulter Epics Elite) with an argon laser (488 nm), collecting green fluorescence emissions at 525 nm and collecting red fluorescence emissions at 610 nm. Gate forward and side scatter parameters to count particles that correspond to the size of a spermatozoan. Count 10,000 spermatozoa/sample and analyze for their fluorescence. Plot emission data on dot-plot cytograms and draw amorphous regions around individual populations to quantify live (green fluorescence), dead (red fluorescence), or moribund (dual-fluorescence) spermatozoa populations (*see* **Fig. 3**).

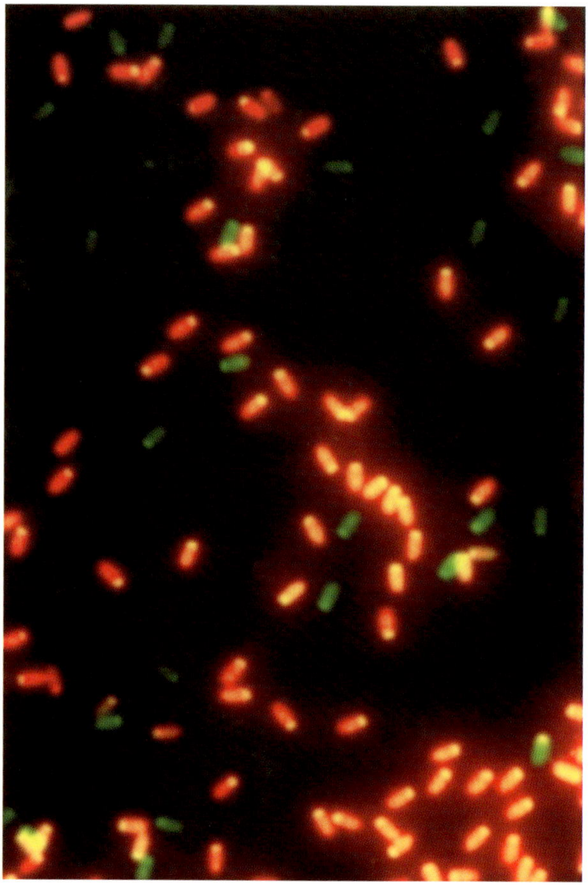

Fig. 2. Stallion sperm stained with SYBR-14 and PI, incubated 15 min (25°C), and viewed with a fluorescence microscope (magnification: 400×). Sperm stained with SYBR-14 appear green and are considered live, whereas sperm stained with PI appear red and are considered dead.

3.3.3. Hoeschst

Incubate fresh or frozen-thawed stallion spermatozoa with the fluorescent dye H33258. Mix sperm suspensions (100 µL; 20 × 10^6 sperm/mL) 1:1 (vol:vol) with prepared H33258 (*see* **Subheading 2.3., step 3**) and incubate in the dark (*see* **Note 2**) for 3 min at room temperature. Determine viability immediately after incubation. One operator counts at least 2 × 100 sperm (*see* **Note 11**) using a fluorescence microscope with excitation at 365 nm and emission at 435 nm; sometimes in order to see unstained sperm, you must first

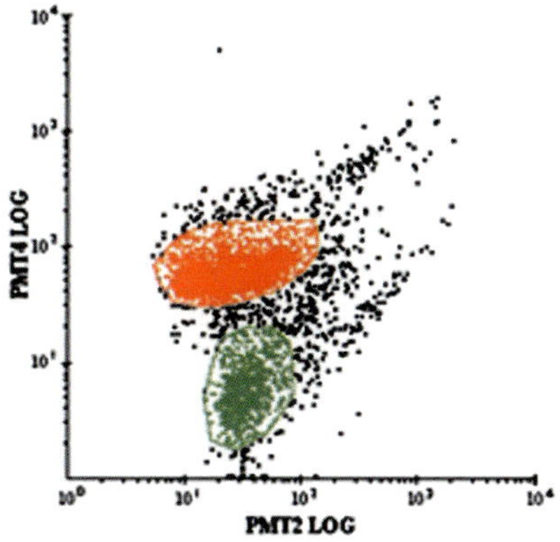

Fig. 3. Flow cytometer dot plot of stallion sperm stained with SYBR-14 and PI. At least 10,000 events were recorded, and the log of fluorescence intensities were plotted. Amorphous regions were drawn to depict dead (red) and live (green) sperm populations.

look at a field under fluorescence and then with bright field. Categorize spermatozoa as live (unstained) or dead (dark) (*see* **Fig. 4**).

4. Notes

1. Eggs should be very fresh for best effect. To minimize the chances of contamination of the extender, sanitize the shells by wiping with 70% alcohol. Air-dry and then separate yolk from white. Complete separation by rolling egg yolk on clean filter paper. Puncture the membrane around the yolk and pour the yolk into a graduated cylinder or other measuring device.
2. Note that SYBR-14, PI, and H33258 are all fluorescent dyes and must be stored and worked with in the dark to prevent photobleaching. Wrapping containers in aluminum foil is an effective way of keeping them dark. When actually working with the fluorescent material, try to work in dim conditions—we usually have a light on in one corner of the room that is away from our working area.
3. Stallions are very individual in their preferences for working with an artificial vagina. When training a stallion to collect, a variety of temperatures can be tested to determine what temperature a particular stallion prefers. The procedure for collection can directly affect the quality of the semen sample, and if the stallion is not completely comfortable with the temperature in the artificial vagina, he may produce a poor quality sample.
4. It is very important to load the chambers of the hemocytometer according to the manufacturer's directions in order to get an accurate count.
5. Straws may be filled in a number of ways. If only freezing a small number of straws, attach a small bore syringe to a short length of plastic tubing, put the other end of the tubing over the cotton end of the straw, and pull back on the syringe plunger to fill the straw, including the cotton plug. If processing a

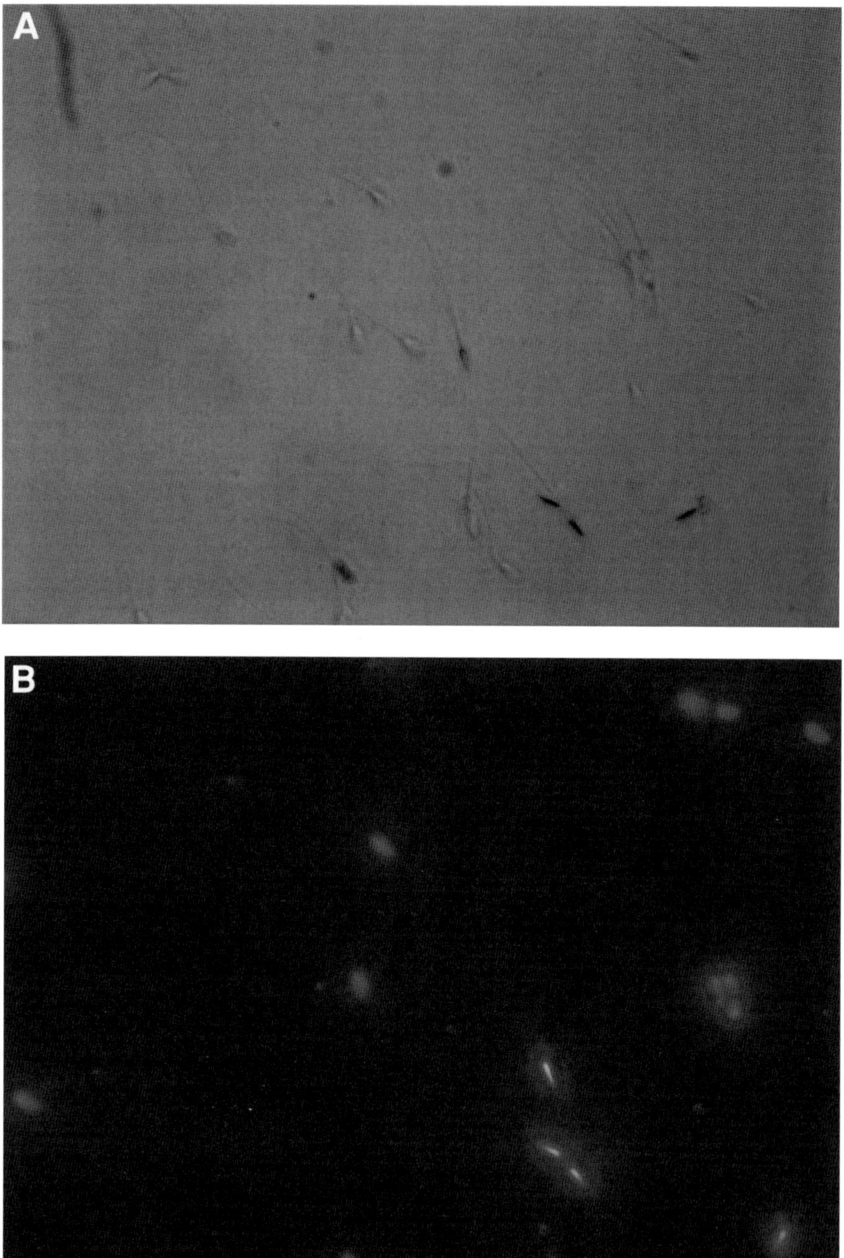

Fig. 4. Stallion sperm stained with H33258, incubated for 3 min (25°C) and viewed with a fluorescence microscope (magnification: 400×) and then with bright field with phase contrast. All sperm can be seen in the upper (phase-contrast) view. Membrane-compromised sperm (dead) fluoresce in the lower (fluorescence) view; live sperm do not fluoresce.

large number of straws, a vacuum pump or an automated straw filler is much more efficient. These may be designed to fill and seal at the same time. If sealing straws manually, dip the open end of straws in polyvinyl chloride powder to form a plug and immerse only the plug end in water (at the temperature of the semen and not above the level of the plug) to seal.

6. Traditionally, stallion semen cryopreserved in 0.5 mL straws has been thawed using a 37°C water bath. However, recent research in our lab has shown an improvement in sperm quality when straws are thawed in a 70°C water bath for 6 s ***(19)***.
7. The best smears are from small drops that give a thin film over most of the length of the slide and should dry quickly. If the sperm are very widely dispersed, double the volume of semen added to the eosin–nigrosin (i.e., 20 µL sperm to 100 µL eosin–nigrosin).
8. Eosin–nigrosin slides that have been prepared under mountant such as Entellan may be stored at –20°C and scored up to 2 wk later. In the absence of a mountant, they should be scored within 24–36 h.
9. A blue filter may help in scoring sperm stained with eosin–nigrosin, particularly when using a milk-based extender such as EZ Mixin. Count sperm in an area where the background is very pale or almost invisible, which is what this technique should produce over most of the slide. When the background is pale, it is easy to determine if the sperm is colored a deep purple-pink (dead) or is pale and white (alive).
10. Acrosomal integrity may be assessed at the same time. Characterization of structural integrity gives an additional quality parameter. It is best to use phase-contrast microscopy at 1000× magnification when assessing acrosomal integrity. Normal cells will appear round and smooth, whereas compromised cells will show evidence of flattening, denting, expansion, or breakdown of acrosomal membranes.
11. Individual people always seem to assess sperm somewhat differently. There is less individual variation with the clearly different red/green differentiation of SYBR-14/PI than with eosin–nigrosin, but, nonetheless, there is almost always operator-to-operator variation. Therefore, to have consistent assessment of viability for all semen samples graded within any one laboratory, or within a series of experiments, or within one veterinary practice, it is advisable to assign the counting to one individual.

References

1. Merkies, K. and Buhr, M. M. (1998) Epididymal maturation affects calcium regulation in equine spermatozoa exposed to heparin and glucose. *Theriogenology* **49,** 683–695.
2. Rousset, H., Chanteloube, P. H., Magistrini, M., et al. (1987) Assessment of fertility and semen evaluation of stallions. *J. Reprod. Fertil.* **35(Suppl.),** 25–31.
3. Pattie, W. A. and Dowsett, K. F. (1982) The repeatability of seminal characteristics of stallions. *J. Reprod. Fertil.* **32(Suppl.),** 9–13.

4. Pickett, B. W. and Amann, R. P. (1993) Cryopreservation of semen, in *Equine Reproduction* (McKinnon, A. O. and Voss, J. L., eds.), Lea and Feibiger, Philadelphia, pp. 769–789.
5. Katila, T. (2001) In vitro evaluation of frozen-thawed stallion semen: a review. *Acta. Vet. Scand.* **42(2),** 199–217.
6. Graham, J. K. (2001) Assessment of sperm quality: a flow cytometric approach. *Anim. Reprod. Sci.* **68(3–4),** 239–247.
7. Heitland, A. V., Jasko, D. J., Graham, J. K., et. (1995) Motility and fertility of stallion spermatozoa cooled and frozen in a modified skim milk extender containing egg yolk and liposome. *Biol. Reprod. Monogr.* **1,** 753–759.
8. Voss, J. L., Pickett, B. W., and Squires, E. L. (1981) Stallion spermatozoal morphology and motility and their relationship to fertility. *J. Am. Vet. Med. Assoc.* **178,** 287–289.
9. Casey, P. J., Hillman, R. B., Robertson, K. R., et al. (1993) Validation of an acrosomal stain for equine sperm that differentiates between living and dead sperm. *J. Androl.* **14,** 289–298.
10. Blach, E. L., Amann, R. P., Bowen, R. A., et al. (1988) Use of a monoclonal antibody to evaluate integrity of the plasma membrane of stallion sperm. *Gamete Res.* **21,** 233–241.
11. Janett, F., Thun, R., Ryhiner, A., et al. (2001) Influence of Eqvalan (ivermectin) on quality and freezability of stallion semen. *Theriogenology* **55(3),** 785–792.
12. Ozaki, T., Takahashi, K., Kanasaki, H., et al. (2002) Evaluation of acrosome reaction and viability of human sperm with two fluorescent dyes. *Arch. Gynecol. Obstet.* **266(2),** 114–177.
13. Pintado, B., de la Fuente, J., and Roldan, R. S. (2000) Permeability of boar and bull spermatozoa to the nucleic acid stains propidium iodide or Hoechst 33258, or to eosin: accuracy in the assessment of cell viability. *J. Reprod. Fertil.* **118,** 145–152.
14. Papaioannou, K. Z., Murphy, R. P., Monks, R. S., et al. (1997) Assessment of viability and mitochondrial function of equine spermatozoa using double staining and flow cytometry. *Theriogenology* **48,** 299–312.
15. Graham, J. K. and Squires, E. L. (1996) Comparison of the fertility of cryopreserved stallion spermatozoa with sperm motion analyses, flow cytometric evaluation, and zona-free hamster oocyte penetration. *Theriogenology* **46,** 559–578.
16. Garner, D. L., Johnson, L. A., Yule, S. T., et al. (1994) Dual DNA staining assessment of bovine sperm viability using SYBR-14 and propidium iodide. *J. Androl.* **15,** 620–629.
17. Barth, A. D. and Oko, R. J. (1989) Abnormal Morphology of Bovine Spermatozoa. Iowa State University Press, Ames, IA.
18. Merkies, K., Chenier, T., Plante, C., et al. (2000) Assessment of stallion spermatozoa viability by flow cytometry and light microscope analysis. *Theriogenology* **54,** 1215–1224.
19. Bradford, L. L. and Buhr, M. M. (2003) Function of cryopreserved horse semen is improved by optimized thawing rates. *Equine Vet. Sci.* **22,** 546–550.

5

Visualization of Sperm Accessory Structures in the Mammalian Spermatids, Spermatozoa, and Zygotes by Immunofluorescence, Confocal, and Immunoelectron Microscopy

Peter Sutovsky

1. Introduction

1.1. Importance of Studying the Fate of Sperm Antigens and Accessory Structures After Fertilization

Mammalian fertilization is traditionally viewed as a process during which the fertilizing spermatozoon penetrates the egg vestments to plant its chromosomes into the fertile environment of oocyte cytoplasm. Much less attention has been paid to the events that occur between gamete fusion and first embryonic cleavage. The interaction of the mammalian spermatozoa with the oocyte during and after gamete fusion is a complex and meticulously orchestrated cascade of events that can come to a halt at any given stage, resulting in pathological conditions. Researching these "late" events of fertilization and pronuclear development is a major challenge, which will be undoubtedly rewarded by new knowledge and, in the long term, by the improved treatments of infertility in humans and improved reproductive performance in farm animals. The fate of various sperm accessory structures came under scrutiny recently, as it became obvious that in addition to the sperm-borne chromosomes, other structures of the fertilizing spermatozoon make important contributions to the mammalian zygote *(1)*. Yet other sperm accessory structures are degraded in an orderly fashion as to not interfere with normal embryo development. These include the sperm proximal centriole, perinuclear theca, sperm mitochondria, and axonemal fibrous sheath and outer dense fibers.

From: *Methods in Molecular Biology, vol. 253: Germ Cell Protocols: Vol. 1 Sperm and Oocyte Analysis*
Edited by: H. Schatten © Humana Press Inc., Totowa, NJ

In most mammals, except rodents ***(2)***, the spermatozoon contains a reduced, inactive form of the *centrosome*, within which one of the two centrioles as well as the entourage of the pericentriolar material are degraded during the final stages of spermiogenesis ***(3)***. Such an incomplete centrosome, consisting of a proximal centriole embedded in the dense mass of sperm tail capitulum, must be released into oocyte cytoplasm at fertilization in order to attract microtubule-nucleating pericentriolar proteins from the surrounding oocyte cytoplasm. Failure of converting the reduced sperm centriole into such active zygotic centrosome may be a reason for postfertilization developmental arrests affecting couples treated at in vitro fertilization (IVF) clinics ***(4)***.

Sperm perinuclear theca (PT) is a skeletal capsule that envelopes the sperm nucleus and provides rigid support for the sperm head structure. In addition, PT may harbor molecules important for early development, such as signaling protein PT32, a candidate component of the sperm-borne oocyte-activating factor, transcription factor Stat 4, and histonelike proteins with a possible role in rebuilding the nuclear scaffold of the male pronucleus after fertilization (reviewed in **ref. *5***). PT is removed from the sperm nucleus at an early stage of sperm incorporation and dissolves in the oocyte cytoplasm concomitantly with oocyte activation ***(6,7)***. Further research is thus likely to provide support for the role of sperm PT in oocyte activation and initiation of normal pronuclear development. After intracytoplasmic sperm injection (ICSI), a technique of assisted fertilization in which the natural steps of sperm penetration, demembranation, and incorporation are bypassed, the redundant sperm PT may interfere with the development of the male pronucleus (PN). This likely lowers the success rate of assisted fertilization ***(8,9)***.

Sperm mitochondria, harboring their own mitochondrial genome, remain at the center of ongoing controversy. Despite the evidence for ubiquitin-dependent degradation of sperm mitochondria after fertilization ***(10,11)***, some authors maintain that paternal mitochondrial DNA (mtDNA) could escape and eventually recombine with maternal mtDNA, thus defying the axiom of strictly maternal mtDNA inheritance in humans/mammals. Because the evidence for such a recombination event has been considerably weakened by recent retractions and contradicting reports (reviewed in **ref. *12***), further research is needed. This knowledge could help redirect mitochondrial inheritance during cloning in farm animals and human-assisted reproduction techniques such as cytoplasm donation. Unlike the sperm-borne paternal mitochondria, the "foreign" mitochondria of somatic and embryonic cells used for nuclear transfer and the mitochondria of donor oocyte cytoplasm are not prone to rapid degradation inside the reconstructed embryo.

Flagellar fibrous sheath (FS) and outer dense fibers (ODFs) are intimately apposed within the sperm flagellum; yet, their postfertilization fates are different. Whereas FS dissolves in the oocyte cytoplasm almost immediately after sperm tail incorporation, ODFs remain intact throughout pronuclear and early embryonic development ***(6)***. The reasons for and significance of such differential processing are not known and warrant further research.

Other developmental and clinical considerations are as follows: Are male and female pronuclei (PN) equal in their protein makeup, developmental programming, and uptake of molecules from zygotic cytoplasm? To resolve such dilemmas, reliable tools are needed to distinguish between male and female PN. This can be accomplished by the tagging of sperm mitochondria with MitoTracker probes prior to IVF (**Subheading 3.1.**), which renders them fluorescent and detectable inside the fertilized oocyte. As the sperm mitochondria remain associated with the male PN throughout PN development, it is easy to determine which PN is the male, sperm-derived one, whether any of the examined zygotes are parthenogenetic, or if the multiple PN seen inside some zygotes result from polyspermy. Sperm acrosome and plasma membrane do not enter the oocyte cytoplasm during natural fertilization. Are they processed properly and in a timely fashion when an intact spermatozoon is injected into oocyte cytoplasm during the ICSI procedure? Detection of acrosomal antigens, sperm surface proteins, and PT components will help to answer these questions.

The following protocols are designed to provide a detailed description of how to perform a parallel, multichannel immunodetection and imaging of sperm and egg components by epifluorescence or confocal microscopy (*see* **Note 1**) and how to obtain corresponding detection at the ultrastructural level by colloidal gold immunocytochemistry. The chapter focuses on the immunocytochemical techniques rather than on the methods for in vitro fertilization and imaging, which are described in the literature. Therefore, **Subheading 3.1.** will introduce readers to a simple technique for vital labeling of the sperm mitochondria with a vital, fixable, mitochondrion-specific fluorescent probe, MitoTracker. This procedure renders the sperm mitochondria detectable inside the fertilized eggs, as the MitoTracker probes are relatively stable and resistant to permeabilization or extraction. In **Subheading 3.2.**, processing of spermatozoa and spermatogenic cells will be introduced, and the detection of sperm structures inside the fertilized egg will be described in **Subheading 3.3.** Finally, **Subheading 3.4.** will demonstrate how to adapt Protocols 3.2 and 3.3 for pre-embedding colloidal gold labeling, detected by electron microscopy. Images collected by epifluorescence, confocal, or electron microscopy can be processed using common image-editing software. Adobe Photoshop (version 5.0 or later; Adobe Systems, Mountain View, CA) is an excellent choice. It is to be emphasized that after image processing, the images should remain faithful to the original, unedited raw data, and such (raw) data should be stored and made available for peer review.

2. Materials

2.1. Media/Chemicals

1. Sperm TL or TALP-HEPES.
2. KMT medium: 100 m*M* KCl, 2 m*M* $MgCl_2$, 10 m*M* Tris-HCl; adjust pH to 7.0.
3. Dimethyl sulfoxide (DMSO; Sigma).
4. Normal goat serum (NGS).
5. Phosphate-buffered saline (PBS): 8 g NaCl, 0.2 g KCl, 0.26 g K_2HPO_4, 0.5 g NaN_3 in 1 L of ultrapure water, pH 7.2.
6. Formaldehyde (ultrapure, methanol-free, EM grade, 10% solution).
7. Paraformaldehyde (EM grade 4%).
8. Glutaraldehyde (EM grade).
9. Triton X-100.
10. Poly-L-lysine (MW 300,000; Sigma).
11. VectaShield (Vector Labs, Burlingame, CA) or similar mounting medium.
12. Hyaluronidase (Sigma).
13. Pronase/protease (Sigma).
14. Polyvinyl pyrrolidone (PVP; MW 40,000; Sigma).
15. MitoTracker Green FM (Molecular Probes, Eugene, OR).
16. Erythrocyte-lysing buffer: 155 m*M* NH_4Cl, 10 m*M* $KHCO_3$, 2 m*M* EDTA, pH 7.2, filtered into a sterile bottle.

2.2. Antibodies

1. Goat anti-rabbit immunoglobulins (IgGs) conjugated with FITC, TRITC, or Cy5.
2. Goat anti-mouse IgGs and IgMs, conjugated with FITC, TRITC, or Cy5.
3. DAPI or Hoechst 33258; DNA stains.
4. Colloidal gold conjugated anti-mouse IgGs.

2.3. Consumables

1. Regular and four-well Petri dishes (both 10 cm).
2. Microscopy slides.
3. Microscopy coverslips (22 × 22 mm; 18 × 18 mm).
4. 15-mL centrifuge tubes (Falcon).
5. Clear nail polish.
6. 53-μm Spectra mesh (Spectrum Labs).
7. 1.5-mL Eppendorf tubes.
8. Lint-free lens paper (Ross Optical Lens Tissue; A. Rosmarin Co., Plano, TX).

2.4. Equipment

1. Benchtop clinical centrifuge.
2. Slide warmer.
3. Epifluorescence microscope.
4. 20-, 200-, and 1000-μL pipetters (Gilson or similar).

5. Timers.
6. Vortex or similar shaker/mixer.
7. Incubator (39°C, 5% CO_2).

2.5. Tools

1. Scalpels and blades.
2. Fine tweezers.
3. Scissors.
4. Sterile glassware and funnels.
5. Nine-well glass plates (Pyrex brand; Fischer cat. no. 13-748B, or similar).

3. Methods

3.1. MitoTracker Labeling of Sperm Mitochondria

Bull sperm mitochondria are incubated with a vital, fixable, mitochondrion-specific fluorescent probe MitoTracker Green FM. The labeled spermatozoa can be fixed, permeabilized, and immunostained without loss of MitoTracker labeling, as MitoTracker dyes are incorporated mainly into the detergent-resistant mitochondrial membranes. Such prelabeling allows for easy colocalization of sperm proteins with the mitochondrial sheath in the isolated spermatozoa (**Subheading 3.2.**) or for the studies of interactions between the sperm mitochondria and oocyte-derived proteins and organelles after fertilization (**Subheading 3.3.**) in bovine and other mammalian species (*see* **Note 2**).

1. Prepare 1 m*M* primary stock of green fluorescent MitoTracker Green FM by adding 74 μL of DMSO in an original 50 μg vial of MitoTracker. Stock can be kept in dark at 4°C for several months with repeated freeze–thawing. For MitoTracker CMTM Ros or other variants, the amount of DMSO must be recalculated, as their respective molecular weights vary from that of MitoTracker Green FM.
2. Prepare the sperm pellet by simple centrifugation in sperm medium [e.g., Sperm TL *(13)*]. Sperm can also be prepared by gradient separation on Percoll or similar separation medium, or by a swim-up technique with subsequent centrifugation in Sperm TL or similar sperm preparation medium.
3. Prepare the secondary stock of MitoTracker by diluting 5 μL of 1 m*M* MitoTracker stock in 45 μL of KMT medium in 1-mL Eppendorf tube, vortex for 10 s, and wrap in aluminum foil to prevent fading. Use the secondary stock as soon as possible. Do not store and reuse this stock.
4. Dilute the sperm pellet in 1 mL of sperm TL or similar HEPES-buffered sperm prep medium; then, add the desired amount of MitoTracker to the final concentration of 100–500 n*M* (e.g., 4 μL of MitoTracker Green FM stock in 1 mL of Sperm TL to obtain 400 n*M* concentration). Incubate for 10 min at 37°C; then, mix with 10 mL of SpermTL and wash by centrifugation.
5. Wash the sperm by repeated pelleting and resuspending in Sperm TL.
6. (*Optional*) Determine sperm count and add the labeled sperm into the drops of fertilization medium with oocytes, to the desired sperm concentration. At the

desired time-point after insemination, fix the eggs for 30 min in 2% formaldehyde in PBS. Other aldehyde fixatives such as 4% paraformaldehyde in PBS are also suitable. Permeabilize cells with 0.1% Triton X-100 in PBS and process with an antibody and/or a DNA dye of your choice (*see* **Subheading 3.3.**). Sperm alone can be fixed and processed for immunofluorescence as described below or observed alive without fixation in epifluorescence (vital dyes such as Hoechst 33342 can be used to stain live sperm's DNA).

3.2. Multichannel Immunofluorescence Imaging of Sperm Accessory Structures and Proteins in the Spermatids and Spermatozoa

Isolated, fixed, and permeabilized (optional, not necessary for the detection of sperm surface antigens) mammalian spermatozoa and spermatogenic cells are processed with specific antibodies and appropriate fluorescent conjugates for localization of proteins in the sperm accessory structure. Cells are attached to poly-L-lysine-coated coverslips for easy handling, fixed, and permeabilized to facilitate the penetration of antibodies and their binding to intracellular epitopes. Nonspecific antibody-binding sites resulting from residual aldehyde groups are blocked by preincubation with concentrated 50–100 μg/mL goat, rabbit, or bovine serum. Multichannel imaging (*see* **Fig. 2A**) will be described in general terms, without specifying primary antibodies. It is assumed that: A=mouse monoclonal antibody against protein "A" (visualized by the red fluorescent secondary antibody, anti-mouse IgG–TRITC), B=rabbit polyclonal antibody against protein "B" (visualized by far-red fluorescent secondary antibody, anti-rabbit IgG–Cy5), C=MitoTracker Green FM (green fluorescent), D=DAPI or Hoechst 33342 (blue fluorescent DNA stains). Epifluorescence microscopes with appropriate filter sets or confocal microscopes with appropriate lasers are used for imaging. Combinations of primary antibodies and fluorescent conjugates can be changed according to available microscope filter sets. As some confocal systems are not equipped with ultraviolet (UV) lasers for excitation of blue fluorescent dyes (DAPI, H33342, H33258), alternative green/red fluorescent DNA stain such as propidium iodide (PI) or SYTO (Molecular Probes, Eugene, OR) can be used.

3.2.1. Isolation and Fixation of Bovine Spermatogenic Cells

1. Using a sterile scalpel, make a 3-cm-deep incision on the lateral side of the testis and remove a piece of testicular tissue without tunica (*see* **Note 3**), of approximate size 2 × 2 × 1 cm.
2. Gently mince the tissue with two pairs of fine tweezers in a large (10-cm) Petri dish filled with a 37°C warm Sperm TL or TALP-HEPES medium ***(14)***. This should be performed on a 37°C warm plate/slide warmer.

3. Discard the minced tissue, collect the medium with cells and filter it through sterile 53-μm Spectra/Mesh (Spectrum Labs, Rancho Dominguez, CA). *Optional:* Contaminating red blood cells can be lysed by a 10-min incubation of the pellet in 5 mL of erythrocyte-lysing buffer. Pellet the cells by a 5-min-long centrifugation in a clinical desktop centrifuge at approx 500*g*.
4. Remove the supernatant and dilute the pellet in 500 μL of SpermTL or similar sperm medium.
5. Overlay microscopy coverslips (22 × 22 mm; precleaned with ethanol) with 500 μL of 1% poly-L-lysine (MW > 300,000; e.g., Sigma cat. no. P-1524) in ultrapure water; incubate for 5 min on a slide warmer (37°C) or overnight in a cuvet (4°C). Shake off lysine solution, blot the coated face of the coverslips with a lint-free lens paper, and let dry for 5 min on a slide warmer, lysine-coated face up ***(15)***.
6. Place each slide on a cross of a four-well Petri dish or six-well Costar/Falcon assay plate on a slide warmer (37ºC), coated face up, and overlay it with 500 μL of warm KMT medium (37°C; pH 7.0–7.3). Depose 5 μL of cell suspension onto each slide and incubate for 5 min to allow the cells to settle onto the poly-L-lysine layer.
7. Shake off the medium with excess cells and submerge the coverslip in 2% formaldehyde in PBS (pH 7.2–7.3) for 40 min (*see* **Note 3**). Best preservation of cell structure is achieved when the coverslips with cells are carefully transferred, coated face up, into a 10-cm Petri dish with 8 mL of 37°C warm PBS; 2 mL of 10% formaldehyde is slowly added to obtain final formaldehyde concentration of 2%.
8. Remove the coverslips with cells from the formaldehyde solution using fine pointed tweezers, wash briefly in PBS, and store in pure PBS or in PBS containing 0.1% Triton X-100 (Sigma; permeabilization agent).

3.2.2. Fixation of Ejaculated Spermatozoa

1. Thaw a straw of frozen bull semen and mix the content with 15 mL of SpermTL medium in a 15-mL centrifugation tube. *Optional:* Motile sperm fraction can be isolated by centrifugation through a two-layer, 45/90% Percoll gradient ***(13)***.
2. Remove supernatant and dilute the pellet in 100 μL of SpermTL.
3. Prepare the coverslips as described in **Subheading 3.2.**, overlay them with warm KMT, and add 3–5 μL of sperm pellet onto each coverslip.
4. Fix sperm as described in **Subheading 3.2.**

3.2.3. Immunofluorescence Processing

1. Permeabilize the cells on coverslip with MitoTracker tagged cells for 40 min at room temperature (RT) or overnight at 4°C in 0.1% Triton X-100 in 0.1 *M* PBS.
2. Place coverslips on the cross of a four-well Petri dish (**Fig. 1C**). Block nonspecific antibody binding to residual free aldehyde groups by 25–40 min incubation in 0.1 *M* PBS containing 5% NGS (Sigma) and 0.1% Triton X-100.
3. Prepare the labeling buffer for washing and antibody dilution by supplementing 0.1 *M* PBS with 0.1% Triton X-100 and 1% NGS. This buffer is used for all remaining steps of this protocol.

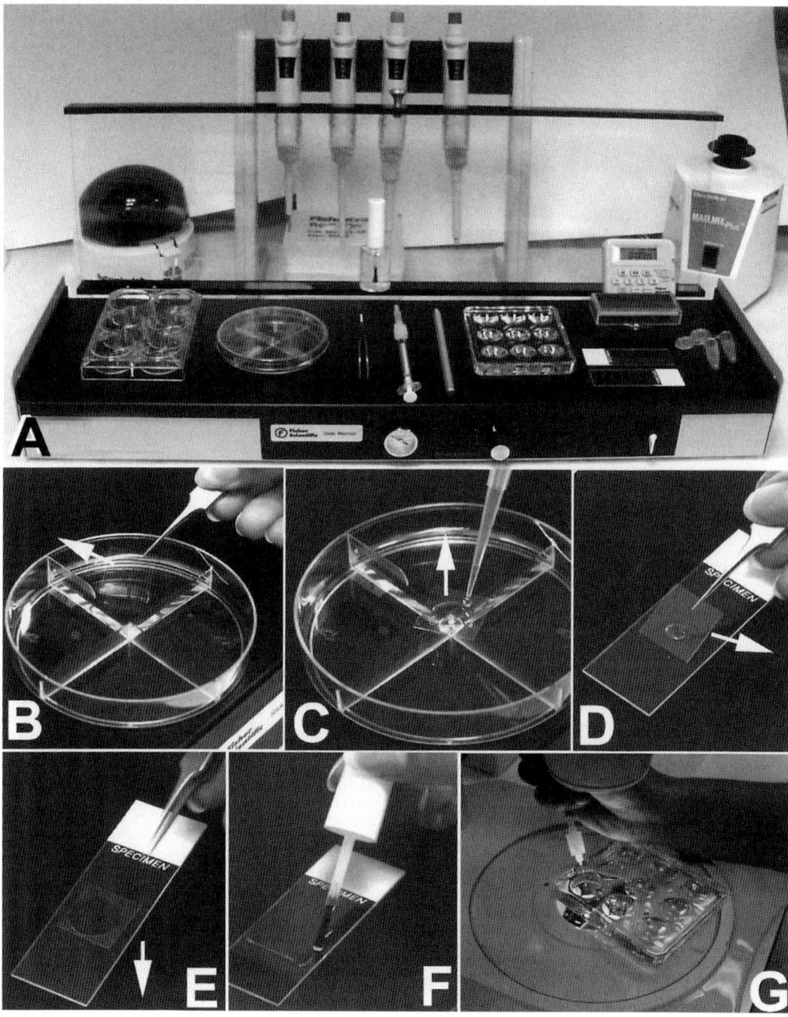

Fig. 1. Immunofluorescence processing of spermatozoa and oocytes. **(A)** Workstation with tools and supplies for immunofluorescence processing of mammalian oocytes. Left to right on top of the slide warmer: six-well assay dish, four-well Petri dish, fine forceps, 1-cm^3 syringe with Unapette pipet, diamond pen, nine-well glass plate, microscopy slides, coverslips, timer, and Eppendorf tubes. Behind the slide warmer, left to right: table-top microcentrifuge, pipetters, and vortex. **(B)** Washing coverslips in the buffer; arrow indicates the cell-coated side up. **(C)** Overlaying the coverslip (coated side up) with antibody solution. **(D,E)** Mounting the processed coverslip (arrow indicates the coated-side down) on the slide with VectaShield mounting medium. **(F)** Sealing the coverslip with clear nail polish. **(G)** Transferring the oocytes from well to well on a glass plate under the dissecting microscope.

4. Dilute antibodies A and B to desired concentration (determined for each antibody by serial dilution experiments) in 200 µL of labeling buffer. *Example:* If antibody A contains 1 mg/mL of immunoglobulin and works best at final concentration (in labeling buffer) of 5 µg/mL, and antibody B has 2 mg/mL of immunoglobulin and works best at 20 µg/mL, you need to combine 1 µL of antibody A with 2 µL of antibody B and 197 µL of the labeling buffer. The dilutions will therefore be as follows: antibody A, 1/200, antibody B, 1/100. Mix by vortexing in a 1.5-mL Eppendorf tube.
5. Shake and blot off the blocking solution. Place the coverslip back on the cross of the four-well Petri dish, overlay it with 200 µL of antibody solution, and incubate for 40 min at RT or on a slide warmer (raising temperature to 37°C may improve antibody binding to PT epitopes) or overnight at 4°C.
6. Wash coverslip by submerging it into pure labeling buffer in the first well of a four-well Petri dish for 5 min (*see* **Fig. 1B**). Prepare the secondary antibody solution, composed of red fluorescent, FITC-conjugated anti-mouse IgG (dilution: 1/100–1/50) and a far-red fluorescent, Cy5-conjugated anti-rabbit IgG. The colors can be switched or omitted if only one antibody is used. If the cells are not tagged with MitoTracker, both antibodies (A and B) can be detected by fluorescent conjugates emitting in the visible-light spectrum: The wavelength of Cy5 emission (>630 nm) is, in most part, beyond the capability of the human eye and can only be captured by a digital camera or on the film. It follows that a Cy5 conjugate should be used as a second probe, so that the appropriate focal plane can be determined by focusing on the labeling in the visible-light spectrum.
7. Add 2.5 µL (1/80 dilution) of 200 µg/mL DAPI stock (4',6-diamidino-2-phenylidone; blue fluorescent DNA stain from Sigma or Molecular Probes) into secondary antibody solution. DAPI stock of 200 µg/mL is prepared in ultrapure water, as PBS would cause precipitation. A tube of DAPI stock can be kept for several months at 4°C, shielded from light by wrapping in aluminum foil, and used as needed.
8. Overlay the washed coverslip with 200 µL of secondary antibody solution and incubate for 40 min at RT or on a slide warmer.
9. Wash coverslips in labeling solution; blot excess buffer with lens paper. Be sure not to smear the coated face of the coverslip.
10. Mount the coverslip on a standard microscopy slide in a 10 µL drop of VectaShield or similar mounting medium (*see* **Fig. 1D,E**). Seal all four sides of the coverslip with clear nail polish (available at local drug stores) to prevent drying (*see* **Fig. 1F**).

3.3. Detection of Sperm Proteins and Accessory Structures Inside the Fertilized Zygote

Mammalian oocytes are fertilized with the MitoTracker-tagged spermatozoa, fixed, permeabilized, and processed with desired specific antibodies and appropriate fluorescently conjugated secondary, species-specific immunoglobulins (*see* **Note 4**). The resultant multichannel labeling (**Fig. 2B**) detects the sperm tail mitochondria inside the egg simultaneously with the detection of

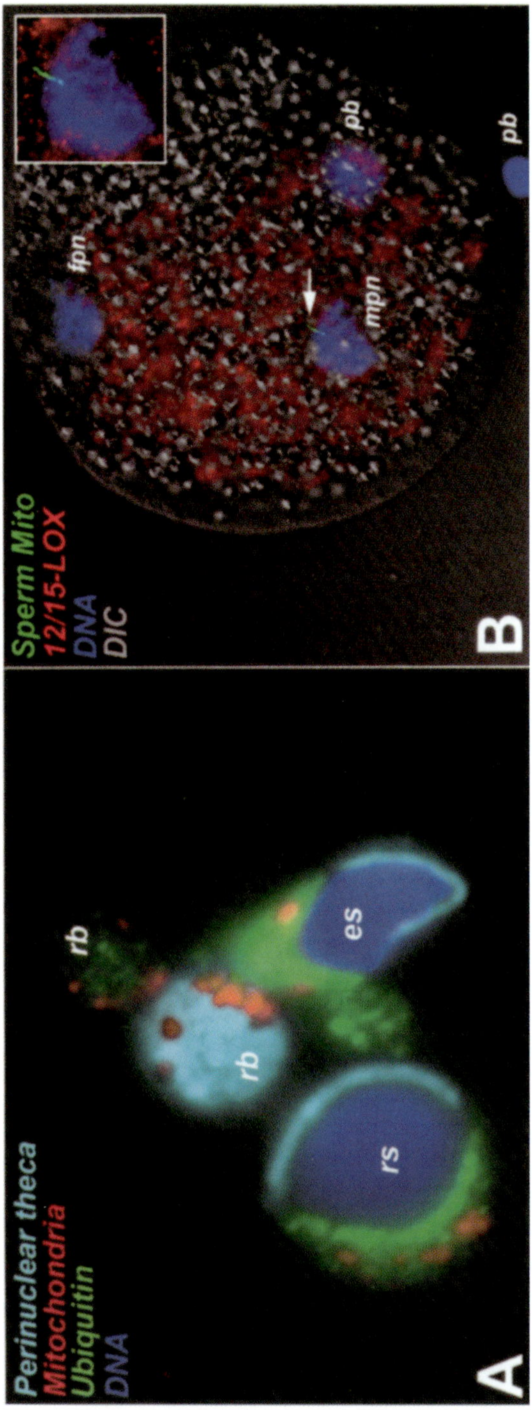
Perinuclear theca
Mitochondria
Ubiquitin
DNA
A
rb
rb
rs
es
Sperm Mito
12/15-LOX
DNA
DIC
B
fpn
mpn
pb
pb

Fig. 2.

DNA and one or two sperm/oocyte antigens. In vitro fertilization techniques and timing of early development after fertilization vary from species to species and are not covered here. MitoTracker-labeled spermatozoa have been used successfully for fertilization in cattle, pig, rhesus monkey, mouse, and humans. Timing of fixation depends on the scope of particular project and on the species. For example, when studying early stages of fertilization and sperm incorporation into bovine oocyte cytoplasm, zygotes are fixed 6–10 h after insemination.

Fig. 2. *(previous page)* Detection of proteins, DNA and sperm organelles in the mammalian spermatogenic cells and zygotes. **(A)** Multichannel labeling of rhesus monkey spermatogenic cells (rs = round spermatid, es = elongated spermatid, rb = residual body). Testicular cell sample was isolated, prelabeled with MitoTracker CMTM Ros (red; Molecular Probes, cat. no. M-7510) and processed as described in **Subheading 3.2.** Green fluorescence is the labeling of ubiquitin, a proteolytic marker peptide that accumulates in the cytoplasmic lobe of spermatids (adjacent to distal end of the spermatid nucleus) and in the residual bodies, representing the redundant cytoplasm rejected by the spermatid during its elongation. Anti-ubiquitin KM691 (dilution 1/100; Kamiya Biomedical Company, Seattle, WA; cat. no. MC-033) was followed by goat–anti-mouse IgM–FITC (dilution 1/80; Zymed Labs; S San Francisco, CA; cat. no. 62-6811). Perinuclear theca (turquoise; far-red fluorescence), the cytoskeletal capsule of the differentiating sperm nucleus, was labeled with antibody PT427 (dilution 1/200; rabbit serum kindly provided by Dr. Richard Oko, Queens University, Kingston, ON), followed by goat-anti rabbit–Cy5 (1/50; Jackson Immunochemicals, West Grove, PA; cat./code no. 111-175-144). DNA (blue) was counterstained with DAPI (2.5 μg/mL in secondary antibody mix; Molecular Probes; D-3571). Each channel (blue, green, red, far-red) was recorded as a gray-scale image on Nikon Eclipse 800 microscope with CoolSnap CCD camera and MetaMorph 4.6 software (Universal Imaging Corp., Downington, PA) in the identical focal plane. Images were pseudocolored and overlapped using Adobe Photoshop 5.5 software (Adobe Systems, Mountain View, CA). **(B)** Multichannel labeling of pig zygote fertilized with MitoTracker Green FM–tagged spermatozoon (green). Oocytes were fixed in 2% formaldehyde, permeabilized with 0.1% Triton X-100, blocked with 5% NGS, and processed sequentially with antibody against 12/15-lipoxygenase (red, 12/15 LOX) and goat–anti-rabbit IgG–TRITC (1/80; Zymed). DNA (blue) was counterstained with 2.5 μg/mL DAPI. Fluorescent channels were recorded as described in **(A)**, merged, and superimposed directly by MetaMorph. Resultant three-channel RGB image was superimposed onto the corresponding differential interference contrast (DIC) image in Adobe Photoshop. Inset shows detail of the male pronucleus (mpn) with the sperm tail mitochondria (green). Antibody against C-terminus of 12/15-LOX, an enzyme known to participate in the degradation of the mitochondria inside the differentiating reticulocytes, was kindly provided by Dr. Klaus Van Leyen, (Sloan Kettering Memorial Institute, New York, NY). fpn = female pronucleus, pb = polar body. One of the polar bodies is out of focus, thus resembling a pronucleus.

For the pronuclear apposition stage, fixation is recommended at 10–16 h after gamete mixing. For the late stages (S-phase, first meiotic prophase), zygotes are fixed 16–24 h after insemination, for two-cell embryos, 24–30 h after insemination, and for blastocysts, d 5–6.

3.3.1. Fixation of Oocytes and Embryos for Immunofluorescence

1. Warm up 1 mL of isotonic PBS in an Eppendorf tube. Prewarm the glass plate with nine cavities on a slide warmer set to 37°C.
2. Using a 1-mL syringe with a Unopette glass pipet, remove oocytes from the fertilization drops at desired time-points, wash in TL-HEPES, and strip cumulus cells by repeated pipetting of the eggs. For unfertilized oocytes, use 5–10 min incubation with 0.5% hyaluronidase (Sigma), added directly to maturation/fertilization drops, prior to stripping of cumulus cells.
3. Remove zona pellucida by short (1–2 min) incubation in TALP-HEPES with 0.5% pronase. Wash oocytes in a serum-free TALP-HEPES medium supplemented with 0.5% PVP (MW 40,000; Sigma).
4. Fill up the first well of a warmed-up, nine-well glass plate with 400 μL of 37°C warm PBS. Carefully transfer the oocytes into the PBS-filled well using a Unopette pipet (**Fig. 1G**).
5. Using a pipetter, slowly add 100 μL of 10% formaldehyde (*see* **Note 4**) into the well with oocytes. This will reduce the concentration of formaldehyde to 2%. Fix for 40 min at RT.
6. After fixation, carefully remove the oocytes and wash them twice in PBS (room temperature) on the same plate used for fixation. Transfer washed oocytes into 400 μL of pure PBS in a clean nine-well plate and seal the plate tightly with Saran Wrap (available at drug/grocery stores) or similar transparent, adhesive kitchen wrap. Store oocytes at 4°C until the day of use. It is best to use (process) the fixed oocyte within 5–10 d, if not immediately, after fixation.

3.3.2. Immunofluorescence Labeling of Oocytes and Zygotes

1. Incubate the oocytes in PBS with 0.1% Triton X-100 for 40 min at RT. You can use the same nine-well plate used for oocytes storage. This will permeabilize the oocytes and make the epitopes inside of them accessible to antibodies. Triton X-100 will also be included in all of the following steps of labeling, to improve the penetration of antibodies. Processing is done by transferring oocytes using a Unopette pipet (Becton Dickinson, cat. no. 365878) from well to well on a nine-well glass plate. Mouth pipetters can also be used.
2. Place the oocytes in the second well with 400 μL of 0.1 *M* PBS containing 5% NGS and 0.1% Triton X-100 to block nonspecific antibody binding to free aldehyde groups. Incubate for 25–40 min.
3. Prepare the labeling buffer by supplementing 0.1 *M* PBS with 0.1% Triton X-100 and 1% NGS. This labeling buffer will be used for all remaining steps of this protocol.

4. Dilute the first antibodies in labeling buffer as described for sperm. Typical dilutions range from 1/500 to 1/10. Transfer 200 μL of first antibody (100 μL if a rare antibody is used) solution into a third well.
5. Transfer oocytes into the third well and incubate for 40 min at RT, on a slide warmer (temperature of 37°C may improve antibody binding to PT epitopes), or overnight at 4°C (overnight incubation allows for higher dilution, i.e., less antibody usage). Mix this and all other solutions using Vortex or similar device.
6. Wash oocytes by transferring them into 400 μL of labeling buffer for 5 min while preparing the second antibody. For 200 μL of antibody solution, include 2.5 μL (1/80 dilution) of 200-μg/mL stock of DAPI (blue fluorescent DNA stain from Sigma; stock is prepared in ultrapure water and buffers will precipitate DAPI) into the secondary antibody solution.
7. Place 200 μL of secondary antibody solution in the next well of a nine-well plate. Transfer the oocytes/embryos and incubate for 40 min at RT or on a slide warmer.
8. Wash oocytes in labeling buffer in the next well.
9. Mount the oocytes/embryos on a standard microscopy slide in a 10-μL drop of VectaShield (Vector Labs) or similar mounting medium and cover them very carefully with a standard microscopy coverslip (18 × 18 or 22 × 22 mm). It is not necessary to use a spacer between the slide and the coverslip. Seal all four sides of the coverslip with clear nail polish to prevent drying.

3.4. Pre-embedding Colloidal Gold Labeling of Sperm Accessory Structures in the Isolated Spermatozoa and Inside the Zygote

In this simple, feasible method of colloidal gold labeling, subcellular detection of sperm proteins is achieved by a modification of **Subheading 3.2.**, whereby the secondary, fluorescently conjugated antibody is replaced by the one conjugated to 5- to 15-nm colloidal gold particles. At a superior ultrastructural level of resolution, such labeling mirrors the data obtained by immunofluorescence and is an excellent, higher-resolution complement to it (**Fig. 3**). Multiple proteins (*see* **Note 5**) can be labeled by using primary antibodies from different species (e.g., a mouse IgG and a rabbit polyclonal) and secondary antibodies conjugated to colloidal gold of distinguishably different size (e.g., anti-rabbit IgG–5 nm gold and anti-mouse IgG–10 nm gold). Instead of attaching cells to lysine-coated coverslips; these are incubated with antibodies and sera in 15-mL Falcon tubes and collected/washed by centrifugation and resuspension in labeling buffer, using a hinged rotor benchtop centrifuge (e.g., Centra CL-2 from Fischer). After the processing, spermatozoa are collected by centrifugation, embedded in agar, and processed for electron microscopy as described previously ***(6,8)***. Briefly, 0.5–1 mL of hot agar is poured in the tip of a Falcon tube with the sperm pellet. After cooling down, the conical agar pellet is flushed out using PBS and a glass pipet or a metal needle necessary for

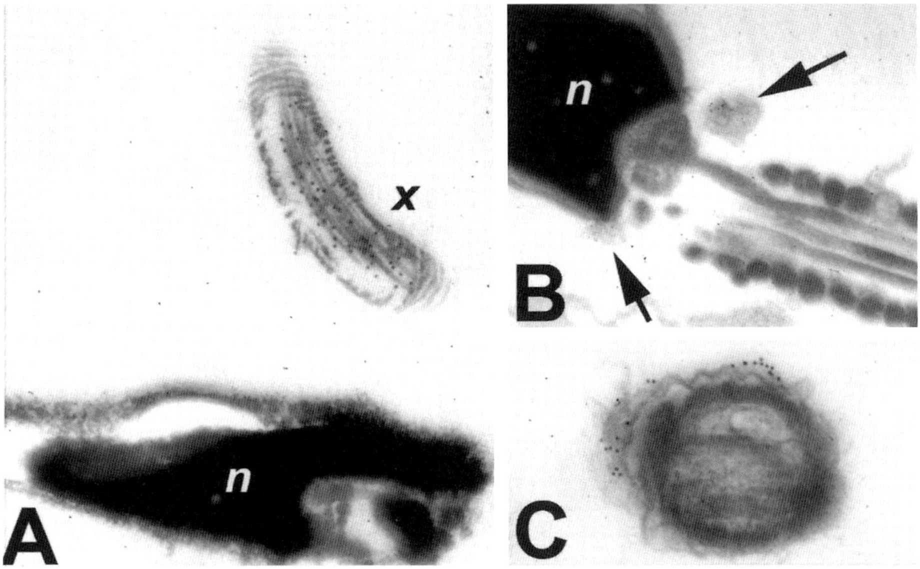

Fig. 3. Pre-embedding colloidal gold labeling of mammalian spermatozoa. **(A)** Formaldehyde-fixed and Triton-permeabilized bull spermatozoon. Primary antibody, anti-α-tubulin mouse IgG (dilution 1/200; Developmental Studies Hybridoma Bank, University of Iowa; cat. no./code E7); secondary antibody: goat anti-mouse IgG–15 nm gold (British BioCell/Ted Pella). Note that the microtubules in the principal piece of the sperm axoneme (x) are heavily decorated with gold particles, whereas there is virtually no labeling on the adjacent sperm head (n = nucleus), which does not contain tubulin. **(B,C)** Formaldehyde-fixed, unpermeabilized stallion spermatozoon (**B**; longitudinal section of the sperm head and sperm tail midpiece) and an unfixed stallion spermatozoon (**C**; cross-section of the sperm tail midpiece) were labeled with mouse IgM raised against human recombinant ubiquitin (dilution 1/100; KM-691, Kamiya Biomedical Company, Seattle, WA). Secondary antibody: goat anti-mouse IgM–12 nm gold (dilution 1/10; Jackson Immunochemicals, West Grove, PA). Stallion spermatozoa were collected and cryopreserved by Dr. Regina Turner, University of Pennsylvania.

detaching the agar from the tip of tube. The agar-embedded pellet is then postfixed in osmium tetroxide (optional, may obscure gold particles in some cases, especially if small 5-nm gold is used), washed, and dehydrated by an ascending ethanol series. The dehydrated pellet is perfused with a mixture of PolyBed 812 and propylene oxide, cut into small pieces and embedded in PolyBed 812, cured at 60°C. Ultrathin sections are prepared and examined in an electron microscope after staining with uranyl acetate (lead citrate staining used in conventional EM processing may be omitted so as to not obscure the colloidal gold particles). This part of the processing can be done by an experienced technician at any electron microscopy core facility and is thus not described in detail. The pre-embedding procedure is less tedious than the postembedding method of colloidal gold labeling, involving the labeling of ultrathin sections on EM grids, which have to be meticulously prescreened to contain suitable sections (i.e., those with the right part of the cell, which can be a problem when localizing very small structures such as centrosomes inside the fertilized egg). For oocyte labeling, the processing is identical to immunofluorescence, except for the substitution of fluorescent secondary antibodies by the colloidal-gold-conjugated ones.

1. Collect spermatozoa or spermatogenic cells as described in **Subheading 3.2.**
2. Fix cells by resuspending the collected pellet in 10 mL of 4% paraformaldehyde or 2% formaldehyde in a 15-mL Falcon tube. After 40 min of fixation, collect the cell pellet by a 5-min centrifugation at 300*g* to 500*g*. Supernatants are removed/discarded using a glass Pasteur pipet with a rubber bulb.
3. Wash cells in 10 mL of PBS, added to the tube with cell pellet. Collect pellet by a 5-min centrifugation.
4. Permeabilize the cells for 40 min at RT or overnight at 4°C in 0.1% Triton X-100 in 0.1 *M* PBS.
5. Remove permeabilization buffer by centrifugation. Resuspend pellet in 1 mL of 5% NGS in PBS with 0.1% Triton X-100 to prevent nonspecific antibody binding to residual, free-aldehyde groups. Incubate for 25–40 min and remove NGS by centrifugation.
6. Prepare labeling buffer by supplementing 0.1 *M* PBS with 0.1% Triton X-100 and 1% NGS. Dilute the primary antibody to desired concentration in 200–500 μL of labeling buffer and add it to the tube with the sperm pellet. Mix well by vortexing or repeated pipetting. Incubate for 40 min at RT or at 37°C, or overnight at 4°C. Gently shake the tube every 5–10 min to prevent sedimentation of the cell pellet and facilitate penetration of antibodies. This can also be achieved by placing the tube on an electric rocker set to a low speed.
7. Wash by adding 10 mL of labeling buffer and collect cells by centrifugation.
8. Prepare the appropriate colloidal-gold-conjugated secondary antibody (e.g., if the primary antibody is a mouse IgG, secondary will be a colloidal-gold-conjugated anti-mouse IgG, most often raised in the goat). Such antibodies are carried by a number of biomedical companies (e.g., Jackson Immunochemicals, British BioCell, Polyscience, Ted Pella Inc.). Whereas typical dilution of colloidal gold conjugates for postembedding immunolabeling is 1/60 to 1/30, the antibodies for pre-embedding may be used at a much higher concentration, diluted to 1/5-1/20. This gives stronger labeling while the background is very low as a result of a thorough washing in PBS after the labeling.
9. Incubation of sperm is performed in a Falcon tube for 1–2 h at RT or overnight at 4°C, with occasional gentle shaking. Incubation of oocytes can be performed in a nine-well glass plate as described for immunofluorescence.
10. Wash sperm by adding 10 mL of labeling buffer and centrifuge.

11. Fix cells with EM fixative [2.5% paraformaldehyde and 0.6% glutaraldehyde in cacodylate buffer, or similar glutaraldehyde based fixative ***(6,8)***].
12. Wash by resuspension in 10 mL of cacodylate buffer. Postfixation with osmium tetroxide, frequently used in EM protocols to improve membrane structure preservation, is optional, as it may obscure the colloidal gold particles.
13. Collect the pellet and embed it in agar. This step is not necessary for oocyte processing.
14. Electron microscopy processing is routinely performed by core facilities at most research institutions and therefore is not described in further detail.

4. Notes

1. The described techniques allow for the rapid detection of proteins and organelles in the differentiating and mature spermatogenic cells, as well as in the oocytes, zygotes, and preimplantation embryos up to the blastocyst stage. In addition to epifluorescence and confocal microscopy, these techniques can be adapted for electron microscopy and for flow cytometry. We had good success in combining the immunofluorescence technique with the detection of apoptotic cells with TUNEL assay of DNA fragmentation ***(16)***. In such assay, cells were fixed in formaldehyde and permeabilized with Triton X-100 as described here. The 1-h TUNEL processing was applied according to the manufacturer's protocol (enclosed in the package; In Vitro Apoptosis Kit; Boehringer); cells were washed in PBS and blocked and processed with antibodies as described here. Details of a similar technique are described elsewhere in this book. Our fluorescent techniques can be combined with any other organelle-specific, fixable probes (many are available from Molecular Probes, Eugene, OR). Readers are welcome to contact author with any questions or comments at SutovskyP@missouri.edu.
2. **Subheading 3.1.** has been successfully adapted for the labeling of bull, rhesus monkey, human, boar, and mouse spermatozoa. The only substantial modification is the use of sperm-processing medium appropriate for a given species. For IVF, optimal concentration of MitoTracker (typically 100–500 n*M*) has to be determined experimentally, as some types of MitoTracker conjugates may affect sperm motility and reduce fertilization rates.
3. In **Subheading 3.2.**, it is very important to remember that aldehydes are carcinogenic; thus, fixation must be performed under a fume hood or with a face mask with filters specifically formulated for protection against aldehyde vapors. It is crucial to always keep the coverslip with the cell-coated side up. Similar procedure can be used for the isolation of epididymal spermatozoa, wherein lesser mincing is necessary to release the spermatozoa into collection medium. An alternative to formaldehyde/paraformaldehyde fixation is fixation in 100% methanol for 10 min at –20°C, after which the coverslips are stored in pure PBS, and Triton permeabilization is not necessary. Acetone fixation is also permeabilizing and may be used in the same manner as methanol. However, these non-cross-linking fixations may reduce immunoreactivity of some sperm antigens and extract some cytosolic and membrane antigens completely.

The primary antibodies are often available in very limited quantities, in which case the amount of antibody solution can be reduced to 100 µL. An alternative to Triton permeabilization of aldehyde-fixed cells is saponin [5 mg/mL in PBS/labeling buffer ***(17)***], which may result in better preservation of cell structure. For blocking, 2% (v/v) nonfat, dry milk or 10% teleost fish gelatine (from Sigma) in PBS can be used instead of goat serum. Brief centrifugation in a benchtop microcentrifuge can be used to clear and concentrate the antibody stocks in the Eppendorf tubes. Negative controls include the omission of the first antibody (labeling buffer alone is used for the first antibody incubation) and the replacement of the first antibody with appropriate (i.e., rabbit with rabbit, rat with rat, goat with goat) nonimmune serum. Ideally, this "preimmune" serum should come from the same animal as the immune serum, being isolated prior to the first immunization. If such "normal" serum is not available, commercial nonimmune sera can be purchased. If a recombinant or purified protein, which was used as immunogen, is available, it can be used to saturate the antibody to provide an excellent negative control. By processing the cells in a solution, instead of attaching them to a coverslip, this technique can be easily adapted for flow cytometry ***(16,18)*** or for electron microscopy/colloidal gold labeling (*see* **Subheading 3.4.**).

4. In **Subheading 3.3.**, it is very important to remember that aldehydes are carcinogenic; fixation must be performed under a fume hood or with a face mask with filters specially formulated for protection against aldehyde vapors. An alternative for zona removal is incubation in the acid Tyrode (pH 2.0). Oocytes can be individually attached to the lysine-coated coverslips ***(15)*** in a manner similar to the processing of spermatozoa. This will facilitate the handling of large numbers of oocytes, wherein these can be moved from solution to solution very rapidly. Four-well or six-well dishes are used for such processing in a fashion similar to sperm processing. Placing individual zona-free oocytes on the lysine-coated surface of a coverslip, however, requires substantial skill, the oocytes may detach and the lysine layer may bind impurities from labeling solutions. Detachment of oocytes can be reduced by substituting 0.5% PVP for serum albumin in TALP-HEPES used for the oocyte wash. Increased background fluorescence may be caused by the binding of antibodies and debris to lysine layer. Permeabilization with saponin and fixation with ice-cold ethanol or acetone can be substituted as described for sperm. These nonaldehyde fixatives, however, may diminish the immunoreactivity of certain epitopes or remove it completely as they extract proteins from the cytosol.
5. Multiple types of labeling can be performed on the same sample. For example, one antibody can be a mouse IgG, visualized with green fluorescent anti-mouse IgG–FITC; the other antibody can be a rabbit polyclonal, visualized by a red fluorescent goat anti-rabbit–TRITC. Blue fluorescent DNA signal will be rendered by DAPI. Plates with oocytes can be covered by aluminum foil during the second antibody incubation to prevent photobleaching of fluorescent probes. Unapette pipets can be pulled over a burner and trimmed with a diamond pen to obtain a narrow tip with an opening only slightly larger than that of an oocyte.

This will help minimize the carryover of solutions during processing and mounting.

In **Subheading 3.4.**, electron microscopic processing of cells that were previously fixed and permeabilized for immunofluorescence may result in the loss of some antigens. Membrane structures on the surface and inside the cell may be destabilized by permeabilization and thus more prone to extraction during the dehydration by ethanol series, as used for EM. If the examined antigens are on the sperm surface, permeabilization (**step 4**) can be omitted (*see* **Fig. 3B**), thus preserving excellent ultrastructure. In some cases, both fixation and permeabilization (**steps 2–4**) can be omitted (*see* **Fig. 3C**) and live cells can be labeled with colloidal gold. In such case, the processing buffer is substituted by sperm-handling medium and the processing of live cells is done at 37°C. *See* **Note 3** for negative controls.

Acknowledgments

The protocols described here were developed during author's training in the laboratories of Dr. Jan Motlík, Dr. Jacques E Fléchon, and Dr. Gerald Schatten and are based on the protocols used in the laboratories of these respected scientists. The author is currently supported by grants from NIH/NIOSH, USDA, and F21C Program of the University of Missouri–Columbia. He wishes to thank his wife/assistant Miriam and current and former colleagues at the University of Missouri–Columbia (Columbia, MO), Institute of Animal Physiology (Libéchov, Czech Republic), INRA (Jouy-en-Josas, France), University of Wisconsin (Madison, WI), and Oregon Regional Primate Research Center (OHSU, Beaverton/Portland, OR). Special thanks belong to collaborators Dr. Regina Turner (University of Pennsylvania), Dr. Richard Oko (Queens University, Kingston, ON), and Dr. Klaus Van Leyen (Sloan Kettering Memorial Institute, New York, NY) and to many other collaborators in United States, Argentina, Australia, Canada, Japan, New Zealand, Slovakia, and United Kingdom.

References

1. Sutovsky, P. and Schatten, G. (2000) Paternal contributions to the mammalian zygote: fertilization after sperm-egg fusion. *Int. Rev. Cytol.* **195,** 1–65.
2. Schatten, H., Schatten, G., Mazia, D., et al. (1986) Behavior of centrosomes during fertilization and cell division in mouse oocytes and sea urchin eggs. *Proc. Natl. Acad. Sci. USA* **83,** 105–109.
3. Sutovsky, P., Manandhar, G., and Schatten, G. (1999) Biogenesis of the centrosome during mammalian gametogenesis and fertilization. *Protoplasma* **206,** 249–262.
4. Asch, R., Simerly, C. S., Ord, T., et al. (1995) The stages at which human fertilization arrests: microtubule and chromosome configurations in inseminated oocytes

which fail to complete fertilization and development in humans. *Hum. Reprod.* **10,** 1897–1906.

5. Oko, R., Aul, R., Wu, A., et al. (2001) The sperm head skeleton, in *Andrology in the 21st Century* (Robaire, B., Chemes, H., and Morales, C. R., eds.), Medimond, Englewood, NJ, pp. 37–51.
6. Sutovsky, P., Navara, C. S., and Schatten, G. (1996) The fate of the sperm mitochondria, and the incorporation, conversion and disassembly of the sperm tail structures during bovine fertilization in vitro. *Biol. Reprod.* **55,** 1195–1205.
7. Sutovsky, P., Oko, R., Hewitson, L., et al. (1997) The removal of the sperm perinuclear theca and its association with the bovine oocyte surface during fertilization. *Dev. Biol.* **188,** 75–84.
8. Sutovsky, P., Hewitson, L., Simerly, C. R., et al. (1996) Intracytoplasmic sperm injection (ICSI) for Rhesus monkey fertilization results in unusual chromatin, cytoskeletal, and membrane events, but eventually leads to pronuclear development and sperm aster assembly. *Hum. Reprod.* **11,** 1703–1712.
9. Hewitson, L., Dominko, T., Takahashi, D., et al. (1999) Births of ICSI monkeys reveal checkpoints unique to the first cell cycle of fertilization. *Nature Med.* **5,** 431–433.
10. Sutovsky, P., Moreno, R., Ramalho-Santos, J., et al. (1999) Ubiquitin tag for sperm mitochondria. *Nature* **402,** 371–372.
11. Sutovsky, P., Moreno, R., Ramalho-Santos, J., et al. (2000) Ubiquitinated sperm mitochondria, selective proteolysis and the regulation of mitochondrial inheritance in mammalian embryos. *Biol. Reprod.* **63,** 582–590.
12. Cummins, J. M. (2001) Mitochondria: potential roles in embryogenesis and nucleocytoplasmic transfer. *Hum. Reprod. Update* **7,** 217–228.
13. Parrish, J. J., Susko-Parrish, J. L., Leibfreid-Rutledge, M. L., et al. (1986) Bovine in vitro fertilization with frozen-thawed semen. *Theriogenology* **25,** 591–600.
14. Boatman, D. E. (1987) In vitro growth of non-human primate pre- and peri-implantation embryos, in *The Mammalian Preimplantation Embryo* (Bavister, B. D., ed.), Plenum, New York, pp. 273–308.
15. Simerly, C. and Schatten, G. (1993) Techniques for localization of specific molecules in oocytes and embryos. *Methods Enzymol.* **225,** 516–552.
16. Sutovsky, P., Neuber, E., and Schatten, G. (2002) Ubiquitin-dependent, sperm quality control mechanism recognizes spermatozoa with DNA defects, as revealed by dual ubiquitin-TUNEL assay. *Mol. Reprod. Dev.* **61,** 406–413.
17. Sutovsky, P., Fléchon, J. E., Fléchon, B., et al. (1993) Dynamic changes of gap junctions and cytoskeleton during in vitro culture of cattle oocyte cumulus complexes. *Biol. Reprod.* **49,** 1277–1287.
18. Sutovsky, P., Terada, Y., and Schatten, G. (2001) Ubiquitin-based sperm assay for the diagnosis of male factor infertility. *Hum. Reprod.* **16,** 250–258.

6

Localization of Zona Pellucida Receptors on Live Sperm by Fluorophore-Conjugated Solubilized Zona Pellucida Proteins

Heather R. Burkin, Ana Paula Alves-Vieira, and David J. Miller

1. Introduction

The zona pellucida is an extracellular glycoprotein coat that surrounds the oocyte and, in many mammals, provides the first barrier sperm must penetrate for fertilization to occur. The molecular interactions between the sperm and zona pellucida appear to be relatively species-specific ***(1,2)*** and have been studied in most detail in the mouse. The mouse zona pellucida is composed of three glycoproteins termed ZP1, ZP2, and ZP3 in order of decreasing molecular weight ***(3)***.

Because of the relative ease of obtaining gametes required for zona pellucida isolation and fertilization studies, the pig is a useful organism for comparing various steps in the fertilization process to the mouse. Like the mouse zona, the porcine zona pellucida is also composed of three glycoproteins, which we will refer to as ssZPA (molecular weight [MW] = 70 kDa), ssZPB (MW = 59 kDa), and ssZPC (MW = 46 kDa) according to proposed nomenclature ***(4,5)***. Based on sequence data, these glycoproteins appear to be homologs of mouse ZP2, ZP1, and ZP3, respectively ***(4)***. ZP3 is the primary receptor for mouse sperm ***(3)***. The identity of the receptor for porcine sperm is more equivocal. The extensive glycosylation of porcine zona proteins makes them difficult to separate biochemically; however, recent data suggest that ssZPB and ssZPC are primarily responsible for sperm binding ***(6)***. Direct knowledge of the location of zona pellucida receptors on boar sperm will be useful in determining the potential roles of various zona-binding proteins that have been identified in the boar ***(7–12)***.

From: *Methods in Molecular Biology, vol. 253: Germ Cell Protocols: Vol. 1 Sperm and Oocyte Analysis*
Edited by: H. Schatten © Humana Press Inc., Totowa, NJ

Although binding and penetration through the zona pellucida is essential for successful fertilization, the exact regions of sperm that are capable of binding the zona pellucida have remained controversial ***(13–17)***. Sperm are highly differentiated cells with distinct and specialized plasma membrane domains that restrict movement of receptors. Herein, we describe a fluorescence microscopy technique used to localize solubilized zona pellucida proteins on live acrosome-intact and acrosome-reacted boar sperm (*see* **Note 1** and **ref. *18***). This technique uses zona pellucida proteins labeled directly with a bright fluorophore. Using labeled zona proteins avoids multiple sperm centrifugation steps, maintains sperm viability, and reduces the opportunity for zona pellucida proteins to induce changes in sperm membranes and the acrosome reaction. In addition to identifying the region of sperm that binds the zona pellucida, this technique is also useful in determining the developmental stage at which sperm begin to display receptors for the zona pellucida ***(18)***. This technique has been applied recently to mouse sperm ***(19)***. In mice, ZP3 is responsible for binding sperm and triggering the acrosome reaction. Following the release of the acrosome, sperm gain affinity for ZP2. Therefore, it is likely that different membrane domains of sperm contain unique receptors for ZP2 and ZP3. This technique has been used to help resolve this issue ***(19)***.

This chapter discusses methods to purify porcine zona pellucida proteins, conjugation of fluorophore to zona proteins, confirmation of zona protein biological activity, collection and preparation of sperm, and labeling of sperm with fluorophore-conjugated zona proteins. Some information given here has previously been published ***(18)***.

2. Materials

1. Special equipment: ultracentrifuge with rotor and appropriate centrifuge tubes, microfuge, Polytron with standard generator, Hamilton Beach Chef Prep food processor (six cup capacity with S blade) or a gang of razor blades, microscope with epifluorescence capabilities and camera to detect zona protein bound to sperm, inverted microscope with phase contrast to count sperm bound to oocytes, dissecting microscope, computer with image-analysis software.
2. Alexa labeling kit from Molecular Probes (Eugene OR) containing 1 *M* $NaHCO_3$.
3. 1.5 *M* Hydroxylamine, pH 8.9, made fresh.
4. Phosphate-buffered saline (PBS): 137 m*M* NaCl, 2.7 m*M* KCl, 10 m*M* Na_2HPO_4, 2 m*M* KH_2PO_4, pH to 7.4. This should be autoclaved and stored at room temperature.
5. Make dmTALP medium by removing the proper volumes from stock solutions to arrive at the final concentration of salts. The stock solutions are 2 *M* NaCl; 1 *M* KCl; 1 *M* $CaCl_2$; 1 *M* $MgSO_4$; 1 *M* KH_2PO_4. These stock solutions are filter-sterilized and can be stored at 5°C for several months. The remaining ingredients are added directly to the medium to give a final composition of 2.1 m*M* $CaCl_2$, 3.1 m*M* KCl,

1.5 m*M* $MgCl_2$, 100 m*M* NaCl, 0.29 m*M* KH_2PO_4, 0.36% lactic acid, 26 m*M* $NaHCO_3$, 0.6% bovine serum albumin (BSA) (Fraction V), 1 m*M* pyruvic acid, 20 m*M* HEPES, pH 7.3, 10 units/mL penicillin, and 10 μg/mL streptomycin. The medium should be sterilized by filtration and can be stored for 1 wk at 5°C.

6. We use either dmTALP or dmKRBT for zona preparations; both work equally well. To make dmKRBT, remove the proper volumes from stock solutions to arrive at the final concentration of salts. The stock solutions are 2 *M* NaCl; 1 *M* KCl; 1 *M* $CaCl_2$; 1 *M* $MgSO_4$; 1 *M* NaH_2PO_4, which are filter-sterilized and stored at 5°C for several months. The remaining ingredients are added directly to the medium to give a final composition of 120 m*M* NaCl, 2 m*M* KCl, 2 m*M* $CaCl_2$, 10 m*M* $NaHCO_3$, 1.2 m*M* $MgSO_4$, 0.36 m*M* NaH_2PO_4, 5.6 m*M* glucose, 1.1 m*M* pyruvic acid, 25 m*M* TAPSO, 18.5 m*M* sucrose, 0.6% BSA (Fraction V), 10 units/mL penicillin, 10 μg/mL streptomycin, pH 7.3. The medium should be sterilized by filtration and can be stored for up to 1 wk at 5°C.
7. Protease inhibitors: Leupeptin from 1 m*M* stock in distilled water, phenylmethyl sulfonylfluoride from 100 m*M* stock in isopropanol, pepstatin from 1 m*M* stock in ethanol. These are added to dmKRBT or dmTALP immediately before homogenization, as the phenylmethyl sulfonylfluoride degrades quickly in aqueous solutions.
8. Hyaluronidase (from bovine testis) and DNase I are added fresh before homogenization.
9. 10X HBS: 1.3 *M* NaCl, 40 m*M* KCl, 10 m*M* $CaCl_2$, 5 m*M* $MgCl_2$. After sterilization by filtration, this can be stored at 5°C for several weeks.
10. Stock of 20% NP-40 (can be stored at room temperature) and 20% deoxycholate (made fresh).
11. Protein assay kit such as the BCA protein assay (Pierce, Rockford, IL).
12. Equipment and supplies for sodium dodecyl sulfate-polyacrylamide gel electrophoresis (SDS-PAGE).
13. Dialysis tubing or microdialysis chamber like Pierce Slide-A-Lyzer with a 10,000 MW cutoff.
14. Medium B: 127 m*M* NaCl, 5.3 m*M* KCl, 18.2 m*M* HEPES, pH 7.2. This should be filter-sterilized.
15. 4% Paraformaldehyde: Paraformaldehyde is dissolved to 40 mg/mL in a solution of 15.5 mg/mL Na_2HPO_4. Paraformaldehyde dissolves most quickly at 60°C with occasional agitation. After dissolution, the solution is neutralized by adding NaH_2PO_4 to a final concentration of 2.98 mg/mL.
16. 1.5% Formaldehyde: 1.5 mL formaldehyde in 100 mL of medium B.
17. Coomassie G-250 stain: 0.22% Coomassie blue G-250, 50% methanol, 10% acetic acid, and 40% water. This should be stored in a sealed container at room temperature.
18. Calcium ionophore A23187 dissolved in dimethyl sulfoxide as a 10-m*M* stock and kept in a lightproof container at 5°C.
19. 0.5 mg/mL Propidium iodide dissolved in PBS and stored in the dark at 5°C.
20. 0.1 *M* Ammonium acetate in water pH 9.0 (can be stored at 5°C for several months).

3. Methods

3.1. Zona Pellucida Protein Purification and Solubilization

1. Collect porcine ovaries from an abattoir (ovaries can be stored at room temperature for up to 8 h) and then disrupt ovaries with razor blades in dmKRBT medium to free oocytes from follicles. Using this method, repeatedly crush the ovaries individually using the razor blades. Collect the fluid containing oocytes. Alternatively, place the ovaries in groups of 50 in 10 mL of dmKRBT in a six-cup capacity food processor with an S blade. Give the ovaries two 10-s pulses in the food processor to release the oocytes. The food processor method works well and is much easier and safer to perform than using razor blades and usually provides higher yields of zona pellucida protein. After chopping in the food processor, pour the material through a standard food strainer and collect the liquid that contains the oocytes.
2. Transfer the liquid to 50-mL conical centrifuge tubes. Use one 50-mL tube for liquid from about 50 ovaries. Centrifuge the material removed from the ovaries at 800*g* for 10 min, followed by one wash with dmKRBT to move oocytes to the bottom of the tube. If the food processor is used, transfer the pellets containing oocytes to a clean tube before homogenization, leaving the congealed pellet with erythrocytes behind. The pellet volume should be from 5 to 10 mL. Add dmKRBT to bring the total volume to 20 mL.
3. Add the following, in some cases from stock solutions (described in **Subheading 2.**), to arrive at the given final concentration: 1 μg/mL leupeptin, 1 μg/mL pepstatin, 10 μg/mL phenylmethyl sulfonylfluoride, 0.3 mg/mL DNase 1, and 0.3 mg/mL hyaluronidase.
4. Homogenize the pellets with three bursts lasting 10 s each at 5/6 maximum speed using a Polytron. After adding 1 mL of NP-40 from a 20% stock (1% final concentration), give three more 7-s homogenizations with a Polytron, as above. Finally, after the addition of 1 mL of deoxycholic acid from a 20% stock (1% final concentration), give three more bursts with the Polytron. Each Polytron homogenization step produces foam, so provide several minutes between each homogenization. Between homogenizations, place the tube of homogenate on ice.
5. Add an equal volume of Percoll to the homogenate and transfer to 25 × 89-mm Quickseal polyallomer tubes. After heat-sealing the tubes, centrifuge the homogenates at 44,000*g* for 70 min at 3°C. The zona pellucida fragments are recovered near the top of the Percoll gradient as a clean white band and added to another 50% Percoll gradient. After centrifugation under the same conditions, remove the zona fragments with a pipet and transfer to a 15-mL conical centrifuge tube. Dilute the zona solution at least 10-fold with dmKRBT and resuspend the zona by vortexing. Wash the pellet three times with dmKRBT and check the zona fragments for purity under a microscope. The fragments should be free of particulates and appear like broken egg shells. Finally, wash the zona proteins at least four times in 10 m*M* Tris pH 9.5, 100 m*M* NaCl and solubilize them in 0.5 mL of the same buffer by heating for 20 min at 73°C. If 200–300 ovaries are used, the total yield should be 1–2 mg protein.

6. Determine the protein concentrations by a protein assay, such as the BCA assay (Pierce, Rockford, IL) using BSA as a standard. The protein concentration should be approx 2 mg/mL for subsequent labeling with Alexa Fluor 488.
7. Confirm protein purity by standard SDS-PAGE (*see* **Note 2**).

3.2. Conjugation of Fluorophore to Zona Pellucida Proteins

1. Isolate porcine zona pellucida proteins as described in **Subheading 3.1.** Irrelevant proteins such as transferrin and BSA should be used as controls for sperm-binding specificity. Dialyze each protein into PBS, pH 7.4.
2. To label each protein, adjust the concentration of zona pellucida or control protein to 2 mg/mL in 0.5 mL PBS and add 50 µL of 1 *M* bicarbonate to increase the pH. Transfer the protein solution to a vial of reactive Alexa Fluor 488 dye (Molecular Probes, Eugene, OR) and stir for 1 h at room temperature using the mini-stir bar in the reaction tube. Stop the reaction with the addition of 100 µL of 1.5 *M* hydroxylamine solution for 1 h at room temperature.
3. Remove the Alexa-conjugated hydroxylamine by extensive dialysis into PBS at 4°C.

3.3. Examination of Biological Activity of Labeled Zona Pellucida Proteins

3.3.1. Sperm–Egg Binding Inhibition Assays

To confirm that labeling of zona proteins does not affect their biological activity, several assays can be performed that test the ability of zona proteins to bind sperm and induce the acrosome reaction. An indirect, competition assay is commonly used to assess the ability of zona proteins to bind sperm. This is based on the principle that bioactive soluble zona proteins bind to sperm and prevent sperm from binding to oocytes (*see* **Note 3**).

1. Aspirate porcine oocytes with an 18-gauge needle from follicles greater than 2 mm in diameter and centrifuge the aspirate at 600*g* for 10 min. Remove the follicular fluid and replace it with about half the volume of medium B.
2. Locate individual oocytes in the medium under a dissecting microscope and transfer them to fresh medium B using a mouth pipet having an internal diameter just larger than the oocytes. Mouth pipets are made by heating Pasteur pipets or capillary tubes and then pulling by hand. Commercial pipet pullers are also available. The pipet is linked to plastic tubing that is coupled to a mouthpiece. This enables the user to pipet small volumes containing oocytes.
3. Vortex oocytes in 100 µL of medium B for 5 min in a 1.5-mL microfuge tube to remove cumulus cells.
4. Isolate oocytes from cumulus cells by consecutive transfer through three culture dishes of medium B by mouth pipet. Fix the washed oocytes in a culture dish with medium B containing 1.5% formaldehyde for 10 min at room temperature. The timing is critical because oocytes are more likely to stick to the dish with longer fixation times. If oocytes adhere strongly to the dish, the fixation time can

be reduced slightly. Transfer the oocytes through three culture dishes of medium B and then to a drop of dmTALP covered with mineral oil. Oocytes can be stored at 4°C for up to 2 mo. Before use, wash the oocytes through three drops of fresh dmTALP.

5. Collect and wash sperm as described in **Subheading 3.4.** and adjust the concentration to 2 × 10^6 sperm/mL with dmTALP, following a hemocytometer count. Incubate sperm in the capacitating conditions described in **Subheading 3.4.** for 4 h.
6. During the last 30 min of capacitation, add 90 µL sperm to 10 µL of 2 mg/mL protein (unlabeled zona protein, Alexa-labeled zona protein, or Alexa-labeled control protein) or 10 µL of buffer control. Add 20 µL of protein-treated sperm to 20 µL dmTALP droplets containing 10 oocytes each. Allow sperm to bind oocytes at 39°C for 15 min. Wash oocytes through one drop of dmTALP to remove loosely bound sperm and transfer them to 20-µL droplets of 4% paraformaldehyde.
7. Count bound sperm using an inverted microscope under phase-contrast optics.

3.3.2. Induction of the Acrosome Reaction with Alexa-Zona Proteins

The biological activity of the Alexa-conjugated zona proteins can be tested by the ability to induce the acrosome reaction (*see* **Note 3**).

1. Collect, wash, and capacitate boar sperm as described in **Subheading 3.4.** For capacitation, adjust the sperm concentration to 2 × 10^6 sperm/mL. Ninety microliters of sperm are added to 10 µL buffer control or 2 mg/mL zona protein (unlabeled or Alexa conjugated) and incubated 1 h at 39°C in 1-mL microfuge tubes.
2. Fix sperm by adding 1 mL of 4% paraformaldehyde and gently mix fixed sperm for 10 min at room temperature ***(20)***. Centrifuge fixed sperm at 5000*g* for 1 min and wash sperm three times in 1 mL of 0.1 *M* ammonium acetate, pH 9.0. Resuspend the final sperm pellet in 50 µL of 0.1 *M* ammonium acetate, pH 9.0, and dry sperm on a microscope slide at room temperature.
3. Stain sperm acrosomes by immersion in Coomassie G-250 stain for 1 min. Rinse the slides by two 10-s immersions in water and cover with Permount and a cover slip. Allow slides to dry for more than 2 h.
4. Count the acrosomes of at least 200 sperm per slide under a microscope at 400×. Intact acrosomes are stained bright blue, particularly at the apical ridge of the sperm acrosome ***(20)***.

3.4. Preparation of Sperm for Labeling

3.4.1. Collection of Acrosome-Intact Sperm

1. Collect ejaculated semen samples from mature boars by allowing the boars to mount a dummy and providing pressure manually on the glans penis.
2. Either process the semen immediately or dilute it in an appropriate extender and store at 16–18°C for 24–48 h. To remove the seminal fluids, layer up to 1 mL of the semen samples over a Percoll solution containing 5.4 mL of Percoll, 0.6 mL of 10X HBS and 4 mL of dmTALP in a 15-mL conical tube.

3. Centrifuge the samples at 800*g* for 10 min, aspirate the supernatant, and resuspend the pellet in 15 mL dmTALP. Centrifuge the resuspended pellet for 5 min at 600*g*. After assessing the concentration using a hemocytometer, resuspend the sperm pellet to a concentration of 2×10^7/mL in dmTALP and incubate the sperm for 4 h at 39°C to capacitate them. If sperm stick together and form thick aggregates, substituting essentially fatty-acid-free BSA for the Fraction V BSA helps reduce the clumps.
4. Discard samples with less than 60% motility after this step.

3.4.2. Collection of Acrosome-Reacted Sperm

To produce a population of sperm that are largely acrosome reacted, the calcium ionophore A23187 is added to capacitated sperm.

1. Add calcium ionophore A23187 from the 10-m*M* stock to sperm in dmKRBT to a final concentration of 10 μ*M*. Incubate sperm for 1 h at 39°C.
2. Confirm the acrosome reaction has been induced by Coomassie staining (*see* **Subheading 3.3.2.**).

3.4.3. Collection of Sperm from the Epididymis

1. To obtain epididymal sperm, remove the testes and epididymides from anesthetized boars. Dissect epididymides immediately or store them for 16 h in dmTALP at room temperature prior to dissection. In our experience, storage did not affect viability, motility, or zona protein binding of recovered epididymal sperm.
2. Remove connective tissue from each epididymis and then divide it into caput, corpus, and cauda sections as defined ***(21)***. Place individual sections in Petri dishes containing dmTALP and macerate the epididymal sections with razor blades to release sperm. Fluid containing sperm is removed from the macerated epididymis and centrifuged for 5 min at 600*g*. The pellets containing sperm from each epididymal region are washed three times in dmTALP.
3. Resuspend the washed epididymal sperm to a concentration of 2×10^7/mL and incubate for 4 h in capacitating conditions described in **Subheading 3.4.1.**

3.5. Labeling of Live Sperm with Fluorophore-Conjugated Solubilized Zona Pellucida Proteins

3.5.1. Fluorescent Zona Protein Binding to Live Sperm

1. Add Alexa-zona protein conjugate to 0.4 mL sperm to yield final concentrations of 10, 50, or 100 μg/mL Alexa-zona and 2×10^7 sperm/mL. After approx 30-min incubations, acquire fluorescence images at 400× or 630× magnification. We generally perform incubations for between 15 and 60 min; however, the fluorescent signal can be observed after as little as 5 min ***(18)***.
2. With most microscopes, it is very difficult to count swimming sperm under fluorescence optics, particularly those that have minimal fluorescence signal [e.g., caput epididymal sperm ***(18)***]. Instead, first identify and count sperm under

phase-contrast microscopy in a small area and then switch to fluorescence optics to detect zona protein binding.

3. Analyze fluorescence intensity using the NIH Image program (available at http://rsb.info.nih.gov/nih-image). Measure integrated densities for at least three sperm per frame and average them. Take measurements on multiple frames to assess the effects of any treatments.
4. The 10-µg/mL Alexa-zona protein concentration yields an intense signal on the porcine sperm head with minimal background fluorescence; 50 µg/mL yields maximal fluorescence and higher concentrations produced increased background fluorescence (*see* **Note 4**). Add 1 µL of 0.5 mg/mL propidium iodide to each sample to detect dead or membrane-compromised sperm.
5. If binding of zona proteins to sperm is specific, it should be saturable. To determine if binding is saturable, a 100-fold excess of unlabeled zona pellucida protein is added with the labeled zona protein to sperm. Binding of Alexa-zona should be diminished markedly (*see* **Note 4**).
6. Incubation with Alexa-zona proteins under these conditions are unlikely to have an effect on porcine sperm (*see* **Note 5**); however, sperm from other species may be more readily altered by centrifugation or incubation with zona pellucida proteins.

4. Notes

1. Incubating sperm with directly labeled zona pellucida proteins avoids multiple sperm centrifugation steps, maintains sperm viability, and reduces the opportunity for zona pellucida proteins to induce changes in sperm membranes and the acrosome reaction. Therefore, we use zona proteins conjugated directly to the bright Alexa 488 fluorophore. The 10-µg/mL concentration of zona pellucida proteins used for this detection system is far lower than the 125 µg/mL required to induce the acrosome reaction in boar sperm *(22)*. Furthermore, even at high zona protein concentrations, only approx 15% of boar sperm typically undergo the acrosome reaction *(22)*. For these reasons, the Alexa-zona conjugate is unlikely to induce physiological changes in live sperm under the conditions used. Alexa-zona proteins are added to live sperm, along with propidium iodide, which stains the nuclei of dead or membrane-compromised sperm. We observe three populations of sperm. The first sperm population stains strongly with propidium iodide and exhibits little Alexa-zona labeling. This population probably consists of dead sperm that have lost their acrosomes *(18)*. The second population stains with propidium iodide and binds zona proteins over the acrosomal vesicle, similar to the labeling pattern observed in fixed sperm *(18)*. These sperm may have died without losing their acrosomal contents. A third population of sperm does not stain with propidium iodide and exhibit concentrated Alexa-zona fluorescence over the acrosomal ridge (*see* **Fig. 1A**). This appears to be the region of live, acrosome-intact sperm that initiates zona pellucida binding, as all motile sperm belong to this third population. We have not found differences in Alexa-zona binding between capacitated and uncapacitated boar sperm samples or in

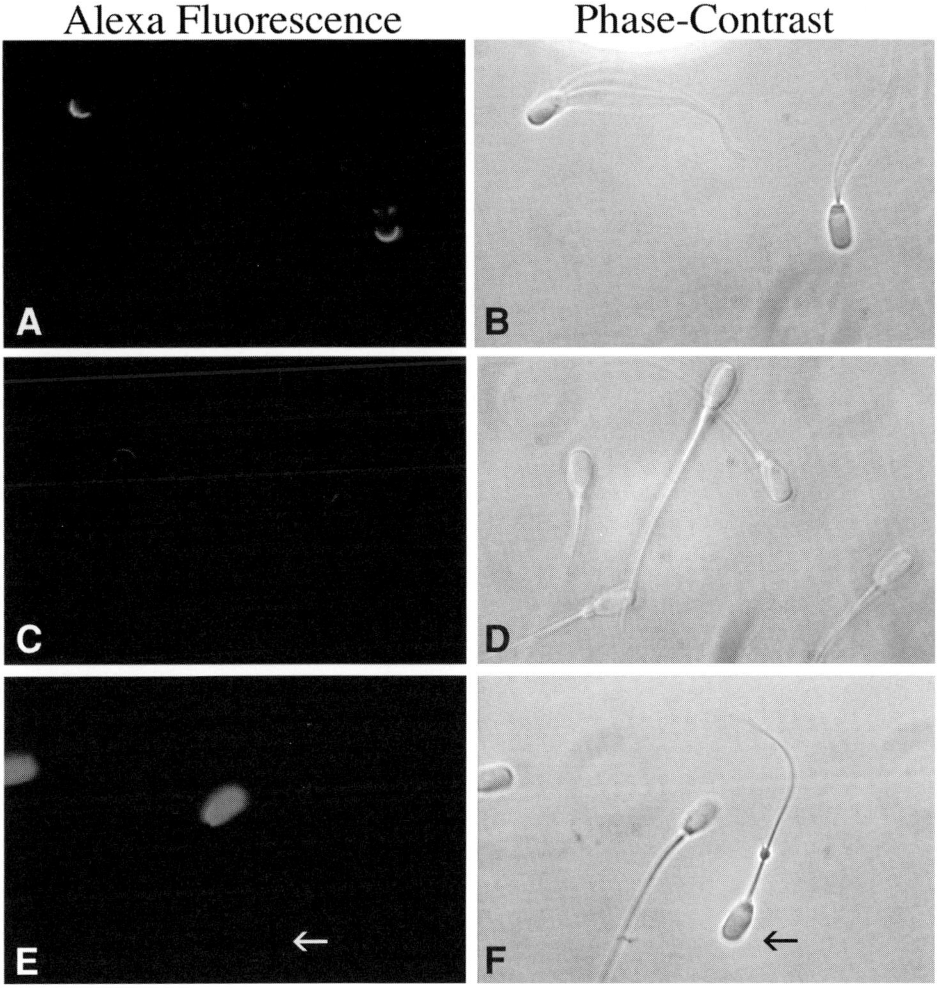

Fig. 1. An example of localization of solubilized zona pellucida proteins on live sperm. Alexa-zona proteins bound live, acrosome-intact sperm over the anterior head region concentrated over the acrosomal ridge **(A)**. The addition of a 100-fold excess of unlabeled zona proteins displaced the signal **(C)**. The Alexa-labeled control glycoprotein, transferrin, did not bind to live sperm indicated by the arrow in **(E)**. Dead sperm stained with propidium iodide are visible because fluorescence images are captured using a filter set that allowed detection of both red and green fluorochromes simultaneously **(A, C, E)**. Two moribund sperm whose entire nuclei are stained in **(E)** are stained with propidium iodide. Corresponding phase-contrast images are shown **(B, D, F)**.

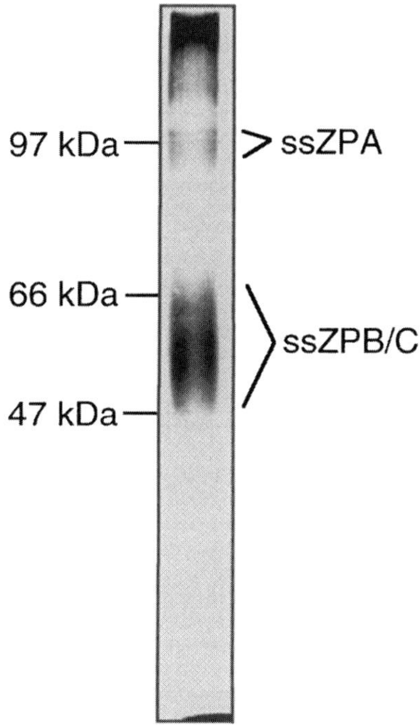

Fig. 2. To confirm purity, zona proteins are separated by 10% SDS-PAGE under nonreducing conditions. In this example, the porcine zona pellucida proteins were detected by silver staining. The numbers on the left correspond to the migration of 97-kDa, 66-kDa, and 47-kDa standards.

samples diluted in the extender ***(18)***. Using mouse sperm, an increase in ZP3 binding during capacitation is noted ***(19)***.

2. The purity of zona pellucida proteins can be assessed by SDS-PAGE using a 10% acrylamide gel. One to ten micrograms of porcine zona protein are loaded onto the gel. After electrophoresis, the gel is silver stained. The reason for using high amounts of protein is that zona proteins do not stain intensely with silver. The major bands are ssZPB and ssZPC that migrate as a broad smear from 45 to 66 kDa and ssZPA that migrates as a 90- to 100-kDa protein (*see* **Fig. 2**).
3. It is important to perform experiments demonstrating that the Alexa-conjugated zona proteins display the same biological activity as unlabeled zona proteins. First, capacitated sperm are preincubated with unlabeled zona protein, Alexa-zona protein, Alexa-transferrin, or a buffer control. At 0.2 mg/mL, Alexa-zona protein inhibited sperm binding to the intact zona pellucida to the same extent as unlabeled zona proteins (*see* **Fig. 3A**). An average of 23.8 ± 2.23 control sperm are bound to each oocyte. Preincubation of sperm with either unlabeled zona protein or Alexa-zona protein inhibited sperm–egg binding (9.67 ± 3.0 bound sperm/oocyte and 2.73 ± 1.15 sperm/oocyte bound, respectively). Preincubation

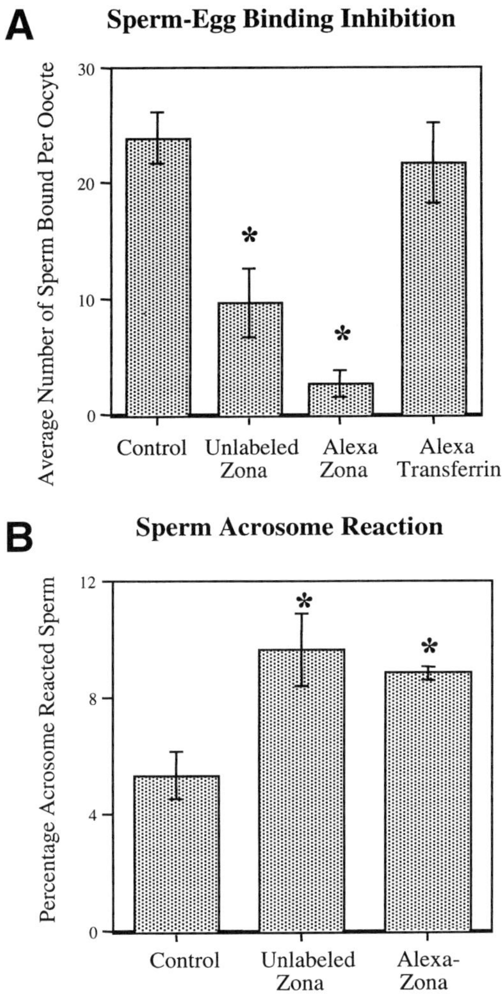

Fig. 3. Alexa-conjugated zona proteins retain biological activity. **(A)** Capacitated sperm are preincubated with 0.2 mg/mL protein or buffer control and then allowed to bind porcine oocytes. The average number of sperm bound per oocyte was determined for triplicate droplets containing 10 oocytes each. Alexa-zona proteins inhibited sperm–egg binding to the same degree as unlabeled zona proteins. Alexa-transferrin did not inhibit sperm binding to the intact zona pellucida. **(B)** Triplicate capacitated sperm samples are incubated with 0.2 mg/mL protein or buffer control for 1 h and acrosomal status determined by Coomassie staining ***(20)***. Alexa-zona protein induced the sperm acrosome reaction to the same extent as unlabeled zona protein. Asterisks indicate samples that are significantly different from the control ($p < 0.05$).

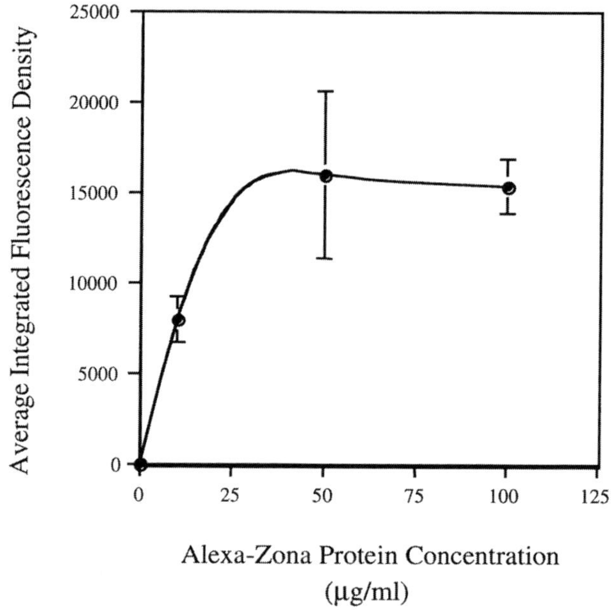

Fig. 4. Saturability of the Alexa-zona fluorescence signal. Capacitated boar sperm were incubated with 10, 50, or 100 μg/mL Alexa-zona protein for 30 min. Sperm fluorescence images are captured and analyzed with NIH Image software. Measurements of integrated fluorescence density are taken for three sperm per image. The average sperm intensity for three images at each concentration was plotted. The Alexa-zona fluorescence signal reached saturation at approx 50 μg/mL.

with Alexa-transferrin did not inhibit sperm–egg binding (21.7 ± 3.42 bound sperm/oocyte). In the second series of experiments, we determined whether Alexa-labeled zona proteins could induce the sperm acrosome reaction to the same extent as unlabeled zona protein (*see* **Fig. 3B**). At a concentration of 0.2 mg/mL, Alexa-labeled zona protein induced the acrosome reaction in the same percentage of sperm as unlabeled zona proteins (8.8 ± 0.21□□% and 9±7 1.2%, respectively). Only 5.3 ± 0.82% of control sperm appeared acrosome reacted. These results are typical of boar sperm, which acrosome react relatively poorly in response to soluble zona protein ***(22)***. Because the Alexa-zona proteins are bound to sperm, inhibited sperm–egg binding and induced the sperm acrosome reaction, we concluded that zona proteins retained biological activity when conjugated to Alexa.

4. Several experiments are performed to show that binding of Alexa-zona to sperm is saturable and specific. First, the signal was blocked with the addition of a 100-fold excess of unlabeled zona proteins (*see* **Fig. 1B**) and no Alexa signal was detected when two labeled control proteins (transferrin and BSA) are incubated with live sperm ***(18)***. Second, we incubated sperm with different concen-

trations of Alexa-zona proteins and quantitated the average intensity of the fluorescent signal per sperm. The fluorescence pattern observed was the same at all concentrations tested. Although the fluorescence intensity between individual sperm was variable, the average fluorescence intensity reached a plateau at approx 50 μg/mL (*see* **Fig. 4**). Therefore, live boar sperm with uncompromised membranes bound labeled zona proteins with a saturable and specific pattern.

5. Although it is theoretically possible that the addition of labeled zona proteins to live sperm could cause the acrosome reaction or other changes in the boar sperm membrane, the experimental conditions we used make this possibility unlikely ***(18)***. First, the concentration of zona proteins used is an order of magnitude lower than the concentration required to induce the acrosome reaction in boar sperm ***(22)***. Second, the labeling times used are as short as 5 min, which does not approach the time required for the boar sperm acrosome reaction ***(13,22)***. Finally, boar sperm generally undergo low rates of soluble zona-induced acrosome reactions (<15%) compared to sperm from other species ***(22)***.

Acknowledgments

The authors thank Excel, Inc for providing porcine ovaries, Russ Wischover at Moorman Research Farm for collecting semen, and other members of the Miller laboratory for helpful comments in preparing the manuscript. Research in the Miller laboratory was supported by the Illinois Council for Food and Agricultural Research, the University of Illinois Agricultural Experiment Station as part of Hatch Project ILLU-35-0335, the University of Illinois Research Board, and the United States Department of Agriculture National Research Initiative (99-35203). HRB was supported by United States Public Health Service Fellowship 5 T32 HD07028. APA-V was supported by a fellowship from the Lalor Foundation.

References

1. Moller, C. C., Bleil, J. D., Kinloch, R. A., et al. (1990) Structural and functional relationships between mouse and hamster zona pellucida glycoproteins. *Dev. Biol.* **137,** 276–286.
2. Peterson, R. N., Russell, L., Bundman, D., et al. (1980) Sperm–egg interaction: evidence for boar sperm plasma membrane receptors for porcine zona pellucida. *Science* **207,** 73–74.
3. Wassarman, P. M., Jovine, L., and Litscher, E. S. (2001) A profile of fertilization in mammals. *Nature Cell Biol.* **3,** E59–E64.
4. Harris, J. D., Hibler, D. W., Fontenot, G. K., et al. (1994) Cloning and characterization of zona pellucida genes and cDNAs from a variety of mammalian species: the ZPA, ZPB and ZPC gene families. *DNA Seq.* **4,** 361–393.
5. Hedrick, J. L. (1996) Comparative structural and antigenic properties of zona pellucida glycoproteins. *J. Reprod. Fertil.* **50(Suppl.),** 9–17.

6. Yurewicz, E. C., Sacco, A. G., Gupta, S. K., et al. (1998) Hetero-oligomerization-dependent binding of pig oocyte zona pellucida glycoproteins ZPB and ZPC to boar sperm membrane vesicles. *J. Biol. Chem.* **273,** 7488–7494.
7. Ensslin, M., Vogel, T., Calvete, J. J., et al. (1998) Molecular cloning and characterization of P47, a novel boar sperm- associated zona pellucida-binding protein homologous to a family of mammalian secretory proteins. *Biol. Reprod.* **58,** 1057–1064.
8. Geng, J. G., Raub, T. J., Baker, C. A., et al. (1997) Expression of a P-selectin ligand in zona pellucida of porcine oocytes and P-selectin on acrosomal membrane of porcine sperm cells. Potential implications for their involvement in sperm-egg interactions. *J. Cell Biol.* **137,** 743–754.
9. Hardy, D. M. and Garbers, D. L. (1995) A sperm membrane protein that binds in a species-specific manner to the egg extracellular matrix is homologous to von Willebrand factor. *J. Biol. Chem.* **270,** 26,025–26,028.
10. Hickox, J. R., Bi, M., and Hardy, D. M. (2001) Heterogeneous processing and zona pellucida-binding activity of pig zonadhesin. *J. Biol. Chem.* **276,** 41,502–41,509.
11. Mori, E., Kashiwabara, S., Baba, T., et al. (1995) Amino acid sequences of porcine Sp38 and proacrosin required for binding to the zona pellucida. *Dev. Biol.* **168,** 575–583.
12. Rebeiz, M. and Miller, D. J. (1999) Porcine sperm surface β1,4galactosyltransferase binds to the zona pellucida but is not necessary or sufficient to mediate sperm-zona pellucida binding. *Mol. Reprod. Dev.* **54,** 379–387.
13. Fazeli, A. R., Holt, C., Steenweg, W., et al. (1995) Development of a sperm hemizona binding assay for boar semen. *Theriogenology* **44,** 17–27.
14. Fazeli, A., Hage, W. J., Cheng, F. P., et al. (1997) Acrosome-intact boar spermatozoa initiate binding to the homologous zona pellucida in vitro. *Biol. Reprod.* **56,** 430–438.
15. Jones, R. (1991) Interaction of zona pellucida glycoproteins, sulphated carbohydrates and synthetic polymers with proacrosin, the putative egg-binding protein from mammalian spermatozoa. *Development* **111,** 1155–1163.
16. Yonezawa, N., Hatanaka, Y., Takeyama, H., et al. (1995) Binding of pig sperm receptor in the zona pellucida to the boar sperm acrosome. *J. Reprod. Fertil.* **103,** 1–8.
17. Yurewicz, E. C., Pack, B. A., Armant, D. R., et al. (1993) Porcine zona pellucida ZP3 alpha glycoprotein mediates binding of the biotin-labeled M(r) 55,000 family (ZP3) to boar sperm membrane vesicles. *Mol. Reprod. Dev.* **36,** 382–389.
18. Burkin, H. R. and Miller, D. J. (2000) Zona pellucida protein binding ability of porcine sperm during epididymal maturation and the acrosome reaction. *Dev. Biol.* **222,** 99–109.
19. Kerr, C. L., Hanna, W. F., Shaper, J. H., et al. (2002) Characterization of zona pellucida glycoprotein 3 (ZP3) and ZP2 binding sites on acrosome-intact mouse sperm. *Biol. Reprod.* **66,** 1585–1595.

20. Larson, J. L. and Miller, D. J. (1999) Simple histochemical stain for acrosomes on sperm from several species. *Mol. Reprod. Dev.* **52,** 445–449.
21. Okamura, N., Dacheaux, F., Venien, A., et al. (1992) Localization of a maturation-dependent epididymal aperm surface antigen recognized by a monoclonal antibody raised against a 135-kilodalton protein in porcine epididymal fluid. *Biol. Reprod.* **47,** 1040–1052.
22. Berger, T., Turner, K. O., Meizel, S., et al. (1989) Zona pellucida-induced acrosome reaction in boar sperm. *Biol. Reprod.* **40,** 525–530.

7

Detection of Human Papillomavirus DNA in Sperm Using Polymerase Chain Reaction

Olufemi A. Olatunbosun, Allison M. Case, and Harry G. Deneer

1. Introduction

Human papillomavirus (HPV) is a highly prevalent sexually transmitted virus that causes genital warts in men and women and is associated with an increased risk of genital tract neoplasia in women *(1,2)*. Nucleic acid amplification techniques such as the polymerase chain reaction (PCR) have enabled the sensitive and specific detection of HPV DNA in many different bodily fluids and tissues *(3–5)*. Although HPVs were first detected in semen in 1986 *(6)*, there have been limited attempts to develop and refine protocols for their detection in this fluid *(7,8)*. Diagnostic methods that enable the detection of HPV in semen *(7,9–11)* offer the promise of reducing the risk of transmission of the virus to the recipients of donor semen during artificial insemination.

Although the use of molecular biology techniques such as PCR is gaining wider acceptance in clinical practice, the majority of testing is still undertaken in the research setting using protocols that are unsuitable for clinical diagnostic laboratories. This poses a challenge for routine HPV DNA testing because it may be performed by technicians not specifically trained in virology or molecular biology, yet there still is a need to produce consistently reliable results in this setting. This is only realistically possible with a standardized and quality-controlled, commercially produced HPV DNA test. Although there are some commercially available HPV assays (e.g., the Digene Hybrid Capture II HPV DNA test), these assays have not been validated for use on semen specimens. PCR has been used successfully in a number of research settings for the detection of HPV DNA in semen *(4,5,7–11)*, although each laboratory has developed its own methodology, making interlaboratory comparisons difficult.

From: *Methods in Molecular Biology, vol. 253: Germ Cell Protocols: Vol. 1 Sperm and Oocyte Analysis*
Edited by: H. Schatten © Humana Press Inc., Totowa, NJ

We have adapted an in-house-developed HPV-specific PCR assay, originally designed for the examination of genital specimens and used it to analyze HPV in semen samples from a cohort of presumed healthy semen donors and volunteers with known genital HPV lesions. The technological basis of this assay, together with available performance data, is reviewed in this chapter and the clinical utility of the test is discussed.

1.1. Specimen Collection and Preparation

Semen samples are collected on site in a private collection room adjacent to the laboratory and maintained at room temperature until complete seminal liquefaction. After liquefaction, total volume, sperm concentration, motility, and progressivity are determined using a 200× phase-contrast microscope. Approximately 0.5 mL of raw (unfractionated) semen is then taken for HPV DNA testing. Alternatively, sperm cells may be separated by Percoll gradient centrifugation, washed, and then subjected to DNA extraction and HPV testing. Semen or sperm specimens can be stored at –20°C if DNA extraction cannot be performed immediately. DNA is extracted and purified following a modified protocol using the MasterPure DNA Extraction Kit (described in **Subheading 3.1.**), which we have found to yield good quality amplifiable DNA.

1.2. PCR Amplification

Detection of HPV DNA in semen employs a PCR amplification protocol using broad-range consensus primers that target the highly conserved L1 gene encoding the capsid protein of HPV *(12)*. The primers are degenerate in several positions, allowing for the amplification of many, if not all, subtypes of HPV with equal efficiency. The PCR method is not intended, however, to discriminate between subtypes of HPV, nor will it provide a quantitative measure of viral load. PCR products are visualized by ultraviolet (UV) transillumination following agarose gel electrophoresis. A specimen positive for HPV will produce an amplification product of approx 450 basepairs (between 448 and 454 bp depending on the HPV subtype). Specimens that are negative for HPV DNA after the first amplification reaction are confirmed using a seminested PCR protocol, which reamplifies a small portion of the first-round reaction and thereby has increased sensitivity. In practice, very few specimens require nested PCR reamplification—a specimen that is a true positive for HPV DNA will generally appear as such after the first round of PCR amplification.

1.3. Quality Control and Contamination Prevention

Because of the inherent sensitivity of PCR amplification, prevention of false-positive results resulting from contamination with previously amplified DNA is of utmost importance. We have not employed uracil DNA glycosylase for biochemical contamination control because of the need to occasionally perform nested PCR on negative specimens. Instead, stringent laboratory techniques, as outlined by Kwok ***(13)*** and as further described in the Notes subheading, serve to minimize problems resulting from contamination. For example, gloves and gowns are worn at all times and must be changed frequently. Aerosol-resistant pipet tips are always used and pipettors and work surfaces are frequently decontaminated with 10% bleach. At a minimum, two physically separated laboratory rooms are required: one room for extraction of specimens and assembly of amplification reaction tubes and a second room for amplification and gel analysis of PCR products. Persons leaving the second room must always change gowns and gloves before returning to the extraction room. Also, PCR reagents should be kept stored in a separate location, aliquoted in small volumes, and discarded and replaced if contamination is noted. Finally, the frequent use of negative controls, introduced at the DNA extraction stage, is critical.

1.4. Performance Characteristics

We have found that the PCR technique described in this chapter is capable of reliably detecting between 10 and 20 copies of HPV DNA when spiked into a 150-µL semen specimen. With this technique, we have tested the semen of 45 donors with historical or clinical evidence of HPV infection and 40 HPV lesion-negative donors ***(7)***. Of the men with past or current genital HPV infections, 24 (53%) were also positive for HPV DNA in their semen. In addition, we found that three (8%) of the apparently uninfected men tested positive for HPV DNA. Overall, PCR detected HPV in 21 of 32 subjects with identifiable lesions and in 6 of 53 subjects without lesions. Swim-up washing of these 27 positive sperm specimens reduced HPV DNA to below levels detectable by PCR in only two cases ***(7)***.

1.5. Conclusions and Clinical Perspectives of Sperm HPV DNA Detection by PCR

The major problem in evaluating the diagnostic performance of DNA amplification methods is that there is no clinical or microbiological reference standard for HPV infection in semen. The evaluated diagnostic performance will, therefore, always be a reflection of the chosen comparator(s). Although HPV DNA tests have improved greatly in recent years, assessing their sensitivity and specificity is not straightforward, as these measures depend on the knowledge of a gold standard that should reflect the true situation ***(14,15)***. Despite this, it seems that DNA amplification methods for detecting HPV are more sensitive than are techniques based on immunology. Nevertheless, there is a need for continuous monitoring of and attempts at improving the reproducibility, sensitivity, and specificity of this testing method.

During fertilization, male and female gametes are exposed to HPV and other infectious agents in the reproductive tract. The effects of these agents on embryogenesis and early embryos are poorly understood, but if these could be evaluated, it could have enormous implications for the prevention of vertical transmission of infectious agents to gamete recipients and their offspring ***(16–19)***. The improved sensitivity afforded by DNA amplification methods allows the use of semen material in which the content of HPV has been reduced or eliminated by sperm washing, as has been shown to be possible for HIV ***(20)***. As well, because transport conditions seem less critical for test performance, individuals themselves can obtain specimens in the privacy of the home and these, subsequently, can be mailed directly to the diagnostic laboratory. Such a strategy for testing would encourage compliance with universal screening. Tests with an optimal diagnostic performance have not yet been reached, but a milestone for new strategies in detection and preventing genital HPV and other sexually transmitted infections has been reached with the availability of DNA amplification techniques.

2. Materials

2.1. Containers and Equipment

1. 1.5-mL Screw-cap microcentrifuge tubes (e.g., Sarstead, cat. no. 72.692.105).
2. Thermal cycler (Applied Biosystems GeneAmp 9700 or equivalent), preferably equipped with a heated lid.
3. Horizontal agarose gel electrophoresis system.
4. Microcentrifuge (capable of 16,000*g*).
5. Digital gel photodocumentation system (e.g., Bio-Rad Gel-Doc system). A standard ultraviolet (UV) light transilluminator equipped with a Polaroid photography system may be used as an alternative.
6. Adjustable pipettors (200–1000 μL, 20–200 μL, 10–100 μL, and 0.2–2 μL) with aerosol-resistant pipet tips.
7. Thin-walled PCR tubes, 0.2 or 0.5 mL (size will depend on the type of thermal cycler being used).

2.2. Reagents

1. MasterPure DNA Purification Kit (Epicentre, Madison, WI., product no. MCD85201).
2. Pellet Paint NF co-precipitant (Novagen Inc., Madison, WI., product no. 70748-3).
3. Isopropyl alcohol (isopropanol).
4. TE buffer: 10 m*M* Tris-HCl (pH 8.6), 1 m*M* EDTA.
5. 70% Ethanol.
6. Electrophoresis-grade agarose.
7. Distilled, deionized water (sterile).
8. 10X PCR reaction buffer: 200 m*M* Tris-HCl (pH 8.4), 500 m*M* KCl, 15 m*M* $MgCl_2$.

Table 1
Sequences of Oligonucleotide Primers

Name	Nucleotide Sequence, 5' to 3'
MY11	GC(AC)CAGGG(AT)CATAA(CT)AATGG
MY09	CGTCC(AC)A(AG)(AG)GGA(AT)ACTGATC
MY11-IN	ATGG(TC)(GA)TTTG(CT)TGG(CG)(AG)(TCA)AA(TC)CA

Source: **ref. *12***.

9. Deoxynucleotide triphosphate mix: Starting from 100-m*M* stocks (available from Amersham Pharmacia Biotech, product no. 27-2035-01), prepare a mixture in water that contains 2.5 m*M* dATP, 2.5 m*M* dCTP, 2.5 m*M* dGTP, and 2.5 m*M* dTTP. Store at –20°C.
10. HPV-specific oligonucleotide primers (*see* **Table 1**). Each primer is dissolved in TE buffer to a stock concentration of 100 μ*M*. Working stocks diluted to 20 μ*M* in TE buffer are then prepared. Stocks are stored at –20°C and are stable for at least 1 yr. Multiple freeze–thaw cycles should be avoided.
11. Platinum *Taq* DNA polymerase, 5 units/μL (Invitrogen). Other "hot-start" enzymes may also be used, although the 10X PCR reaction buffer composition may require modifications.
12. 1X TBE–ethidium bromide buffer: 89 m*M* Tris base, 89 m*M* boric acid, 2 m*M* EDTA, 0.5 μg/mL ethidium bromide.
13. Gel-loading buffer: 2.5 mg/mL xylene cyanol, 65% sucrose in TE buffer.
14. PCR master mix (sufficient for 30 reactions: *see* **Note 1**): Mix the following components in a 1.5-mL screw-capped tube; store at 4°C:

Sterile water	858 μL
10X PCR reaction buffer	150 μL
Deoxynucleotide triphosphate mix	120 μL
Primer MY11 (20 μ*M* stock)	30 μL
Primer MY09 (20 μ*M* stock)	30 μL
Platinum Taq DNA polymerase	12 μL

3. Methods

3.1. Extraction and Purification of DNA from Semen Specimens

The following is a modified protocol using the MasterPure DNA Purification Kit. All indicated reagents, with the exception of isopropanol, ethanol, and Pellet Paint NF coprecipitant solution, are included with the kit. In order to avoid contamination, all manipulations, with the exception of the incubation and centrifugation steps, should be performed in a laminar-flow biohazard hood (*see also* **Note 11**).

1. To 150 μL of raw semen or washed sperm in a sterile 1.5-mL screw-capped microcentrifuge tube, add 150 μL of 2X tissue and cell lysis solution, and 3 μL of 50 μg/mL proteinase K. Mix thoroughly by vortexing and incubate at 65°C for 1 h, vortexing every 15 min (*see* **Note 2**).
2. Place the tubes into a boiling water bath for 5 min, followed by chilling on ice for 5 min.
3. Add 170 μL of MPC protein precipitation reagent to each tube and vortex vigorously for 10 s. Place the tubes on ice for 5 min.
4. Pellet the debris by centrifugation for 10 min at 16,000*g*.
5. Pour the supernatant into a new sterile 1.5-mL screw-capped microcentrifuge tube and discard the pellet.
6. Add 1 μL of Pellet Paint NF coprecipitant to the supernatant (*see* **Note 3**).
7. Add 500 μL of isopropanol to the supernatant. Vortex the tube for 10 s.
8. Pellet the DNA by centrifugation for 12 min at 16,000*g*.
9. Carefully pour off the isopropanol without dislodging the pellet. The pellet itself should be clearly visible as a dark blue "button" at the bottom of the tube.
10. Add 900 μL of 70% ethanol to the tube. Vortex for 10 s and centrifuge for 5 min to return the pellet to the bottom of the tube.
11. Pour off and discard the ethanol. Close the tube and centrifuge for 5 s.
12. Without touching the pellet, carefully remove any residual ethanol using a pipet tip.
13. With the tube uncapped, allow the DNA pellet to air-dry for 3–5 min.
14. Add 40 μL of TE buffer to the pellet. Resuspend the DNA by gentle vortexing until there is no longer a blue-colored pellet visible in the bottom of the tube. The DNA solution can now be placed at –20°C for long-term storage or used immediately for PCR.

3.2. PCR Amplification (see Note 4)

1. Set out and label the required number of 0.2- or 0.5-mL thin-walled PCR tubes. For each extracted specimen, two separate reaction tubes must be set up. In addition, extra tubes for positive and negative controls are required.
2. Add 40 μL of PCR master mix to each tube. Leave the tubes uncapped.
3. Add 10 μL of extracted DNA preparation to each tube. Each patient specimen is tested in duplicate, with one reaction tube containing 10 μL of undiluted extracted DNA and the other containing 10 μL of a 1 in 5 dilution of the extracted DNA (*see* **Note 5**). The total volume in the reaction tubes will be 50 μL.
4. Prepare the control tubes as in **steps 2** and **3**. For negative controls, add 10 μL of water that has been extracted in parallel with the specimens to 40 μL of PCR master mix (it is not necessary to test a 1 in 5 dilution of the negative control). For positive controls, add 10 μL of extracted material (*see* **Note 6**) to 40 μL of PCR master mix.
5. Cap the tubes, mix by brief vortexing, and transfer to the thermal cycler (*see* **Note 7**).

6. Begin a thermocycling program as follows:
   a. Initial denaturation and *Taq* activation, 95°C for 5 min.
   b. Denaturation, 94°C for 35 s.
   c. Annealing, 55°C for 35 s.
   d. Extension, 72°C for 40 s.
   e. Repeat **steps b–d** 40 more times (41 total cycles of amplification).
   f. Final extension, 72°C for 5 min.
   g. Cool down to 4°C indefinitely.

3.3. PCR Product Detection and Interpretation (see *Note 8)*

1. Prepare a horizontal 1% agarose gel using 1X TBE–ethidium bromide buffer.
2. Remove 20 µL of the reaction mix after amplification and mix with 4 µL of gel-loading buffer.
3. Load the entire 24-µL volume into one well of the agarose gel. Include a separate lane containing a DNA marker (100-bp or 123-bp DNA ladder).
4. Electrophorese at 100–110 V until the tracking dye has migrated 1.5–2 cm into the gel.
5. Photograph the gel under UV transillumination (*see* **Note 9**).
6. A specimen positive for the presence of HPV will be indicated by the presence of an approx 450 bp amplified DNA fragment (between 448 and 454 bp, depending on the HPV subtype).
7. Specimens negative for HPV DNA after the initial amplification reaction should be confirmed by a seminested PCR protocol.

3.4. Seminested PCR Amplification

1. Assemble a PCR master mix as described in **Subheading 2.2.**, but substitute a 20-µ*M* stock of the MY11-IN primer (*see* **Table 1**) in place of the MY11 primer.
2. Remove 2 µL of the reaction products from the original amplification reaction and add it to 40 µL of the MY11-IN master mix in a thin-wall PCR tube (*see* **Note 10**). Add 8 µL of sterile water to bring the total volume to 50 µL.
3. Begin the same thermocycling program as above (*see* **Subheading 3.2.**).
4. Analyze the reaction products on a 1% agarose gel as above (*see* **Subheading 3.3.**).
5. A positive result is indicated by the presence of a DNA fragment of 430 bp.

4. Notes

1. For convenience, a PCR "master mix," which contains all of the reaction components with the exception of the extracted specimen itself, may be prepared in advance and is stable for at least 3 wk if stored at 4°C. The master mix recipe described is sufficient for 30 individual PCR reactions but may be made in smaller amounts if desired.
2. In addition to the semen specimens, include at least one tube containing 150 µL of water as a negative control. Ideally, one negative control tube is included for every six clinical specimens to be extracted. These controls are then carried through the extraction procedure in parallel with the semen samples.

3. The addition of Pellet Paint coprecipitant is optional, but is recommended. It does not affect the yield or quality of the extracted DNA, but because of its blue-black color, it helps to give a DNA pellet that is readily visible during subsequent steps of the extraction protocol.
4. To help avoid contamination, it is recommended that all manipulations involving PCR tubes (i.e., labeling, addition of master mix, addition of extracted specimen) be done in a separate, dedicated "dead-air" hood or biohazard hood. This hood must be different from the one in which semen specimens are extracted (although it may be in the same room) and a separate, dedicated set of pipettors must be used. If a dedicated hood is not available, then PCR reaction tubes can be set up on a bench that is physically removed from the area in which specimens were extracted.
5. Dilution of the extracted specimen prior to PCR amplification sometimes results in the appearance of an amplification product, whereas the same specimen, if tested undiluted, will not give a product or else will give only a very weak band. This may be because of PCR inhibitory material that is copurified with the DNA or could be the result of the presence of large amounts of human cellular DNA that is extracted with the HPV DNA. Dilution of the specimen should be done in sterile water.
6. Positive controls are set up only at the PCR amplification stage and are not extracted in parallel with clinical specimens. Positive controls may consist of DNA purified (using the MasterPure kit as described) from a cultured cell line known to harbor HPV. For example, the SiHa cell line (ATCC no. HBT35) carries HPV type 16, whereas the common HeLa cell line carries a copy of HPV type 18 stably inserted into the genome. Alternatively, positive controls may be purified plasmids containing the cloned HPV L1 gene or may be DNA extracted from a known positive clinical specimen. In all cases, the positive control DNA should be titrated such that if 10 μL is added to a PCR reaction tube, only a weakly positive signal will result.
7. If the thermal cycler does not have a heated lid (to prevent condensation in the lids of the reaction tubes), then the contents of the PCR tubes must be overlain with 50 μL of sterile paraffin oil, which will act as an evaporation barrier.
8. The electrophoresis equipment used to visualize PCR products must be set up in a room physically removed from the area used for DNA extraction and reaction setup. Ideally, the thermal cycler can be set up in the electrophoresis room as well. This arrangement avoids the possibility of amplified PCR products inadvertently being carried over into new clinical specimens and giving rise to false-positive results.
9. Gels can be photographed using a Polaroid camera and a standard UV transilluminator. However, we have found that the ability to visualize very faint DNA bands is greatly improved by using a digital photoimaging system consisting of a UV transilluminator, a charge-coupled device (CCD) camera, and a computer interface.

10. The potential for contamination at this stage is great because the reaction tubes from the first round of amplification may contain amplified HPV DNA. Care should be taken to avoid aerosols when opening PCR tubes. Pipettors and working surfaces should be wiped with 10% bleach after the seminested reactions have been set up.
11. Good general laboratory technique is essential for obtaining reliable results and for avoiding problems resulting from carryover contamination. In addition to some of the considerations addressed, the following points should be emphasized:
    - Follow techniques that ensure accurate pipetting, especially when small volumes are involved.
    - Perform monthly calibration checks of pipettors and other equipment.
    - Decontaminate pipettors and work surfaces with 10% bleach after each use.
    - Gowns and gloves must be worn at all times and changed frequently.
    - Aerosol-resistant pipet tips are used for all pipetting steps.
    - A minimum of two physically separated laboratory rooms are required: one room for extraction of specimens and assembly of reaction tubes and a second room for amplification and gel electrophoresis. Dedicated pipettors are required for each room.
    - Store critical reagents (e.g., primers, buffers) as small volumes in multiple tubes. This avoids the deleterious effects of multiple freeze–thaw cycles and ensures that if contamination does occur, large stocks of reagents are not ruined.

References

1. Melbye, M. and Frisch, M. (1998) The role of human papillomaviruses in anogenital cancers. *Semin. Cancer Biol.* **8(4),** 307–313.
2. Kyo, S., Inoue, M., Koyama, M., et al. (1994) Detection of high-risk human papillomavirus in the cervix and semen of sex partners. *J. Infect. Dis.* **170(3),** 682–685.
3. van den Brule, A. J., Meijer, C. J., Bakels, V., et al. (1990) Rapid detection of human papillomavirus in cervical scrapes by combined general primer mediated and type specific polymerase chain reaction. *J. Clin. Microbiol.* **28,** 2739–2743.
4. Lai, Y. M., Yang, F. P., and Pao, C. C. (1996) Human papillomavirus deoxyribonucleic acid and ribonucleic acid in seminal plasma and sperm cells. *Fertil. Steril.* **65(5),** 1026–1030.
5. Astori, G., Pipan, C., Muffato, G., et al. (1995) Detection of HPV-DNA in semen, urine and urethral samples by dot blot and PCR. *New Microbiol.* **18(2),** 143–149.
6. Ostrow, R. S., Zachow, K. R., Nimura, M., et al. (1986) Detection of papillomavirus DNA in human semen. *Science* **231,** 731–733.
7. Olatunbosun, O., Deneer, H., and Pierson, R. (2001) Human papillomavirus DNA detection in sperm using polymerase chain reaction. *Obstet. Gynecol.* **97(3),** 357–360.
8. Green, J., Monteiro, E., Bolton, V. N., et al. (1991) Detection of human papillomavirus DNA by PCR in semen from patients with and without penile warts. *Genitourin. Med.* **67(3),** 207–210.

9. Tanaka, H., Karube, A., Kodama, H., et al. (2000) Mass screening for human papillomavirus type 16 infection in infertile couples. *J. Reprod. Med.* **45(11),** 907–911.
10. Rintala, M. A., Pollanen, P. P., Nikkanen, V. P., et al. (2002) Human papillomavirus DNA is found in the vas deferens. *J. Infect. Dis.* **185(11),** 1664–1667.
11. Chan, P. J., Su, B. C., Kalugdan, T., et al. (1994) Human papillomavirus gene sequences in washed human sperm deoxyribonucleic acid. *Fertil. Steril.* **61,** 982–985.
12. Bauer, H. and Manos, M. (1993) PCR detection of genital human papillomavirus, in *Diagnostic Molecular Microbiology, Principles and Applications* (Persing, D., Smith, T., Tenover, F., et al., eds.), American Society for Microbiology, Washington, D.C., pp. 407–413.
13. Kwok, S. (1990) Procedures to minimize PCR product carryover, in *PCR Protocols: A Guide to Methods and Applications* (Innis, M., Gelfand, D., Sninsky, J., et al., eds.), Academic, San Diego, pp. 142–145.
14. Schiffman, M. H. and Schartzkin, A. (1994) Test reliability is critically important to molecular epidemiology: an example from studies of human papillomavirus infection and cervical neoplasia. *Cancer Res.* **54,** 1944–1947.
15. Jacobs, M. V., Snijders, P. J. F., Voorhorst, F. J., et al. (1999) Reliable high risk HPV DNA testing by polymerase chain reaction: an intermethod and intramethod comparison. *J. Clin. Pathol.* **52,** 498–503.
16. CFAS (2001) *Canadian Fertility and Andrology Society Guidelines on Artificial Insemination*, CFAS, Montreal, pp. 1–31.
17. American Fertility Society (1991) Revised guidelines for the use of semen donor insemination. *Fertil. Steril.* **56,** 394–401.
18. Lavitrano, M., Camainoni, A., Fazio, V. M., et al. (1989) Sperm cells as vectors for introducing foreign DNA into eggs: genetic transformation of mice. *Cell* **57,** 717–723.
19. Chan, P. J., Seraj, I. M., Kalugdan, T. H., et al. (1995) Blastocysts exhibit preferential uptake of DNA fragments from the E6–E7 conserved region of the human papillomavirus. *Gynecol. Oncol.* **58,** 194–197.
20. Kim, L. U., Johnson, M. R., Barton, S., et al. (1999) Evaluation of sperm washing as a potential method of reducing HIV transmission in HIV-discordant couples wishing to have children. *AIDS* **13,** 645–651.

8

Spermatogonial Stem Cells

Karim Nayernia, Manyu Li, and Wolfgang Engel

1. Introduction

1.1. Origin and Specification of Spermatogonial Stem Cells

Spermatogenesis is a continuous, highly organized process comprised of sequential steps of cell proliferation and differentiation *(1)*. In male mammals, spermatogenesis proceeds for the reproductive lifetime of the animals. The continuation of this process depends on a pool of spermatogonial stem cells within the testes that undergo asymmetric division to both maintain the stem cell population and give rise to progenitors that will proceed through spermatogenesis to generate mature spermatozoa (*see* **Fig. 1A**). The male germline proceeds through several developmental steps prior to the establishment and initiation of spermatogonial stem cell division in the testis *(2–4)*. Primordial germ cells (PGCs) are the founders of the gametes. Progenitors in mice have been recognized in the proximal epiblast *(5)*. The PGCs proliferate and migrate from their site of origin to the future position of the gonad, where they associate with somatic gonadal precursor cells to form the gonad *(5)*. Once within the gonad, the PGCs differentiate in a sex-specific manner, including a distinct program of proliferation and quiescence (*see* **Fig. 1A**) *(5)*. In the male genital ridge, the PGCs become enclosed by somatic supporting cells, the precursor Sertoli cells. PGCs and Sertoli cells then, together, form solid strands of cells, which are called seminiferous cords. Later during development, these cords form a lumen and become seminiferous tubules. When the PGCs are enclosed in seminiferous cords, they change morphologically and are then called gonocytes. The gonocytes proliferate for a few days and then arrest in the G0/G1 phase of the cell cycle (*see* **Fig. 1A**). In rats and mice, the gonocytes resume proliferation within a few days after birth to give rise to spermatogonial stem

From: *Methods in Molecular Biology, vol. 253: Germ Cell Protocols: Vol. 1 Sperm and Oocyte Analysis*
Edited by: H. Schatten © Humana Press Inc., Totowa, NJ

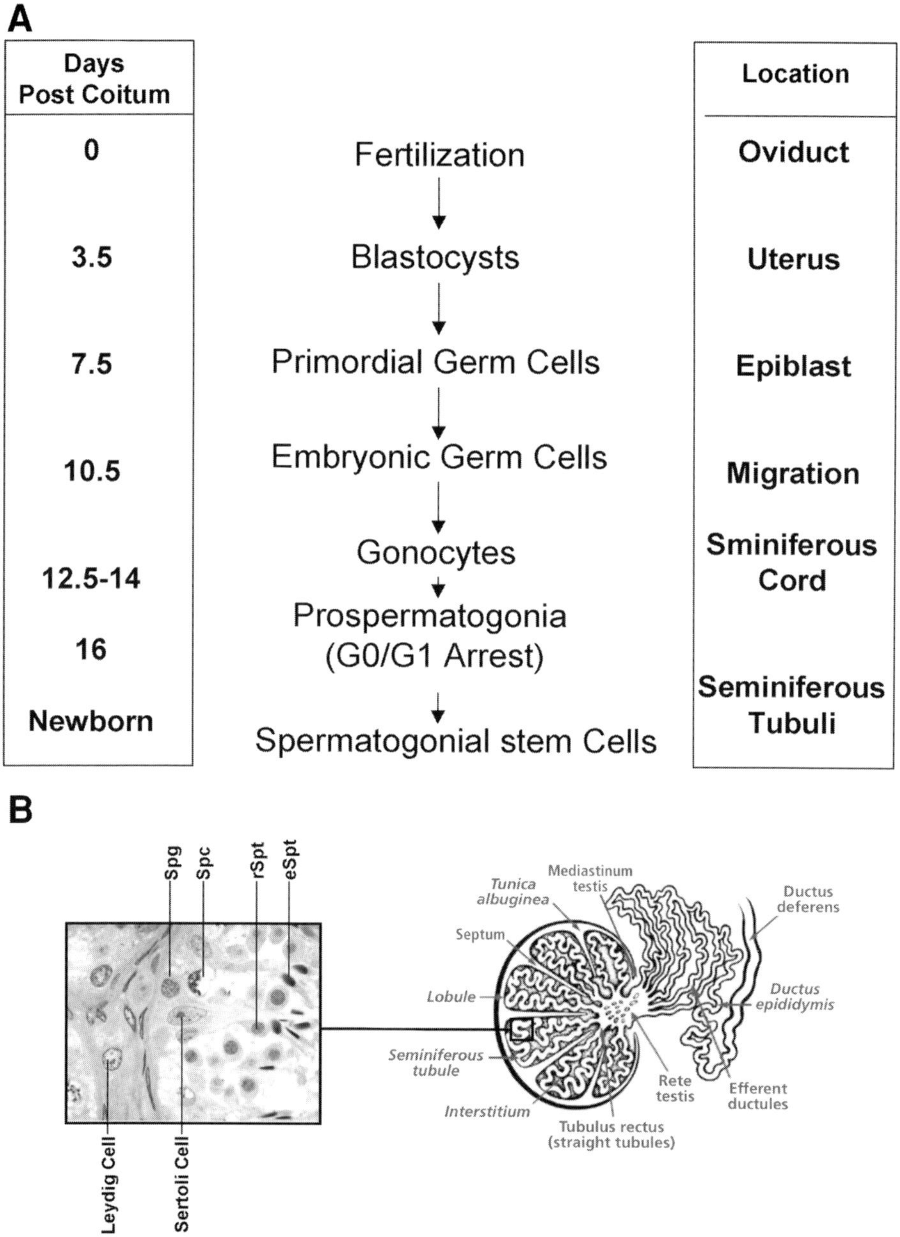

Fig. 1. **(A)** Origin of spermatogonial stem cells in prenatal and newborn mice. **(B)** Location of spermatogonial stem cell (Spg) in testis and differentiation of Spg to further stages: spermatocyte (Spc), round spermatid (rSpt), and elongating spermatid (eSpt).

cells (SSCs) ***(6)***. SSCs initiate the first round of spermatogenesis that will produce sperms at the onset of reproductive age.

1.2. Location of SSCs in the Testis

Mammalian spermatogenesis takes place in the seminiferous epithelium, inside the multiple seminiferous tubules that compose the testis (*see* **Fig. 1B**). The ability to identify spermatogonial stem cells *in situ* is, at present, relatively unique to the male germline, in contrast to the difficulty of unambiguously identifying most other stem cell types in vivo (e.g., hematopoietic stem cells within the bone marrow). Spermatogenesis proceeds in a spatial gradient within the testis (*see* **Fig. 1B**), allowing SSCs actively to be easily visualized as the continual production of differentiating germ cells (*see* **Fig. 1B**). Germ cells move radially inward as spermatogenesis proceeds until sperm are released into the central lumen (*see* **Fig. 1B**). Spermatogenesis occurs in successive waves along the length of the tubules. Spermatogenesis is a cyclic process that, in mice, can be divided into 12 stages (I–XII) ***(7)***.

In stage VIII, SSCs are present. From stage X onward, these cells start to proliferate ***(8)***. Spermatogonial multiplication and stem cell renewal can best be studied in whole mounts of seminiferous tubules, because in this way, the topographical arrangement of the cells is preserved ***(9)***.

1.3. Molecular and Cellular Characteristics of SSCs

The generally accepted scheme of spermatogonial renewal and differentiation has been proposed by Huckins ***(10)*** and Oakberg ***(11)***. In this model, the A_s (A single) spermatogonia are considered to be the stem cells of spermatogenesis. Upon division of A_s spermatogonia, the daughter cells can either separate and become two independent stem cells or they can remain connected by an intercellular bridge and become A_{pr} (pair) spermatogonia. The Apr spermatogonia divide further to form chains of 4, 8, or 16 (aligned) spermatogonia. A_{al} spermatogonia divide and differentiate into type A spermatogonia (A1–A4) that further develop into intermediate (In) and B spermatogonia. Finally, type B spermatogonia differentiate into primary spermatocyte (*see* **Fig. 2**). Numerous cellular characteristics were described for spermatogonial cell types ***(11–13)***. The group comprising A_s, A_{pr}, and A_{al} spermatogonia could be differentiated based primarily on mottling of heterochromatin throughout the nucleus in the absence of heterochromatin lining the nuclear envelope. The A_1 cells display finely granular chromatin throughout the nucleus and virtually no flakes of heterochromatin along the nuclear membrane. The A_2 through A_4 spermatogonia contain progressively more heterochromatin rimming the nucleus. Intermediate-type spermatogonia display flaky or shallow heterochromatin that completely rims the nucleus. Type B spermatogonia show rounded heterochro-

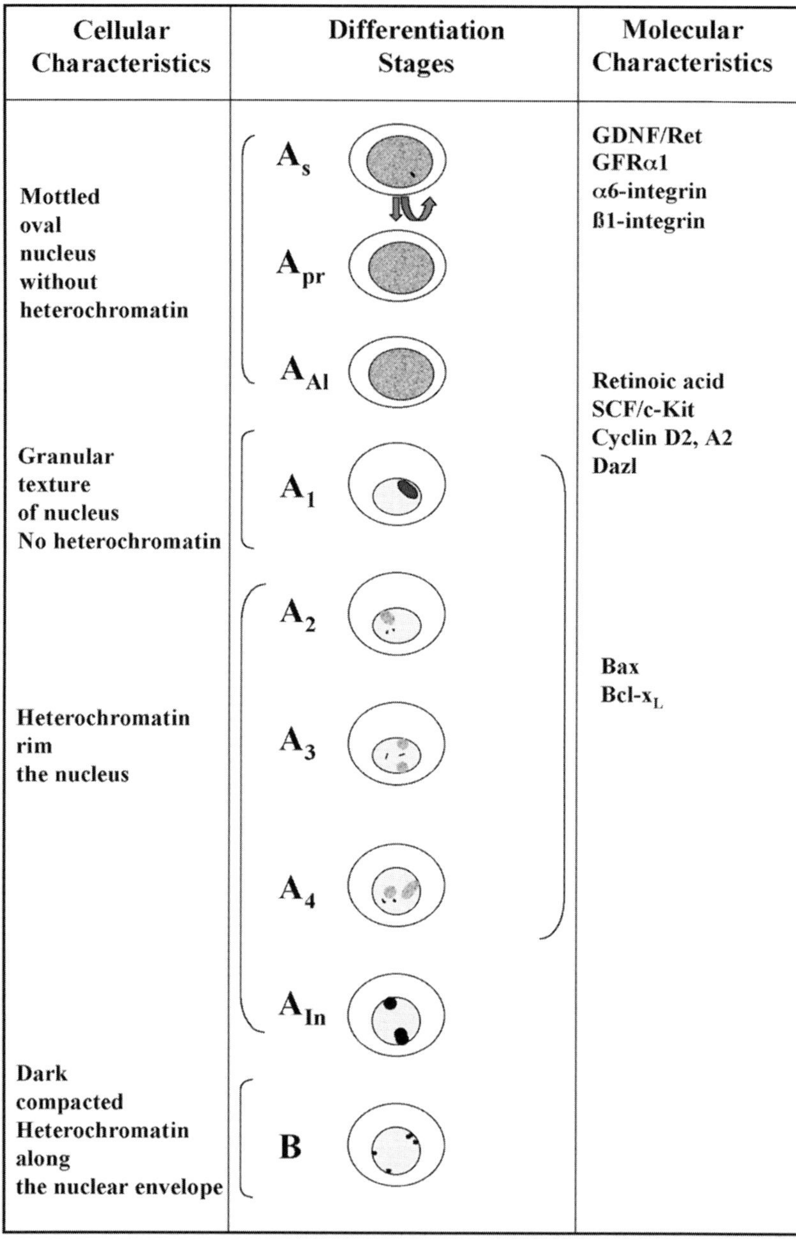

Fig. 2. Cellular and molecular characteristics of spermatogonia. Spermatogonial stem cells (A_s) proliferate and also differentiate to A-type paired spermatogonia (A_{pr}) that are joined together by intercellular cytoplasmic bridges. Further divisions of A_{pr} produce chains of aligned spermatogonia (A_{al}). These differentiate through six mitotic divisions into A1, A2, A3, and A4, intermediate (A_{In}), and B spermatogonia.

matin periodically along the nuclear envelope (*see* **Fig. 2**). Spermatogonial stem cells and their progeny are contained in the germinal epithelium of the semniferous tubules, in close association with the somatic Sertoli cell ***(14)***. Regulatory mechanisms mediated by growth factors induce or inhibit the proliferation and differentiation of SSCs ***(15)***.

There must be regulatory mechanisms controlling the ratio between self-renewal and differentiation of SSCs. In the normal case, the ratio between self-renewal and differentiation of SSCs should be about 1.0. Currently, there is a limited understanding of the molecular basis that controls these mechanisms. Transgenic loss-of-function and overexpression models show that the dosage of glial cell line-derived neurotrophic factor (GDNF) produced by Sertoli cells regulates the cell fate decision of undifferentiated spermatogonial stem cells. At a low GDNF level, spermatogonia favor differentiation, and at a high level, they favor self-renewal ***(16)***. In addition, the receptors for GDNF, Ret, and GFRα1 are expressed by a spermatogonial subset that indicates that this group represents the GDNF-responsive cells and that Sertoli cells can regulate the number of SSCs (*see* **Fig. 2**). Different molecules involved in differentiation of SSCs into A1 spermatogonia have been reported ***(17,18)***. A specific, early point of differentiation in spermatogonia lineage takes place in epithelial stages VII–VIII, during which A_{al} spermatogonia differentiate into A_1 spermatogonia. This differentiation step involves the action of retinoic acid, the SCF (stem cell factor)/c-kit system, cyclins D2 and D3, and the Dazl protein and seems to become disturbed by high concentrations of testosterone (*see* **Fig. 2**) ***(18)***. In adult testis, the tyrosine-kinase receptor c-kit is re-expressed in differentiating type A spermatogonia, but not in spermatogonial stem cells, whereas Sertoli cells under follicle stimulation hormone (FSH) stimulation express its ligand, SCF. SCF/c-kit system upregulates cyclin D3 and promotes cell cycle progression in spermatogonia through a rapamycin-sensitive phosphoinositide 3–kinase/p70 S6 kinase pathway ***(19)***. Mutations in the c-kit and the SCF genes have a variable effect on spermatogenesis, indicating that this system has a role at various stages in the spermatogenetic lineage.

Dazl (deleted in azoospermia-like) appears to be an RNA-binding protein and is unique among the characterized genes in that it is expressed exclusively in spermatogonia, so its effects are not mediated indirectly through the Sertoli cells ***(20)***. In Dazl-deficient mice, the differentiation of A_{al} spermatogonia into A_1 also does not take place ***(20)***. Cyclin-dependent kinases (cdks) and cyclins likely govern both the mitotic and meiotic divisions. Cyclin D2 and D3 and their catalytic partner Cdk4 were found to be expressed in spermatogonia ***(21)***. By contrast, little cyclin D2 and Cdk4 expression was observed later in spermatocytes and spermatids ***(21)***, although cyclin D3 expression was maintained ***(21)***. The finding of null mice for cyclins D2 and D3 and Cdk4 suggest that cyclins D2 and D3 in combination with Cdk4, and possibly Cdk6, may regulate G1 progression and differentiation of spermatogonia *(22,23)*.

There are several other factors that are involved in the differentiation of spermatogonia. For example, in vitamin A–deficient mice and rats, the differentiation of the A_{al} spermatogonia into A_1 spermatogonia is arrested *(24)*. In mice made artificially cryptorchid, spermatogenesis deteriorates to the point at which only actively proliferating A spermatogonia remain that produce few or no B spermatogonia *(25)*. Furthermore, in juvenile spermatogonial depletion (jsd) mutant mice, spermatogenesis starts normally during development but then also declines ultimately to the point at which only proliferating undifferentiated and possibly differentiating type A spermatogonia are left *(26)*. Finally, differentiating germ cells have been reported to be missing in adult mice homozygous for the Steel17H mutation *(27)*. Other molecules also have been found expressed in spermatogonia or Sertoli cells, but their functions remain unknown. For instance, three Notch receptors have been detected on spermatogonia, and Delta 1 and Jagged 1, ligands for Notch, have been detected on Sertoli cells or spermatogonia *(28)*. Notch functions in cell fate decisions in other tissues, so it is conceivable that these signaling pathways have similar roles in spermatogonial decisions *(29)*.

2. Materials

Use the following materials for the isolation, culture, and transplantation of spermatogonial stem cells:

1. Donor male mice and rats, available from the Jackson Laboratory (Bar Harbor, ME, USA).
2. Homozygous Insl3–/– male mice *(30)*.
3. Recipient mice (Jackson Laboratory).
4. Nylon mesh 70-µm pore size (352350; BD Bioscience, Heidelberg, Germany).
5. 15P-1 Sertoli cell line *(31)*.
6. SIM mouse embryo-derived thioguanine- and ouabain-resistent fibroblast cell line (STO) (CRL-1503; American Type Culture Collection, Manassas, VA) *(32)*.
7. 1-mL Plastic syringe barrel (Braun, Melsungen, Germany).
8. Pipet puller (Bachofer, Beutlingen, Germany).
9. Grinding stone (Bachofer).
10. Borosilicate glass pipet (World Precision-Instruments, Berlin, Germany).
11. Microforge (Schütt, Göttingen, Germany).

2.1. Reagents

1. Collagenase type IV (CI889, Sigma-Aldrich, or 17104, GIBCO Invitrogen Corp., Karlsruhe, Germany).
2. DNAse I, D5025 (Sigma-Aldrich).
3. Dispase (17105; GIBCO Invitrogen Corp.)

4. Laminin (L6274; Sigma-Aldrich).
5. Mitomycin C (M4287, Sigma-Aldrich).
6. Busulfan (1,4-butanediol dimethansulfonate) (B2635; Sigma-Aldrich).
7. DMSO (dimethyl sulfoxide) (B2635; Sigma-Aldrich).
8. Ketavet 10% (Parke-Davis, Berlin, Germany).
9. Rumpun 2% (Bayern, Leverkusen, Germany).
10. Trypan blue (T1646, Sigma-Aldrich).

2.2. Solutions

1. Dulbecco's modified Eagle's medium (DMEM), 10829-018; GIBCO Invitrogen Corporation, containing 10% (v/v) fetal bovine serum (FBS), F3018; Sigma-Aldrich, 1X penicillin–streptomycin solution (100X), P0781; Sigma-Aldrich, 1X nonessential amino acids, 11140-019, GIBCO Invitrogen Corporation.
2. DMEM medium, P04-03590; PAN Biotech containing 0.5% Nu Serum type IV, 355504 (BD Biosciences, Heidelberg, Germany).
3. Collagenase/DNAase solution: Hank's balanced salt solution without calcium chloride and magnesium sulfate (H9394, Sigma-Aldrich, Munich, Germany) containing 1 µg/mL collagenase and 300 µg/mL DNAase.
4. Dispase/DNase solution: Hank's balanced salt solution without calcium chloride and magnesium sulfate containing 1.5 U/mL and 400 µg/mL DNAse.
5. Trypsin–EDTA solution (0.25%) (T4049; Sigma-Aldrich).
6. Cell-freezing solution: Dulbecco's modified Eagle's medium (DMEM) containing 10% FBS, 2 m*M* glutamine, 6 m*M* lactate, 0.5 m*M* pyruvate, 30 µg/mL penicillin, and 50 µg/mL streptomycin containing FBS and DMSO (in a ratio of 3:1:1).
7. Cell-culture-freezing medium (11101, GIBCO Invitrogen Corp.)
8. Dulbecco's phosphate-buffered saline (DPBS) (P04-36100; PAN Biotech, Aidenbach, Germany).

3. Methods

3.1. Isolation of Spermatogonial Cell Populations

Spermatogonial stem cells represent a small percentage of cells in the testis of any animal. In the mouse, there are about 10^8 cells in the testis and approx $(2–3) \times 10^4$ are thought to be stem cells ***(33)***. The rat testis contains about 1 stem cell per 504 total cells (418.5×10^6 stem cells per testis) ***(10)***. The rat testis has both a high concentration and a large number of spermatogonial stem cells as compared with the mouse testis ***(34)***. Thus, most germ cells in the seminiferous tubules are mitotic spermatogonia, spermatocytes, and highly differentiated cells such as spermatids and sperm ***(1)***. It could be calculated from cell counts in mouse testis that spermatogonia stem cells (A_s spermatogonia) constitute 10.6% of the undifferentiated spermatogonial, 1.25% of all spermatogonia, and 0.03% of all germ cells ***(35)***. Therefore, it is advantageous to start from testes that do not contain the full complement of germ cells.

To achieve this, several models are developed. Starting from young mice, over 90% pure A spermatogonia could be prepared *(36)*. Cell suspension from young animals will contain undifferentiated type A spermatogonia as well as the more numerous differentiating type A_1–A_4 spermatogonia and probably not more than about 1% stem cells.

In order to purify undifferentiated spermatogonia specifically, one can use animal models in which these cells are enriched. For example, in vitamin A–deficient rats and mice, spermatogenesis becomes arrested at the level of undifferentiated spermatogonia *(24,37)*. Starting from testes of vitamin A–deficient mice and rats, 80–90% pure undifferentiated A spermatogonia can be prepared *(24,37)*. Other models that show an arrest at about the same differentiation stages include cryptorchid mice *(38)* and irradiated rats *(39)*.

In the following method, we used the testes from Insl3–/– mice, which show bilateral cryptorchidism *(30)*. A population of testis cells from Insl3–/– mice was prepared by a modified enzymatic procedure described by Bellvé *(40)*. The stem cell potential of isolated cells was be examined using the transplantation technique (*see* **Subheading 3.4.**).

1. Preparation of collagenase/DNAase solution: Dissolve the nonsterile, lyophilized collagenase in Hank's balanced salt solution (HBSS) in a concentration of 10 mg/mL. Filter-sterilize the solution with a cell-culture-approved filtration unit, aliquot, and store frozen. Thaw stock solution in a refrigerator immediately prior to use and dilute with HBSS to a final concentration of 1 mg/mL. Add 300 µg/mL DNAase to the solution.
2. Preparation of dispase/DNase solution: Dissolve the lyophilized dispase in PBS to 10 mg/mL. Filter-sterilize through a 0.22-µm filter membrane. Further dilute with HBSS to a final concentration of 1.5 U/mL. Add 400 µg/mL DNAse to the solution.
3. Isolate testes from four or five homozygous Insl3$^{-/-}$ mice (*see* **Note 1**), wash in PBS, and put in 10 vol (4–5 mL) of collagenase /HBSS solution from **step 1** in a 60-mm sterile Petri dish.
4. Remove the tunica albuginea from the testes.
5. Incubate at 37°C with gentle agitation in a rocker platform until the tubules separate (*see* **Note 2**).
6. Pour off the tubules using a 10-mL pipet and transfer to a 10-mL cell culture centrifugation tube.
7. Centrifuge at 4°C at 650*g* for 5 min.
8. Aspirate the supernatant, add 4-5 mL of HBSS, and centrifuge again at 4°C at 650*g* for 5 min.
9. Repeat **step 8** three times.
10. Add 2 mL of dispase solution from **step 2** (*see* **Note 3**).
11. Stir slowly at 37°C until the tissue is sufficiently dissolved (30 min). Dispersion of the tubule cells can be hastened by pipetting and gentle agitation. Fresh dispase

solution may be added to the remaining tubule fragment if further disaggregation is required.

12. Following digestion, the cell suspension is separated from large pieces by passing the mixture through a nylon mesh with a 70-μm pore size.
13. Pellet cells by centrifugation at 16°C at 650*g* for 5 min and decant the enzyme solution.
14. Resuspend the pellet in DMEM containing 10% FBS (*see* **Note 4**).

3.2. Enrichment of Spermatogonial Stem Cells

Because α6-and α1-integrins comprise a known receptor for laminin and both integrins were individually identified on mouse ***(41)*** and rat spermatogonial stem cells ***(34)***, these antigens can be used for further enrichment of rat and mouse testis cell populations for spermatogonial stem cells. In the mouse, laminin selection provided 5-fold to 7-fold enrichment ***(38)***, and in the rat, in vitro selection using laminin resulted in 8.5-fold enrichment ***(34)***.

1. Preparation of laminin-coated dishes: Incubate 60-mm Falcon Petri dishes overnight at 37°C with 5 mL Laminin solution (20 μg/mL PBS). Remove solution and wash the dishes three times with 2 mL PBS. Nonspecific binding is prevented by preincubating the dishes with 0.5 mg/mL of bovine serum albumin (BSA)–PBS solution for 1 h at 37°C. Wash the dishes three times with PBS before adding cells.
2. Suspend 4×10^7 cells from **step 14** (*see* **Subheading 3.1.**) in 8 mL DMEM containing 10% FBS.
3. Add 2 mL of cell suspension to each laminin-coated dish.
4. Incubate for 15 min at 37°C and wash the dishes five times with PBS to remove unbound cells.
5. Cover the cells with dispase solution, prewarmed to 37°C. Incubate for 5 min at 37°C.
6. Decant the solution and incubate for an additional 10 min at 37°C.
7. Monitor detachment using a microscope. If necessary, incubate for an additional 15 min until detachment is complete.
8. Wash the cells with culture medium and resuspend the cells in fresh DMEM medium containing 10% FBS.

3.3. Culture of Spermatogonial Stem Cells

Spermatogonial stem cells can be maintained in culture for long periods of time and probably are undergoing cell division ***(42)***. To date, culture systems for spermatogonia of various species have been set up ***(43)***. The objectives varied, but success has been measured by the recapitulation of some specific sequence in the total spermatogenic process (e.g., the initiation of meiosis or spermatid differentiation) or by the initiation of spermatogenesis after transplantation to the testis. For further differentiation, spermatogonial stem cells should be cocultured with Sertoli cells. It was shown that somatic support cells restrict germline stem cell self-renewal and promote differentiation ***(31,44)***. For replication of spermatogonial stem cells and suppression of differentiation, a coculture with STO feeder cells is recommended ***(42)***.

*3.3.1. Coculture with Sertoli Cell Line 15P-1 (see **Note 5**)*

1. Plate Sertoli cells in 50-mm tissue culture dishes in DMEM medium containing 10% FBS and incubate at 32°C and 5% CO_2 for 3 d.
2. Trypsinize and seed the Sertoli cells into Falcon 24-well plates at a concentration of 1×10^5 cells per well in DMEM medium containing 10% FBS and incubate for 1 d.
3. Pour off the medium from the wells and replace it by DMEM medium containing 0.5% Nu serum (*see* **Note 6**).
4. Add 10^5–10^6 isolated germ cells from **step 14** (*see* **Subheading 3.1.**) and incubate the cocultures at 32°C and 5% CO_2 for 3 d.
5. After 3 d, pour off the medium and replace it by fresh DMEM medium containing 10% FBS.
6. Medium is changed two or three times each week, and the cells are cultured for further analysis or are frozen as described in **Subheading 3.4.**

*3.3.2. Coculture with STO Feeder Cells (see **Note 5**)*

1. Prepare gelatinized six-well plates: Add 0.1% gelatine into plate and incubate at 4°C for 2–3 h. After incubation, wash the plates two times with PBS.
2. Prepare STO medium: DMEM + 5% FBS +1% nonessential amino acid.
3. Plate STO cells on 60-mm culture dishes in STO medium and culture the cells for 2–3 d at 37°C and 5% CO_2.
4. Replace the medium on a confluent culture of STO cells with DMEM medium containing mitomycin C (final concentration 8 µg/mL).
5. Return STO cells to incubator for 3 h.
6. After incubation, wash the cells three times with PBS, trypsinize, spin down, pour off medium, and resuspend in STO medium.
7. Count the cells and dilute to 2.5×10^5/mL in STO medium.
8. Plate 2 mL of the STO cell to the six-well plates. These cells can be used as feeders up to 10 d after preparation.
9. Add 10^5–10^6 isolated germ cells from **step 14** (*see* **Subheading 3.1.**) and incubate the cocultures at 37°C and 5% CO_2.
10. The cells are cultured for further analysis or are frozen as described in **Subheading 3.4.**

3.4. Cryopreservation of Testicular Germ Cells

Cryopreservation of various cell suspensions is widely practiced in science and medicine, and rapid freezing, using either demethyl sulfoxide (DMSO) or glycerol as cryoprotectant, provides satisfactory results for most purposes.

Avarbock et al. ***(45)*** were able to establish spermatogenesis in recipient animals by transplanting a frozen-thawed suspension of mouse testicular cells having used DMSO as a cryoprotectant.

1. Prepare cell-freezing medium: DMEM medium containing 10% FBS, 2 m*M* glutamine, 6 m*M* lactate, 0.5 m*M* pyruvate, 30 μg/mL penicillin, and 50 μg/mL streptomycin.
2. Prepare freezing solution: Freezing medium + FBS + DMSO (in a ratio of 3:1:1).
3. Pellet germ cells (from **step 14**, **Subheading 3.1.**) by centrifugation at 650*g* for 5 min and pour off the medium.
4. Add freezing solution slowly by drops to the cells and mix (at a concentration of about $[20–40] \times 10^6$ cell/mL).
5. Allocate cells 1.0 mL per freezing vial and freeze them at –70°C for 12–24 h.
6. Transfer the vials in liquid nitrogen (–196°C) for long-term storage.
7. For thawing, remove cells from storage temperature and quickly thaw in a 37°C water bath.
8. Slowly dilute in 1 mL of frozen cells by drops with 3 mL of freezing medium by swirling and mixing gently.
9. Mix cells gently and pellet cells by centrifugation at 650*g* for 5 min.
10. Remove supernatant and discard. Gently resuspend cells in DMEM medium containing 10% FBS.

3.5. Transplantation of Spermatogonial Stem Cells

Spermatogonial stem cell transplantation is a novel technique in which donor testicular cells are transferred into recipient testes. Spermatogonial stem cells not only colonize the recipient testes, but they also initiate spermatogenesis and produce sperm capable of fertilization ***(45)***. This technique will allow the scientist to investigate the fundamental aspect of spermatogenesis, provide a method to regenerate spermatogenesis in infertile individuals, and genetically manipulate germ cells in vitro to develop transgenic animals.

Isolated germ cells from **step 8** of **Subheading 3.2.** can be injected into the testis by several methods. The most widely used method involves inserting a needle through the efferent ductules outside of the testis and passing the needle into the rete testis, a bundle of tubules that connects the seminiferous tubules to the epididymis. The injected cell mixture contains trypan blue, so that filling of seminiferous tubules can be monitored (*see* **Fig. 3**). The injection is performed without a micromanipulator or pressure injection. The micropipet is hand-held and pressure is created manually using a 1-mL syringe connected to the micropipet. Importantly, to enable the evaluation of the success of the transplantation, the recipient testis should have no or negligible endogenous spermatogenesis. Therefore, busulfan-treated mice or homozygous dominant white spotting mutant (W/W^v) mice, which do not have endogenous spermatogenesis, can be used.

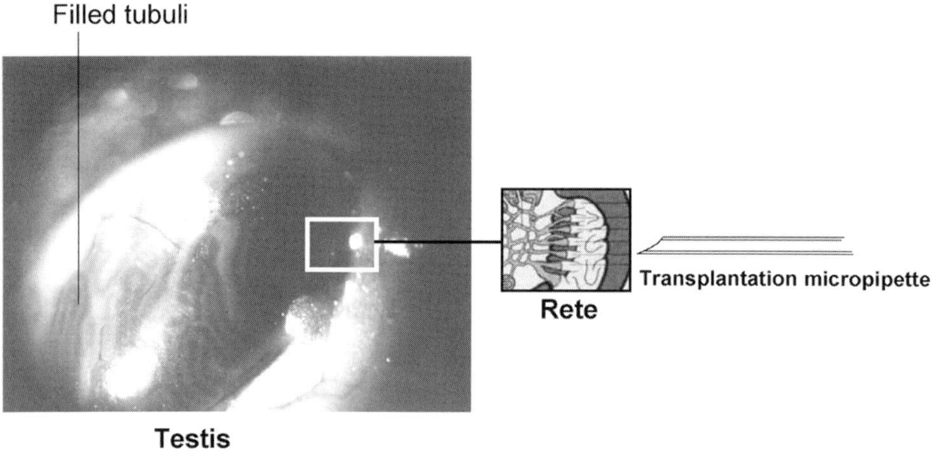

Fig. 3. Transplantation of spermatogonial cell suspension through rete testis. Micropipet tip is inserted directly into the rete. After injection, the tubules of testis are filled.

1. Preparation of busulfan solution: Dissolve busulfan in DMSO and then add an equal volume of sterile distilled water to provide a final concentration of 4.0 mg/mL. Precipitation of busulfan should be delayed by mixing the DMSO and water phase just before use and maintaining the solution at 37°C.
2. Preparation of busulfan-treated mice: It is important that the recipient mice be immunologically tolerant to the donor testis cells. Inject busulfan solution intraperitoneally to the mice (6–8 wk old) at a dose of 0.40 mg/10 g body weight. After busulfan treatment, the male mice are maintained for 1–2 mo before use for transplantation (*see* **Note 7**).
3. Preparation of transplantation micropipet: Draw a borosilicate glass tube (8 cm in length) on a pipet puller to form two pipets with thin, pointed tips. Then, break the tip of the pipet on a glass anvil using a micoforge to give a diameter of 40–50 μm. Then, bevel the pipet tip on a stone-grinding wheel. Rinse the beveled pipet tip several times with distilled water. Sharpen the tip of the pipet by touching it to a microforge filament. After sharpening, rinse pipets repeatedly in distilled water. A micropipet for transplantation is shown in **Fig. 3**.
4. Preparation of testicular germ cells: Resuspend the cells from **step 8** of **Subheading 3.2.**) or from **step 8** of **Subheading 3.2.**) in 400 μL DMEM medium at a concentration of $(2–3) \times 10^7$ cells/mL (*see* **Note 8**). Add 0.4% trypan blue to the cell suspension. Load the cell suspension into a transplantation micropipet.
5. Preparation of anesthetic solution: Add 1 mL Rumpun and 1 mL Ketavet to 8 mL distilled water and mix it by shaking.
6. Preparation of recipient mice: Weigh the busulfan-treated male mice and anesthetize by peritoneally injection with Rumpun/Ketavet solution (100 μL/10 g of body weight). Shave the abdomen and wipe with 70% ethanol.

7. Open the body wall. Cut the skin at a point level with the top of the legs. Both testes can be reached through the one incision.
8. Remove one testis from the body and reflect it laterally. Carefully dissect away surrounding fat and tissue from the area cranial to the vascular pedicle connecting the testis to the testicular cord. The rete lies on the surface of the testis, mostly under the spreading branches of the vessels (*see* **Fig. 3**).
9. Insert the loaded micropipet into the area cranial to the vessels and adjacent to the efferent ducts (*see* **Fig. 3**). The micropipet should be almost parallel to the surface of the testis (*see* **Note 9**).
10. After entering the rete, increase pressure very slowly in the injection tube until the cell suspension fills the rete and flows into the tubules (*see* **Fig. 3**). About 10–20 μL of cell suspension is required to fill one testis (*see* **Note 10**).
11. After the tubules are completely filled, withdraw the micropipet.
12. Return the testis to the abdomen and repeat the injection on the other testis.
13. After replacing the testis in the abdominal cavity, suture the skin with one or two clips. Dissection time is normally 10–15 min, and the tubule is filled in less than 1 min.
14. Recipient male mice are generally maintained 3–5 mo before analysis.

4. Notes

1. The age of mice is between 30 and 60 d. Alternatively, normal male mice between 5 and 10 d can be used.
2. The incubation time depends on different parameters. Normally, 10–15 min of incubation is sufficient. Dispersion of the tubules can be hastened by careful dissection, spreading of the tubules, and intertubular cellular strands when visible.
3. Alternatively, HBSS containing 0.25% trypsin–EDTA solution can be used. In this case, incubate at 37°C for 5 min and terminate the action of trypsin by adding a 20% volume of FBS. However, dispase is recommended for desegregation procedures because it does not damage cell membranes, which are important for cell–cell interaction and signal transduction.
4. If the cells are used immediately for transplantation to testes, resuspend the cells in 400 μL DMEM. The cells can be maintained 1–5 h at 4°C until transplantation. If the cells are cultured, resuspend the cells in 2 mL DMEM medium containing 0.5% Nu serum. If the cells are used for enrichment of SSCs, resuspend the cells in 8 mL DMEM medium containing 10% FBS.
5. If the germ cells are cocultured with Sertoli cells or STO feeder cells, isolation of the germ cells harboring *LacZ* transgene from the transgenic mice line 129-TgR (ROSA26)26S or is recommended for X-gal staining to distinguish the germ cells from the Sertoli or STO feeder cells.
6. Low concentrations of Nu serum, an artificial serum, are used to limit the proliferation of peritubular myoid cells co-isolated with germ cells.
7. Homozygous W/Wv male mice (3–4 mo old) can also be used as recipient mice for transplantation. No busulfan treatment is needed in this case. A treatment of the recipient mice with leuprolide acetate (a gonadotropin-releasing hormone

agonist) at a concentration of 7.6 mg/kg of body weight, 4 wk before transplantation can improve the colonization after spermatogonial transplantation.

8. Concentration above 10^8 cells/mL could be associated with a high level of clumping and plugging of pipets.
9. The pipet should be almost parallel to the surface of the testis, because the depth of the rete is 0.2 mm or less. If the angle is too large or the forward movement of the pipet is too great, the intratesticular boundary of the rete will be penetrated, and the cells will be inserted into the intertubular tissue rather than filling the tubules.
10. A cell suspension of about 3–5 µL can be injected in each testis if W/W^v mice are used. The testis of these mice (approx 15 mg) is smaller than the busulfan-treated wild-type testis (approx 30–40 mg).

Acknowledgments

We would like Angelika Winkler for secretarial help. This work was supported by Forschungsförderungsprogramm 2002 of University Göttingen.

References

1. Russell, L. D., Ettlin, R. A., Sinha Hikim, A. P., et al. (1990) Mammalian spermatogenesis, in *Histological and Histopathological Evaluation of the Testis*, Cache River Press, Clearwater, FL, pp. 1–40.
2. Prigle, M. J. and Page, D. C. (1997) Somatic and germ cell sex determination in developing gonad, in Infertility in the Male (Lipshultz, L. I. and Howards, S. S., eds.), Mosby, St. Louis, MO, pp. 3–22.
3. Saffan, E. E. and Lasko, P. (1999) Germline development in vertebrates and invertebrates. *Cell Mol. Life Sci.* **55,** 1141–1163.
4. Wylie, C. (1999) Germ cells. *Cell* **100,** 157–168.
5. Matsui, Y. (1998). Developmental fates of the mouse germ cell line. *Int. J. Dev. Biol.* **42,** 1037–1042.
6. De Rooij, D. G. and Van Diessel-Emiliani, F. M. F. (1997) Regulation of proliferation and differentiation of stem cells in the male germ line, in Stem Cells (Potten, C. S., ed.), Academic, London, pp. 283–313.
7. Oakberg, E. F. (1956) A description of spermiogenesis in the mouse and its use in analysis of the cycle of the seminiferous epithelium and germ cell renewal. *Am. J. Anat.* **99,** 391–414.
8. De Rooij, D. G. (2001) Proliferation and differentiation of spermatogonial stem cells. *Reproduction* **121,** 347–354.
9. Clermont, Y. and Bustos-Obregon, E. (1968) Re-examination of spermatogonial renewal in the rat by means of seminiferous tubules mounted "in toto." *Am. J. Anat.* **122,** 237–247.
10. Huckins, C. (1971) The spermatogonial stem cell population in adult rats. I. Their morphology, proliferation and maturation. *Anat. Rec.* **169,** 533–557.
11. Oakberg, E. F. (1971) Spermatogonial stem-cell renewal in the mouse. *Anat. Rec.* **169,** 515–531.

12. Chiarini-Garcia, H. and Russel, L. D. (2001) High resolution light microscopic characterization of mouse spermatogonia. *Biol. Reprod.* **85,** 1170–1178.
13. Huckins, C. and Oakberg, E. F. (1978) Morphological and quantitative analysis of spermatogonia in mouse testis using whole mounted seminiferous tubules. I. The normal testis. *Anat. Rec.* **192,** 519–528.
14. Skinner, M. K. (1991) Cell–cell interactions in the testis. *Endocr. Rev.* **12,** 45–77.
15. Jegou, B. (1993) The Sertoli-germ cell communication network in mammals. *Int. Rev. Cytol.* **147,** 25–96.
16. Meng, X., Lindahl, M., Hyvönen, M. E., et al. (2000) Regulation of cell fate decision of undifferentiated spermatogonia by GDNF. *Science* **287,** 1489–1493.
17. de Rooij, D. G. and Grootegoed, J. A. (1998) Spermatogonial stem cells. *Curr. Opin. Cell Biol.* **10,** 694–701.
18. De Rooij, G., Schrans-Stassen, B. H. G. J., von Pelt, A. M. M., et al. (2000) Regulation of the differentiation of the undifferentiated spermatogonia, in *The Testis: From Stem Cell to Sperm Function* (Goldberg, E., ed.), Springer-Verlag, New York, pp. 43–54.
19. Feng, L. X., Ravindranath, N., and Dym, M. (2000) Stem cell factor/c-kit up-regulates cyclin D3 and promotes cell cycle progression via the phosphoinositide 3-kinase/p70 S6 kinase pathway in spermatogonia. *J. Biol. Chem.* **275,** 25,572–25,576.
20. Schrans-Stassen, B. H., Saunders, P. T., Cooke, H. J., et al. (2001) Nature of the spermatogenic arrest in Dazl–/– mice. *Biol. Reprod.* **65,** 771–776.
21. Ravink, S. E., Rhee, K., and Wolgemuth, D. J. (1995) Distinct patterns of expression of the D-type cyclins during testicular development in the mouse. *Dev. Genet.* **16,** 171–178.
22. Sicinski, P., Donaher, I. L., Geng, Y., et al. (1996) Cyclin D2 is an FSH-responsive gene involved in gonadal cell proliferation and oncogenesis. *Nature* **384,** 470–474.
23. Tsutsui, T., Hesabi, B., Moons, D. S., et al. (1999) Targeted disruption of CDK4 delays cell cycle entry with enhanced p27Kip1 activity. *Mol. Cell. Biol.* **19,** 7011–7019.
24. van Pelt, A. M. and de Rooij, D. G. (1990) Synchronization of the seminiferous epithelium after vitamin A replacement in vitamin A-deficient mice. *Biol. Reprod.* **43,** 363–367.
25. Nishimune, Y., Haneji, G., and Aizawa, S. (1981) Testicular DNA synthesis in vivo: changes in DNA synthesis activity following artificial cryptorchidism and its surgical reversal. *Fertil. Steril.* **35,** 359–362.
26. Ohta, H., Yomogida, K., Tadokoro, Y., et al. (2001) Defect in germ cells, not in supporting cells is the cause of male infertility in the jsd mutant mouse: proliferation of spermatogonial stem cells without differentiation. *Int. J. Androl.* **24,** 15–23.
27. de Rooij, D. G., Okabe, M., and Nishimune, Y. (1999) Arrest of spermatogonial differentiation in jsd/jsd, Sl/7H/Sl/7H, and cryptorchid mice. *Biol. Reprod.* **61,** 842–847.
28. Dirami, G., Ravindranath, N., Achi, M. V., et al. (2001) Expression of Notch pathway components in spermatogonia and Sertoli cells of neonatal mice. *J. Androl.* **22,** 944–952.

29. Kojika, S. and Griffin, J. D. (2001) Notch receptors and hematopoiesis. *Exp. Hematol.* **29,** 1041–1052.
30. Zimmermann, S., Steding, G., Emmen, I. M., et al. (1999) Targeted disruption of the Insl3 gene causes bilateral cryptorchidism. *Mol. Endocrinol.* **13,** 681–691.
31. Rassoulzadegan, M., Paquis-Flucklinger, W., Bertino, B., et al. (1993) Transmeiotic differentiation of male germ cells in culture. *Cell* **75,** 997–1006.
32. Robertson, E. J. (1987) Embryo-derived stem cell lines, in *Teratocarcinomas and Embryonic Stem Cells: A Practical Approach* (Robertson, E. J., ed.), IRL, Oxford, pp. 71–112.
33. Meistrich, M. L. and van Beek, M. E. A. B. (1993) Spermatogonial stem cells, in *Cell and Molecular Biology of the Testis* (Desjardins, C. and Ewing, L. L., eds.), Oxford University Press, New York, pp. 266–295.
34. Orwig, K. E., Shinohara, T., Avarbock, M. R., et al. (2002) Functional analysis of stem cells in the adult rat testis. *Biol. Reprod.* **66,** 944–949.
35. Tegelenbosch, R. A. J. and de Rooij, D. G. (1993) A quantitative study of spermatogonial multiplication and stem cell renewal in the C3H/101 F1 hybrid mouse. *Mutat. Res.* **290,** 193–200.
36. Bellve, A. R., Cavicchia, J. C., Millette, C. F., et al. (1977) Spermatogenic cells of the prepuberal mouse. Isolation and morphological characterization. *J. Cell. Biol.* **74,** 68–85.
37. van Pelt, A. M. M., Morena, A. R., van Dissel-Emiliani, F. M. F., et al. (1996) isolation of the synchronized A spermatogonia from adult vitamin A-deficient rat testes. *Biol. Reprod.* **55,** 439–444.
38. Shinohara, T., Orwig, K. E., Avarbock, M. R., et al. (2000) Spermatogonial stem cell enrichment by multiparameter selection of mouse testis cells. *Proc. Natl. Acad. Sci. USA* **97,** 8346–8351.
39. Shuttlesworth, G. A., de Rooij, D. G., Huhtaniemi, I., et al. (2000) Enhancement of A spermatogonial proliferation and differentiation in irradiated rats by GnRH antagonist administration. *Endocrinology* **141,** 37–49.
40. Bellvé, A. R. (1993) Purification, culture, and fractionation of spermatogenic cells, in *Methods in Enzymology, Volume 225* (Wassarman, P. M. and Depamphilis, M. L., eds.), Academic, Oxford, pp. 159–261.
41. Shinohara, T., Avarbock, M. R., and Brinster, R. L. (1999) β1- and α6 integrins are surface markers on mouse spermatogonial stem cells. *Proc. Natl. Acad. Sci. USA* **96,** 5504–5509.
42. Nagano, M., Avarbock, M. R., Leonida, E. B., et al. (1998) Culture of mouse spermatogonial stem cells. *Tissue Cell* **30,** 389–397.
43. Kierszenbaum, A. L. (1994) Mammalian spermatogenesis in vivo and in vitro: a partnership of spermatogenic and somatic cell lineages. *Endocr. Rev.* **15,** 116–134.
44. Kiger, A. A., White-Cooper, H., and Fuller, M. T. (2000) Somatic support cells restrict germline stem cell self-renewal and promote differentiation. *Nature* **407,** 750–753.
45. Ogawa, T., Dobrinski, I., Avarbock, M. R., et al. (2000) Transplantation of male germ cell line stem cells restores fertility in infertile mice. *Nature Med.* **6,** 29–34.

9

Molecular and Cellular Approaches for Establishment of an In Vitro Spermatogenesis System

Karim Nayernia and Wolfgang Engel

1. Introduction

1.1. Spermatogenesis: Cellular Aspects

Spermatogenesis is a complex, well-organized process including proliferation, differentiation, apoptosis, and morphogenesis through which differentiating daughter cells of spermatogonial stem cells develop into mature sperm *(1,2)*. In men, this process requires approx 64 d, 52 d in the rat, and 35 d in the mouse. This process in mammals occurs in a highly coordinated fashion within the seminiferous epithelium inside the seminiferous tubule. Spermatogenesis can be divided into three phases.

The first phase is a cascade of mitosis initiated by the self-renewing division of spermatogonial cells located in the basal layer of the epithelium against the tubule wall (basal lamina). These mitoses eventually lead to the production of numerous primary spermatocytes. The second phase is the meiosis of primary spermatocytes, which leads to the production of haploid round spermatids (*see* **Fig. 1**). The third phase is the transformation of the round spermatids into spermatozoa. This process, called spermiogenesis, comprises a series of drastic cellular reorganization and morphological changes, including acrosomal formation, flagellum development, nuclear condensation, and cytoplasm elimination, resulting in the formation of mature spermatozoa (*see* **Fig. 1**). During spermiogenesis, histone and nonhistone proteins are replaced successively by basic proteins, which are known in round spermatids as transition proteins and in elongating spermatids as protamines *(3)*.

One of the hallmarks of a normal seminiferous epithelium is that spermatogonia, spermatocytes, and spermatids at a specific phase of development

From: *Methods in Molecular Biology, vol. 253: Germ Cell Protocols: Vol. 1 Sperm and Oocyte Analysis*
Edited by: H. Schatten © Humana Press Inc., Totowa, NJ

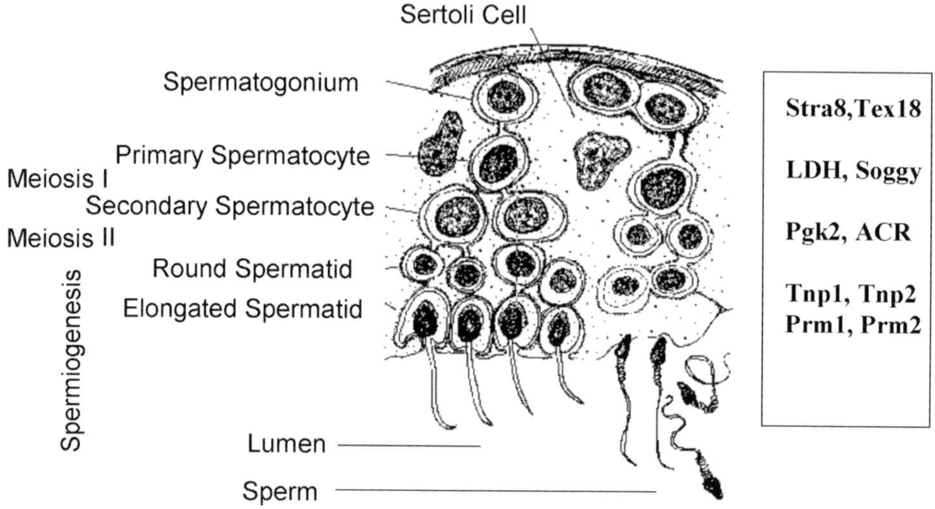

Fig. 1. Differentiation of male germ cells. Examples of genes expressed specifically in different stages of differentiation are shown. LDH: lactate dehydrogenase X; Pgk2: phosphoglycerate kinase 2; ACR: proacrosin; Tnp: transition protein; Prm: protamine.

are always found together. These associations of specific types of germ cells are called the stage of the cycle of the seminiferous epithelium. The number of stages of the cycle varies between species, but this ordered association of spermatogenic cells is realized in all male mammals: human, rat, and mouse have 6, 14, and 12 stages of the cycle, respectively ***(1,2)***. The first stage of the cycle in any mammalian species is defined by the appearance of a new cohort of round spermatids and the last stage by completion of the second meiotic division; over time, the germ cells in any one location in a seminiferous epithelium pass together from one stage of the cycle to the next. In rodents, the stages of the cycle are arrayed linearly along a seminiferous tubule. This phenomenon is called the wave of the seminiferous epithelium. In contrast, the stages in men are arrayed helically around the tubule. Throughout spermatogenesis, the germ cells are entirely dependent on a population of highly specialized somatic cells known as Sertoli cells. In the adult tubule, these large postmitotic epithelial cells span the basal–apical scaffold for the dividing and differentiating germ cells ***(4–6)***. They also generate a specialized microenvironment for each phase of spermatogenesis, which they sustain by the synthesis and delivery of appropriate metabolites ***(6)***. Moreover, the Sertoli cells are uniquely placed to play a major role in spermatogenic regulation. They are active participants in the seminiferous epithelial cycle, undergoing cyclical changes in gene expression, biochemical activity, and morphology as the germ cells develop around them. They are the primary targets of the main hormonal regulators of spermatogenesis, follicle-stimulating hormone (FSH), and testosterone and remain intimately associated with each other and with the developing germ cells through a dynamic repertoire of specialized contacts and junctions ***(6)***. As mediators of both endocrine and paracrine mechanisms of spermatogenic regulation in the adult testis, the Sertoli cells clearly play a central role in the maintenance of the epithelial cycle.

Evidences for the Sertoli–germ cell interactions have been obtained mainly by the primary coculture of germ cells at different stages of development together with Sertoli cells or other somatic testicular cells ***(5,7)***. Germ cells also are able to affect Sertoli cell function. This phenomenon is well illustrated by changes in Sertoli cell gene expression accompanying the progression of the surrounding germ cells through the stages of the cycle of the seminiferous epithelium ***(8)***. The stage-specific changes in gene expression by the Sertoli cell suggest that Sertoli cells detect the developmental stage of the germ cells that surround them and alter the production of their products in response to these germ cells.

1.2. Spermatogenesis: Molecular Aspects

Spermatogenesis is subject to the influence of many genes; the molecular mechanisms involved are beginning to be understood ***(3,9)***. This pathway reflects the expression of unique combinations of facultative (i.e., tissue-specific) and constitutive (i.e., housekeeping genes) in each cell type. In spermatogenesis, commitment is mediated by the mechanism of potentiation whereby specific chromatin domains are selectively opened along each chromosome ***(10)***. Within each open chromatin domain, a unique battery of genes is availed to cell-specific and ubiquitous transacting factors that are necessary to initiate transcription ***(10)***. Only a small percentage of the 1500–2000 different gene transcripts likely to be present in male germ cells have been identified at this time. Some of the gene products are involved in housekeeping functions common to all cell types, some are expressed in distinct developmental patterns ***(11)***, some reflect different isozymic forms of large gene families, and some are expressed uniquely in germ cells ***(3,9)***. At different stages of spermatogenesis, specific expression of many genes are observed (*see* **Fig. 1**). For example, Page et al. ***(12)*** identified several spermatogonia-specific genes; most of them are located on the X chromosome. Fujii et al. ***(13)*** identified several mouse genes whose transcription is up-regulated during spermiogenesis using a stepwise subtraction procedure. Most genes involved in mammalian spermatogenesis function in multiple steps during spermatogenesis and often in other tissues. Their mutants display various effects at multiple stages of spermatogenesis, often with incomplete penetrance. This situation is very similar to most spermatogenic genes in *Drosophila*, whose mutants display specific defects in cytokinesis, mitochondrial morphogenesis, or the shaping of the nucleus ***(14)***. These mutants usually define effector genes that mediate specific morphogenetic processes. The study of these mutants revealed that many of the morphogenic processes proceed via independent pathways. This is one of the two striking features of spermatogenesis.

The second striking feature of spermatogenesis is that an extremely small number of male sterile mutations cause global arrest of spermatogenesis ***(14)***. In these mutants, cell types prior to the blocked stage accumulate and appear to be morphologically normal, whereas cell types that normally arise beyond the arrest point are absent. Such global arrest mutations often identify key regulatory genes controlling transitions between distinct stages of spermatogenesis ***(15)***. Three such control points have been identified: the self-renewing division of spermatogonia, the entry into meiosis, and the switch from the primary spermatocyte program to the completion of meiosis and initiation of spermiogenesis ***(3,14)*** (*see* **Fig. 1**).

In mammals, few genes that display characteristics of a global switch gene have been identified, including *SPRM-1*, *TAK-1*, *ZFY-2*, *OCT-2*, *CREM*, and *ACT* ***(15,16)***.

Intrinsic regulation for spermatogenesis occurs not only at the level of transcription but also at the level of translation and posttranslation ***(17)***. Many transcripts are expressed prior to chromatin condensation at steps 9–12 of spermiogenesis and are stored until being translated at later stages. The delay appears to be regulated by mRNA-binding proteins and 5'- or/and 3'-untranslated regions that alter translational efficiency or prevent translation altogether ***(18)***.

1.3. Experimental Strategies for Establishment of an In Vitro Spermatogenesis System

Establishment of an in vitro spermatogenesis system would greatly enhance our ability to study stage-specific cellular interaction and molecular mechanisms underlying specific gene regulation at different stages of spermatogenesis. These systems are easily accessible for experimental manipulation (e.g., the introduction of DNA, chemical selection, retrovirus treatment, etc.). For these purposes, the system should include the processes of both DNA duplication and differentiation into further stages.

Various in vitro systems have been developed that are based on two main strategies: (1) isolation of male germ cells and coculture with supportive cells and (2) immortalization of male germ cells at discrete stages of differentiation. For example, when mouse male germ cells were cocultured on a Sertoli-like cell line, premeiotic germ cells underwent the meiotic and postmeiotic differentiation into haploid spermatids expressing the protamine gene ***(19)***. When isolated eel (*Anguilla japonica*) germ cells and somatic cells (mainly Sertoli cells) were cocultured, the entire differentiation process from spermatogonia to spermatozoa was completed ***(20)***. These results imply that the initiation or completion of meiosis in normal male germ cells typically requires Sertoli cells or other somatic cells in vertebrates. In contrast with the mouse, primary spermatocytes in lower vertebrates such as amphibians ***(21)*** and fish ***(22)*** differentiated into flagellated spermatids or functional sperm without any supporting cells.

The second strategy is based on immortalization of male germ cells using directed expression of an oncogene in defined differentiation stages. Targeted expression of oncogenes in transgenic mice can immortalize specific cell types to serve as valuable culture model system ***(23)***. However, this technique is very labor intensive and time-consuming. To overcome this problem, germ cells can be isolated and transfected in vitro with oncogene DNA. Although immortalized mouse germ cells undergo meiosis without any supporting cells ***(24,25)***, the immortalization process probably alters the normal course of events. Using this strategy, different germ cell lines have been established ***(24–27)***.

These cell lines may serve as a powerful tool in the elucidation of the molecular mechanisms of spermatogenesis. Furthermore, through genetic modification and transplantation techniques, these male germ cell lines may be used to generate transgenic mice and to develop germ cell gene therapy.

2. Material

Use the following materials for the isolation, culture, and transfection of male germ cells:

1. PBS/PGK-2 Vector ***(26)***.
2. Endo Free Pasmid Maxi Kit (12362; Qiagen, Inc., Valencia, CA).
3. Male mice (18–20 d old) (The Jackson Laboratory, Bar Harbor, ME, USA).
4. Nylon mesh 70-µm pore size (352350; BD Bioscience, Heidelberg, Germany).
5. 15P-1 Sertoli cell line ***(19)***.
6. RNA reagent solution (51583-2; BIOMOL, Hamburg, Germany).
7. Poly A$^+$-RNA isolation kit (70042; Qiagen, Inc., Valencia, CA).
8. RT-PCR kit (27-9555-01; Amersham Bioscience Corp., Freiburg, Germany).

2.1. Reagents

1. Nuclease-free water (P1195; Promega).
2. DNase I (E2215Y; Amersham Bioscience Corp.).
3. Tfx50 (E1811; Promega, Mannheim, Germany).
4. Oligonucleotide primers (Roth, Karsruhe, Germany).
5. G418 (Geneticin) (10131-035; GIBCO Invitrogen Corp.).

2.2. Solutions

1. Dulbecco's modified Eagle's medium (DMEM) (10829-018; GIBCO Invitrogen Corp.) containing 10% fetal bovine serum (FBS) (F3018; Sigma–Aldrich), 1% nonessential amino acids (11140-019; GIBCO Invitrogen Corp.), 1X penicillin–streptomycin solution (100X) (P0781; Sigma–Aldrich).
2. DMEM containing 0.5% Nu serum type IV (355504; BD Bioscience).
3. Collagenase solution: CMRL-1066 medium (11530; GIBCO Invitrogen Corp.) containing 1 mg/mL collagenase type IV (C1889; Sigma–Aldrich, Munich, Germany; or 17104, GIBCO Invitrogen Corp., Karlsruhe, Germany).
4. Dulbecco's phosphate-buffered saline (DPBS) (P04-36100; PAN Biotech, Aidenbach, Germany).
5. Trypsin–EDTA solution (0.25%) (T4049; Sigma–Aldrich).

3. Methods

An overview of the methods described below is shown in **Fig. 2**.

3.1. Generation of Fusion Plasmid and Preparation of pDNA

Cell transfection with a cassette of immortalizing oncogenes controlled by ubiquitous promoters is the conventional method to establish immortalized cell lines *(28)*. However, by using these promoters, all cell types in the culture are potentially immortalized; therefore, predetermination of a defined cell type is not possible. Contamination of the selected cell line with other cells is also a severe problem of this method. Other difficulties in the conventional method may arise from the differences between cell types regarding the expression level of the genes controlled by such promoters. This means that immortalizing a given cell type depends on the promoter activity in that cell type. Previously, we designed a promotor-based selection strategy to specifically select immortalized germ cells at a discrete stage of differentiation. For specific selection of a defined germ cell type, a promoter region is needed that is specifically active in this cell type. For example, we used the promoter region of the human gene for phosphoglycerate kinase 2 (PGK2). The human PGK2 promoter has a specific transcriptional activity in murine male germ cells, and its transcription has been described to start in mouse preloptotene spermatocytes *(29)*. Using the PGK2 promoter region, a fusion vector was generated harboring the SV40 large T antigen and the neomycin phosphotransferase II gene under the control of the SV40 early promoter and PGK2 promoter, respectively *(28)*. A map of this fusion gene is shown in **Fig. 2**.

Plasmid DNA (pDNA) used for the transfection experiment (*see* **Subheading 3.4.**) should be endotoxin-free. Plasmid preparation must be performed in such a manner as to remove most endotoxin. Plasmid DNA preparation procedure can be performed as outlined in the Qiagen protocol by using Qiagen's low-endotoxin kit. This is generally sufficient for transfection of male germ cells.

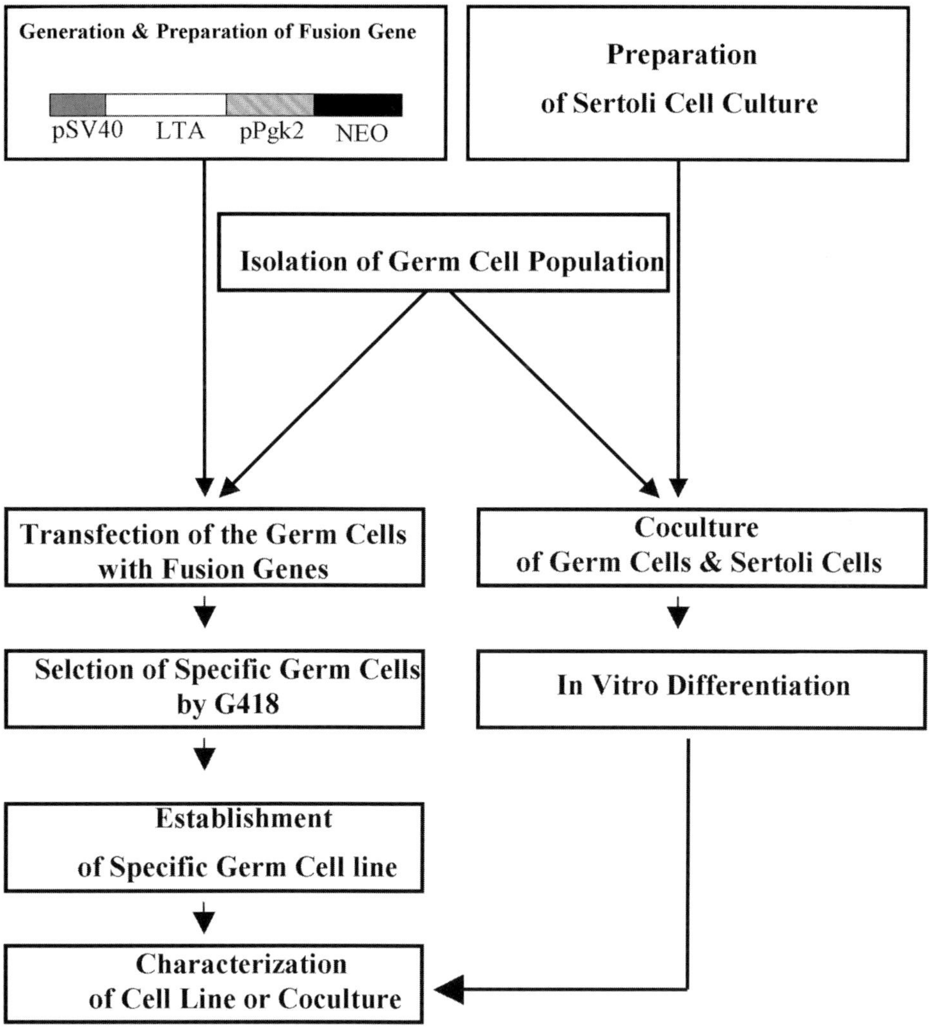

Fig. 2. Outline of strategies used for establishment of an in vitro spermatogenesis system. pSV40: promoter and enhancer of SV 40 virus; LTA: SV40 large T antigen; pPgk2: human promoter of Pgk2; NEO: neomycin phosphotransferase II gene.

3.2. Isolation of Male Germ Cell Populations

The developmental time when each spermatogenic cell type first appears in the mouse testis has been well defined ***(30)***. Thus, for example, primary spermatocytes, pachytene spermatocytes, round spermatids, and elongating spermatids first appear in the mouse testis 10–12, 15–17, 19–21, and 23–25 d after birth, respectively. Therefore, for the study of meiotic process and spermiogenesis, isolation of germ cells from 12- to 18-d-old mice is recommended. For monitoring of differentiation, isolation of the germ cells from transgenic mice harboring a reporter gene (e.g., *LacZ* or *chloramphenicol acetyltransferase*) under the control of male germ cell specific genes (e.g., proacrosin or protamine) can be used.

1. Preparation of collagenase solution: Dissolve the nonsterile, lyophilized collagenase in serum-free CMRL-1066 culture medium in a concentration of 10 mg/mL. Filter-sterilize the solution with a cell-culture-approved filtration unit, aliquot, and store frozen. Thaw stock solution in refrigerator immediately prior to use and dilute with serum-free CMRL-1066 medium to a final concentration of 1 mg/mL.
2. Isolate 8–10 testes from 12- to 13-d-old mice (*see* **Note 1**), wash in PBS, and put in 10 vol (approx 10 mL) of collagenase solution in a 60-mm sterile Petri dish (*see* **Note 2**).
3. Remove the tunica albuginea from the testes.
4. Incubate at 37°C with gentle agitation in a rocker platform until the tubules separate (*see* **Note 3**).
5. Transfer the tubules using a 10-mL pipet to a cell culture centrifugation tube.
6. Centrifuge at 4°C at 650*g* for 5 min.
7. Aspirate the supernatant, add 5 mL of CMRL-1066 medium, and centrifuge again at 4°C at 650*g* for 5 min.
8. Repeat **step 7** three times.
9. Add 1 mL of 0.25% EDTA–trypsin solution and incubate at 37°C for 3–5 min to form a monodisperse population of germ cells (*see* **Note 4**).
10. Add 10 mL CMRL-1066 medium containing 10–20% FBS.
11. Cell suspension is separated from large pieces by passing the mixture through a nylon mesh with a 70-µm pore size.
12. Pellet cells by centrifugation at 16°C at 650*g* for 5 min and decant the enzyme solution.
13. Wash the cells by adding 2 mL PBS and centrifugation at 650*g* for 5 min, three times.
14. Resuspend the pellet in DMEM medium containing 0.5% Nu serum, if coculture with the Sertoli cells and in vitro differentiation will be continued. If establishment of a germ cell is desired (*see* **Subheading 3.3.**), resuspend the cells in serum-free DMEM medium (*see* **Subheading 3.4.**).

3.3. Coculture with Sertoli Cell Line 15P-1

1. Plate Sertoli cells in 50-mm tissue culture dishes in DMEM medium containing 10% FBS and incubate at 32°C and 5% CO_2 for 2–3 d.
2. Trypsinize and seed the Sertoli cells into six-well plates at a concentration of 1×10^6 cells per well in DMEM medium containing 10% FBS and incubate for 1 d.
3. Pour off the medium from the wells and replace it by DMEM medium containing 0.5% Nu serum (*see* **Note 5**).

4. Add 10^5–10^6 isolated germ cells from **step 14** of **Subheading 3.2.** and incubate the coculture at 32°C and 5% CO_2 for 3 d.
5. Pour off the medium and replace it by fresh DMEM medium containing 10% FBS.
6. Medium is changed two or three times each week, and the germ cells are cultured for further analysis (*see* **Note 6**).

3.4. Establishment of a Male Germ Cell Line

1. Suspend the Tfx reagent the day before the transfection and store at –20°C.
2. Determine the density of isolated male germ cells (**step 14** of **Subheading 3.2.**) and dissolve the cells at a concentration of 2×10^6 cells/mL in serum-free DMEM medium.
3. Preparation of DNA/Tfx reagent mixture: Add 5 µg pDNA (*see* **Subheading 3.1.**) and serum-free DMEM medium to a total volume of 500 µL to a sterile tube and vortex.
4. Add 18.0 µL Tfx reagent and vortex immediately.
5. Allow the Tfx reagent/DNA mixture to incubate 15 min at room temperature.
6. While the DNA/Tfx reagent mixtures are incubating, aliquot 500 µL of germ cells (2×10^6) to each well of a six-well plate.
7. Briefly vortex the DNA/Tfx reagent mixture and add to the germ cells (500 µL per well).
8. Return the cells to the incubator for 1 h. During the incubation, warm DMEM medium containing 10% FBS.
9. At the end of the incubation period, add 5 mL of the prewarmed DMEM medium containing 10% FBS to each well. Return the cells to the incubator and continue the incubation for 48 h.
10. After 4 wk, remove medium and replace it by selective medium DMEM containing 10% FBS, 1% nonessential amino acids, and genticin at a concentration of 200 µg/mL.
11. Replace the medium in 3- to 4-d intervals.
12. During the fifth week, monitor the cultures for distinct "islands" of surviving cells.
13. Transfer individual clones to 24-well plates by using a cloning cylinder and continue to culture in selective medium.

3.5. Characterization of Established Cell Lines

Many specific transcripts are produced during male germ cell differentiation. Their expression is regulated in a defined specific stage. These transcripts are generated by using promoters and transcription factors that activate transcription at different start sites upstream or downstream of the usual site, by incorporation of alternative exons, by germ-cell-specific splicing events, and by using alternate initiation sites for polyadenylation ***(11,31)***. Therefore, the expression analysis of defined stage-specific genes is a hint for the differentia-

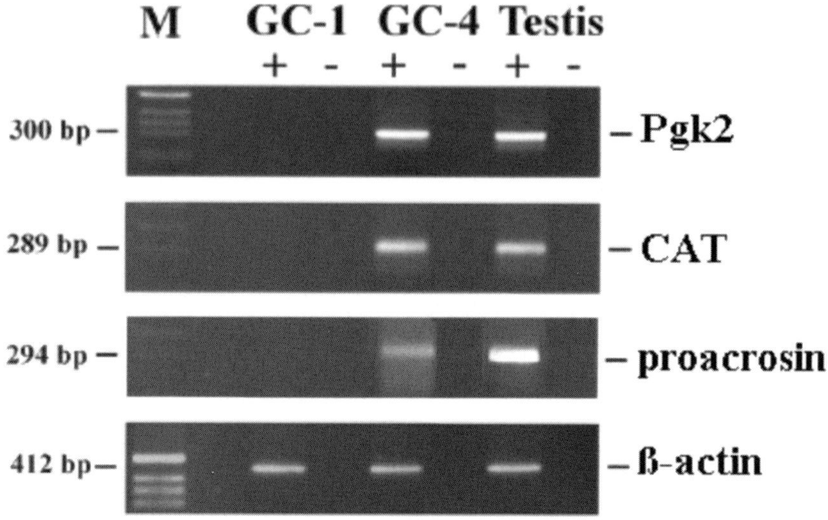

Fig. 3. RT-PCR analysis of RNA isolated from spermatogonia cell line (GC-1), established spermatocyte cell line (GC-4), and testis. Transcripts of spermatocyte specific genes for Pgk2, CAT (under control of proacrosin promoter), and proacrosin were obtained in the RNA of the GC-4 cell line.

tion stage of established cell lines. In the following, reverse transcription–polymerase (RT-PCR) was used to analyze endogenous transcription of Pgk2 (phosphoglycerate kinase 2), proacrosin, and CAT gene. Examples of some genes expressed in defined stages of male germ cell differentiation are shown in **Fig. 3**.

1. Isolation of RNA and poly A^+-RNA can be performed as described in BIOMOL and Qiagen protocols by using RNA reagent solution and the mRNA isolation kit.
2. Add 10 U of DNase RNase-free I to 100 μg total RNA isolated from the germ cells in a reaction volume of 100 μL.
3. Incubate 30 min at 37°C.
4. After DNA digestion, the mRNA is isolated by the mRNA isolation kit.
5. One microgram of poly A^+-RNA is reversed transcribed into cDNA at 42°C for 30 min using the RT-PCR kit.
6. The PCR is performed for 35–40 cycles under the following conditions in a Perkin-Elmer thermocycler for 1 min at 94°C, 1 min at 56°C, and 1 min at 72°C (*see* **Note 7**).
7. The PCR reaction products are visualized by ethidium bromide staining after agarose gel electrophoresis (*see* **Fig. 3**).

4. Notes

1. For monitoring and further characterization of established germ cell lines, it is recommended to use transgenic mice harboring a reporter gene under control of a meiotic or postmeiotic expressed gene. For this study, we used transgenic line TC, which expresses CAT reporter gene in male germ cells under the control of the proacrosin promoter ***(32)***.
2. The DNase treatment should be avoided to prevent DNA damage during the transfection step.
3. Incubation time depends on different parameters. Normally, 10–15 min of incubation is sufficient. Dispersion of the tubules can be hastened by careful dissection, spreading of the tubules, and in the tubular cellular strand when visible.
4. Separation and dispersion of tubule cells can be hastened by pipetting and gentle agitation.
5. Low concentrations of Nu serum, an artificial serum, are used to limit the proliferation of peritubular myoid cells coisolated with germ cells.
6. A transmeiotic differentiation of male germ cells could be observed after 8 d of the start of the coculture.
7. Establish a PCR condition for the genes to be analyzed in germ cells by changing of different parameters, including RNA amount and annealing temperature.

Acknowledgments

We would like Angelika Winkler for secretarial help. This work was supported by Forschungsförderungsprogramm 2002 of University Göttingen.

References

1. Barratt, C. L. R. (1995) Spermatogenesis, in *Gametes—The Spermatozoon* (Grudzinskas, I. G. and Yovich, I. L., eds.), Cambridge University Press, New York, pp. 250–267.
2. Sharpe, R. M. (1993) Regulation of spermatogenesis, in *The Physiology of Reproduction* (Knobil, E. and Neill, J. D., eds.), Raven, New York, pp. 1363–1434.
3. Hecht, N. B. (1998) Molecular mechanisms of male germ cell differentiation. *BioEssays* **20,** 555–561.
4. Skinner, M. K. (1991) Cell–cell interaction in the testis. *Endocrin. Rev.* **12,** 45–77.
5. Syed, V. and Hecht, N. B. (1997) Up-regulation and down-regulation of gene expressed in coculture of rat Sertoli cells and germ cells. *Mol. Reprod. Dev.* **47,** 380–389.
6. Jegon, B. (1993) The Sertoli–germ cell communication network in mammals. *Int. Rev. Cytol.* **147,** 25–96.
7. Nayernia, K., von Meringe, M. H. P., Kraszucka, K., et al. (1999) A novel testicular haploid expressed gene (THEG) involved in mouse spermatid-Sertoli cell interaction. *Biol Reprod.* **60,** 1488–1495.
8. Enders, G. C. (1993) Sertoli–Sertoli and Sertoli–germ cell communications, in *The Sertoli Cell* (Russell, L. D. and Griswold, M. D., eds.), Cache River Press, Clearwater, FL, pp. 447–460.

9. Sha, I., Zhon, Z., Li, I., et al. (2002) Identification of testis development and spermatogenesis-related genes in human and mouse testes using cDNA arrays. *Mol. Hum. Reprod.* **8,** 511–517.
10. Kramer, I. A., McCarrey, I. M., Djakiew, D., et al. (1998) The selective potentiation of chromation domains. *Development* **125,** 4749–4755.
11. Nayernia, K., Adham, I., Kremling, H., et al. (1996) Stage and developmental specific gene expression during mammalian spermatogenesis. *Int. J. Dev. Biol.* **40,** 379–383.
12. Wang, D. J., McCarrey, I. R., Yang, F., et al. (2001) An abundance of X-linked genes expressed in spermatogonia. *Nature Genet.* **27,** 422–426.
13. Fujii, T., Tamura, K., Masai, K., et al. (2002) Use of stepwise subtraction to comprehensively isolate mouse genes whose transcription is up-regulated during spermiogenesis. *EMBO Rep.* **3,** 367–372.
14. Fuller, M. T. (1998) Genetic control of cell proliferation and differentiation in *Drosophila* spermatogenesis. *Semin. Cell Dev. Biol.* **9,** 433–444.
15. Escalier, D. (2001) Impact of genetic engineering on the understanding of spermatogenesis. *Hum. Reprod. Update* **7,** 191–210.
16. Sassone-Corsi, P. (2002) Unique chromatin remodeling and transcriptional regulation inspermatogenesis. *Science* **296,** 2176–2178.
17. Schäfer, M., Nayernia, K., Engel, W., et al. (1995) Translational control in spermatogenesis. *Dev. Biol.* **172,** 344–352.
18. Venables, I. P. and Eperou, I. C. (1999) The roles of RNA-binding proteins in spermatogenesis and male infertility. *Curr. Opin. Genet. Dev.* **9,** 346–354.
19. Rassoulzadegan, M., Paqauis-Flucklinger, W., Bertino, B., et al. (1993) Transmeiotic differentiation of male germ cells in culture. *Cell* **75,** 997–1006.
20. Miura, C., Miura, T., Yamashita, M., et al. (1996) Hormonal induction of all stages of spermatogenesis in germ-somatic cell coculture from immature Japanese eel testis. *Dev. Growth Differ.* **38,** 257–262.
21. Abe, S. I. (1987) Differentiation of spermatogeneic cells from vertebrates in vitro. *Int. Rev. Cytol.* **109,** 159–209.
22. Sakai, N. (2002) Transmeiotic differentiation of zebrafish germ cells into functional sperm in culture. *Development* **29,** 3359–3365.
23. Hanahan, D. (1989) Transgenic mice as probes into complex systems. *Science* **246,** 1265–1275.
24. Hofmann, M. C., Hess, R. A., Goldberg, E., et al. (1994) Immortalized germ cells undergo meiosis in vitro. *Proc. Natl. Acad. Sci. USA* **91,** 5533–5537.
25. Feng, L. X., Chen, Y., Dettin, L., et al. (2002) Generation and in vitro differentiation of a spermatogonial cell line. *Science* **297,** 392–395.
26. Tascou, S., Nayernia, K., Samani, A., et al. (2000) Immortalization of murine male germ cells at a discrete stage of differentiation by a novel directed promoter-based strategy. *Biol. Reprod.* **63,** 1555–1561.
27. Van Pelt, A. M. M., Roepers-Gajadien, H. L., Gademan, I. S., et al. (2002) Establishment of cell lines with rat spermatogonial stem cell characteristics. *Endocrinology* **143,** 1845–1850.

28. Schuermann, M. (1990) An expression vector system for stable expression of oncogenes. *Nucleic Acids Res.* **18,** 4945–4946.
29. Robinson, M. O., McCarrey, I. R., and Simon, M. I. (1989) Transcriptional regulatory regions of testis-specific PGK2 defined in transgenic mice. *Proc. Natl. Acad. Sci. USA* **86,** 8437–8441.
30. Bellve, A. R., Cavicchia, J. C., Millete, C. F., et al. (1977) Spermatogenic cells of the prepuberal mouse. Isolation and morphological characterization. *J. Cell Biol.* **74,** 68–85.
31. McCarrey, J. R. (1998) Spermatogenesis as a model system for developmental analysis of regulatory mechanisms associated with tissue-specific gene expression. *Semin. Cell Dev. Biol.* **9,** 459–466.
32. Nayernia, K., Nieter, S., Kremling, H., et al. (1994) Functional and molecular characterization of the transcriptional regulatory region of the proacrosin gene. *J. Biol. Chem.* **269,** 32,181–32,186.

10

Acrosin Activity Assay for the Evaluation of Mammalian Sperm Acrosome Reaction

Yehudit Lax,* Sara Rubinstein,* and Haim Breitbart

1. Introduction

The sperm acrosome reaction (AR) is an exocytotic event that involves multiple fusions between the outer acrosomal membrane and the overlying plasma membrane. It must occur before the spermatozoon penetrates the zona pellucida (ZP) *(1)*. This exocytotic event is mandatory for fertilization because it enables passage of the sperm through the ZP and its subsequent fusion with the egg oolema. Many artificial stimuli are reported to trigger the AR, either by driving extracellular Ca^{2+} into the sperm cell (Ca^{2+} ionophores) or by acting through intracellular second messengers that are involved in the cascade leading to acrosomal exocytosis *(2,3)*.

The acrosomal components, mostly powerful hydrolytic enzymes, are released by the acrosome reaction *(1)*. Acrosin (E.C. 3.4.21.10), an endoprotease with a trypsinlike substrate specificity, is localized in the acrosomal matrix as an enzymatically inactive zymogen, proacrosin, that is converted into the active form as a consequence of the acrosome reaction ***(4–6)***. At the acrosome reaction, acrosin serves a role by accelerating the dispersal of acrosomal components and further by assisting the sperm penetration of the ZP *(7)*.

Spermatozoa of many mammals (rat, bull, ram, and human) have very thin or lightly fitted acrosomes; hence, their acrosome reactions are difficult to detect *(1)*. Several methods are currently available for monitoring acrosome reactions:

1. Methods using transmission electron microscopy, allowing ultrastructural examination of the acrosome *(8)*

---

*Both contributed equally.

From: *Methods in Molecular Biology, vol. 253: Germ Cell Protocols: Vol. 1 Sperm and Oocyte Analysis*
Edited by: H. Schatten © Humana Press Inc., Totowa, NJ

2. Methods using light microscopy based on staining sperm with specific fluoresceinated plant lectins or various dyes ***(9,10)***
3. Methods using a fluorescence-activated cell sorter and fluoresceinated monocloned antibodies or lectins ***(11)***

In an attempt to overcome most of the disadvantages of the existing methods, such as high cost and expensive equipment, tedious counting of hundreds of sperm on each slide, and not completely objective (observer dependent), we propose here a biochemical method based on measuring the enzymatic activity of released acrosin ***(12–19)***.

1.1. Principle of Method

The method is based on measuring the enzymatic activity of acrosin released during acrosome reaction. Acrosin activity is measured by an esterolytic assay, using BAEE (*N*α – benzoyl-L-arginine ethyl ester) as substrate:

$$\text{BAEE} + H_2O \xrightarrow{\text{Acrosin}} N\alpha - \text{benzoyl-L-arginine} + \text{ethanol}$$

The assay is a continuous spectrophotometric rate measurement calculating the increase in absorbance at 259 nm of *N*α – benzoyl-L-arginine.

1.2. Advantages of Method

The assay has several advantages in comparison to other existing methods:

1. Precise and accurate measurement: The results are objective, do not depend on the observer, and there is no need for tedious slide scoring by microscope.
2. Rapid assay protocol: Acrosin activity determination is immediate, it may be achieved in a very short time (1 min for each sample).
3. Simple method: The assay is very easy to perform; only simple procedures are needed, such as short centrifugation, pH calibration, and spectrophotometric determination.

2. Materials

1. NKM medium, pH 7.4: 110 m*M* NaCl, 5 m*M* KCl, 10 m*M* MOPS (3-[*N*-morpholino] propane sulfuric acid). Dilute to 1000 mL with double-distilled water (ddH_2O); check pH.
2. Glucose-free modified Tyrode's medium (m-TALP), pH 7.4: 100 m*M* NaCl, 3.1 m*M* KCl, 25 m*M* $NaHCO_3$, 0.29 m*M* KH_2PO_4, 21.6 m*M* Na-lactate, 1.5 m*M* $MgCl_2$, 0.1 m*M* Na-pyruvate, 20 m*M* HEPES (*N*-2-hydroxyethylpiperazine-*N*'-2-ethanesulfonic acid), 10 IU/mL penicillin, 3 mg/mL BSA (bovine serum albumin), added just before use. Dilute to 1000 mL with ddH_2O; check pH.
3. BAEE buffer, pH 8.2: 50 m*M* Tris-HCl, 0.2 m*M* $CaCl_2$, 1 m*M* BAEE. Dilute to 100 mL with ddH_2O; check pH.
4. 3 *M* HCl: Prepare 3 *M* HCl by diluting concentrated HCl (10.43 *M*) to 3 *M* with ddH_2O (be careful!).

5. Storage of mediums once prepared:
   a. The NKM medium can be stored in a refrigerator (2–8°C) for at least 2 mo.
   b. The m-TALP medium, devoid of BSA, should be stable for at least 3 mo if stored in the freezer (–20°C) divided into aliquots with an appropriate volume for each experiment. Avoid repeated thawing and freezing cycles of the medium.
   c. The BAEE buffer should be prepared freshly before use and be kept at room temperature.

All chemicals should be of the highest purification available.

3. Methods

3.1. Sperm Preparation

1. Collect ejaculated bovine sperm in an artificial vagina and dilute them (1:2, v:v) in NKM medium (pH 7.4).
2. Wash the cells three times by centrifugation, at room temperature for 10 min at 780*g*.
3. Discard the supernatant and resuspend the pellet in NKM medium to a concentration of (1–3) × 10^9 cells/mL. (See next step for cell-counting procedure.)
4. Count the cells in a Neubauer hemocytometer and calculate number of cells per milliliter.
   a. Cover the Neubauer chamber with a coverslip.
   b. Place a drop (10 µL) of cell suspension (usually diluted 1:10^4; 10 µL of cells in 10 mL ddH_2O) at the edge of the coverslip.
   c. Place the chamber on the stage of the microscope and count all of the cells in the square in the central area.
   d. Total numbers of cells/mL = cells in the central area × dilution × 10^4.*
5. Use only samples exhibiting a minimum of 70% motile sperm.

3.2. Capacitation and Acrosome Reaction Evaluation

1. Dilute washed sperm cells to a concentration of 10^8 cells/mL in glucose-free m-TALP.
2. Evaluate spontaneous acrosin release in uncapacitated cells (zero time). Aliquot 1 mL of sperm cells (10^8 cells/mL) into Eppendorf tubes (1.5 mL) for experimentation. Perform all samples in duplicates. Proceed to **step 5**.
3. Incubate resuspended sperm cells for 4 h at 39°C in m-TALP medium supplemented with heparin (20 µg/mL) and $CaCl_2$ (2 m*M*). After 4 h of incubation, the sperm are considered capacitated cells.
4. Evaluate acrosome reaction in capacitated sperm:
   a. Initiate acrosome reaction by adding various inducers at appropriate concentrations to each tube and mix gently (avoid excess foaming).
   b. Prepare a positive control using Ca^{2+}-ionophore A23187 (10 µ*M*).
   c. Prepare negative control by omitting Ca^{2+} and any other inducer.

---

*10^4 is the camera factor.

   d. Allow reactions to proceed for another 20 min.
   e. Incubate the tubes, loosely capped, in a water bath at 39°C.
5. At the end of the incubation period (*see* **Note 1**), pellet the cells by centrifugation at 4°C for 5 min at 12,930*g*. Transfer the supernatant (800 µL) into new marked parallel Eppendorf tubes (1.5 mL) and discard the sperm cell pellet.
6. The acrosome reaction is determined by measuring the activity of acrosine released into this supernatant.
7. Adjust the supernatant fluid to pH 3.0 with 3 *M* HCl (e.g., add 8 µL of 3 *M* HCl to a 800-µL sample). Check pH with a pH meter or pH-indicator strips (range: 0–6).

At this point samples can be held overnight at 4°C or frozen for up to 3 d if necessary.

3.3. Assessment of Acrosin Activity

1. Measurement of acrosin activity (*see* **Note 2**):
   a. Thaw frozen samples and put them on ice.
   b. Adjust absorption wavelength of the spectrophotometer at 259 nm.
   c. Mix the sample several times by gentle inversion of test tube (*see* **Note 4**).
   d. In a clean quartz cuvet, pipet 900 µL of BAEE buffer.
   e. Add 100 µL of sample to the cuvet (*see* **Note 3**).
   f. Mix several times (*see* **Note 4**).
   g. Place the cuvet in the appropriate chamber of the spectrophotometer.
   h. Record the increase in absorbance at 259 nm ($A_{259\ nm}$) with time. Read each sample to achieve a linear rate line for 1 min ($\Delta A_{259\ nm}$/min). The slope of measurement is directly proportional to the acrosin activity of the sample.
   i. Rinse cuvet with ddH_2O between samples.
2. Data processing, calculation of results:
   a. Calculate acrosin activity units using the following formula:

$$\text{Acrosin activity units} = \frac{\Delta A_{259\ nm} \times df \times 10^6}{1150}$$

   where $\Delta A_{259\ nm}$ is the increase in absorbance at 259 nm for 1 min, df is the dilution factor (100 µL of sample in 1 mL total volume), 1150 $M^{-1}\ cm^{-1}$ is the molar absorption coefficient of BAEE, and 10^6 is the conversion of millimoles to nanomoles.
   b. Express acrosin activity as the rate of BAEE hydrolysis (nmol/10^8 cells × min).
   c. Subtract the spontaneous and nonspecific acrosin release from each value obtained.

4. Notes

The protocol outlined here has been developed in our laboratory to be the optimum and the simplest technique. However, we indicate here several notes that may further help the reader meet specific needs.

1. At the end of incubation period of the acrosome reaction and before centrifugation, avoid incubation of sperm cells on ice; this may be harmful to cells. The sperm cells may release acrosin nonspecifically! Therefore, centrifuge the cells rapidly as possible—immediately at the end of the incubation assay.
2. In cases where samples contain various agonists or antagonists that may affect acrosin activity itself, their effect on acrosin activity has to be tested directly. Furthermore, in cases where samples contain reagents that alter the baseline of the absorbed optical density set reference of the spectrophotometer with the BAEE medium supplemented with the same concentration of these reagents.
3. Because acrosin is a trypsinlike enzyme, it is recommended to use pure trypsin as a substrate for calibrating the enzymatic assay.
4. Excessive foaming of extracts will lead to nonlinearity and nonreproducibility. Therefore, when mixing assay tubes, do gentle inversion to prevent the generation of foam in the samples.

References

1. Yanagimachi, R. (1994) Mammalian fertilization, in *The Physiology of Reproduction* (Knobil, E. and Neil, J. D., eds.), Raven, New York, Vol. 1, pp. 189–317.
2. Zaneveld, L. J. D. and De Jonge, C. J. (1991) Mammalian sperm acrosomal enzymes and the acrosome reaction, in *Mammalian Fertilization* (Dunbar, B. S. and O'Ran, M. G., eds.), Plenum, New York, pp. 63–79.
3. Jaiswal, B. S., Eisenbach, M., and Tur-Kaspa, I. (1999) Detection of partial and complete acrosome reaction in human spermatozoa: which inducers and probes to use? *Mol. Hum. Reprod.* **5,** 214–219.
4. Hedrick, J. L., Urch, U. A., and Hardy, D. M. (1989) In *Biocatalysis in Agricultural Biotechnology* (Whitaker, J. R. and Sonnet, P. E., eds.), American Chemical Society, Washington, DC, pp. 215–229.
5. Klemm, U., Muller-Esterl, W., and Engel, W. (1991) Acrosin, the peculiar sperm-specific serine protease. *Hum. Genet.* **87,** 635–641.
6. Baba, T., Kashiwabara, S., Watanabe, K., et al. (1989) Activation and maturation mechanisms of boar acrosin zymogen based on the deduced primary structure. *J. Biol. Chem.* **264,** 11,920–11,927.
7. Yamagata, K., Murayama, K., Okabe, M., et al. (1998) Acrosin accelerates the dispersal of sperm acrosomal proteins during acrosome reaction. *J. Biol. Chem.* **273,** 10,470–10,474.
8. Kohn, F. M., Mack, S. R., Schill, W. B., et al. (1997) Detection of human sperm acrosome reaction: comparison between methods using double staining, *Pisum sativum* agglutinin, concanavalin A and transmission electron microscopy. *Hum. Reprod.* **12,** 714–721.
9. Cross, N. L. and Meizel, S. (1989) Methods for evaluating the acrosomal status of mammalian sperm. *Biol. Reprod.* **41,** 635–641.
10. Mendoza, C., Carreras, A., Moos, J., et al. (1992) Distinction between true acrosome reaction and degenerative acrosome loss by a one-step staining method using *Pisum sativum* agglutinin. *J. Reprod. Fertil.* **95,** 755–763.

11. Fenichel, P., Hsi, B. L., Farahifar, D., et al. (1989) Evaluation of the human sperm acrosome reaction using a monoclonal antibody, GB24 and fluorescence-activated cell sorter. *J. Reprod. Fertil.* **87,** 699–706.
12. Ben-Av, P., Rubinstein, S., and Breitbart, H. (1988) Induction of acrosomal reaction and calcium uptake in ram spermatozoa by ionophores. *Biochim. Biophys. Acta* **939,** 214–222.
13. Breitbart, H., Lax, Y., Rotem, R., et al. (1992) Role of protein kinase C in the acrosome reaction of mammalian spermatozoa. *Biochem. J.* **281,** 473–476.
14. Dragileva, E., Rubinstein, S., and Breitbart, H. (1999) Intracellular Ca^{2+}-Mg^{2+}-ATPase regulates calcium influx and acrosomal exocytosis in bull and ram spermatozoa. *Biol. Reprod.* **61,** 1226–1234.
15. Gur, Y., Breitbart, H., Lax, Y., et al. (1998) Angiotensin II induces acrosomal exocytosis in bovine spermatozoa. *Am. J. Physiol* **275,** E87–E93.
16. Lax, Y., Rubinstein, S., and Breitbart, H. (1994) Epidermal growth factor induces acrosomal exocytosis in bovine sperm. *FEBS Lett.* **339,** 234–238.
17. Rubinstein, S. and Breitbart, H. (1991) Role of spermine in mammalian sperm capacitation and acrosome reaction. *Biochem. J.* **278,** 25–28.
18. Rubinstein, S., Lax, Y., Shalev, Y., et al. (1995) Dual effect of spermine on acrosomal exocytosis in capacitated bovine spermatozoa. *Biochim. Biophys. Acta* **1266,** 196–200.
19. Spira, B. and Breitbart, H. (1992) The role of anion channels in the mechanism of acrosome reaction in bull spermatozoa. *Biochim. Biophys. Acta* **1109,** 65–73.

11

Isolation of Sea Urchin Sperm Plasma Membranes

Kathryn J. Mengerink and Victor D. Vacquier

1. Introduction

Sea urchin spermatozoa provide an excellent model system for studying the signal transduction events underlying the acrosome reaction ***(1,2)***. Adults can be collected from tide pools, subtidally by SCUBA diving, or purchased from biological supply companies. When adults are injected with 0.5 *M* KCl, the semen comes out of five gonopores on the aboral surface and can be collected with a Pasteur pipet and stored on ice for up to 5 d. The most widely used species in cell research is the California purple sea urchin *Strongylocentrotus purpuratus*. Five milliliters of undiluted semen can be collected from a single male of 60-mL body volume. We have worked with as much as 100 mL of fresh semen for one membrane isolation. Each 1 µL of semen contains 4×10^7 spermatozoa and equals 100 µg of total sperm protein. No other animal model provides such high numbers of spermatozoa and so much plasma membrane at such a low cost in time, labor, and money.

These sperm are terminally differentiated cells that are composed of a head (2.5 µm long × 1 µm wide) containing an acrosomal vesicle and a nucleus, a small midpiece containing a single, large mitochondrion, and a flagellum (50 µm long and 0.2 µm wide) ***(1)***. The plasma membrane surface area of the sperm head is 6 μm^2 and the surface area of the flagellum is 25 μm^2 ***(3)***. Sperm have two main functions: motility and the acrosome reaction; the latter makes them competent to fuse with eggs. Motility is driven by ATP hydrolysis by the flagellar dynein ATPase. The cells are very active in oxidative phosphorylation, which is closely coupled to motility ***(1)***. The mitochondrion provides energy to drive flagellar motility, the ATP traveling down the flagellum by a "phosphocreatine shuttle" mechanism ***(4)***.

From: *Methods in Molecular Biology, vol. 253: Germ Cell Protocols: Vol. 1 Sperm and Oocyte Analysis*
Edited by: H. Schatten © Humana Press Inc., Totowa, NJ

The acrosome reaction involves the exocytosis of the acrosomal vesicle in the anterior of the sperm head and the polymerization of actin to form the 1-μm-long acrosomal process that is coated with the sperm-to-egg adhesion protein bindin ***(5)***. The acrosome reaction involves a net influx of Na^+ and Ca^{2+}, the net efflux of H^+ and K^+, and depolarization of the sperm membrane ***(2)***. The ion channels and exchangers that are up-regulated during swimming and the acrosome reaction are currently under intense study ***(2)***. The ionic changes are activated when sperm contact molecules of the egg jelly coat a fucose sulfate polymer ***(6,7)***, a sialoglycan ***(8)***, and the peptide speract ***(2,9)***. The plasma membrane sperm receptors for these egg-derived signaling molecules are the object of our studies.

To date, more plasma membrane proteins have been cloned from sea urchin sperm than from any other animal model. The list includes a unique 150-kDa creatine kinase ***(10)***, a160-kDa receptor guanylyl cyclase that is regulated by phosphorylation ***(11)***, a 77-kDa accessory protein to the guanylyl cyclase that is the receptor for the egg peptide speract ***(12)***, a GPI-anchored 63-kDa flagellar membrane protein with multiple EGF repeats ***(13)***, a cAMP-modulated pacemaker cation channel ***(14)***, two proteins, suREJ1 ***(15)*** and suREJ3 ***(16)***, both of which share significant homology to human polycystin-1 (the protein mutated in 85% of human autosomal dominant polycystic kidney disease), a K^+-dependent Na^+/Ca^{2+} exchanger ***(17)***, a 190-kDa ABC transporter ***(18)***, and the exocytotic regulatory proteins syntaxin ***(19)*** and SNAP-25 ***(20)***. Within the coming year, the whole genome sequence of this species will be known, permitting a proteomics approach to the identification, isolation, cloning, and expression of the ion channels and receptors that reside in the sperm plasma membrane. Interest in the ion channel events mediating the acrosome reaction stimulated the development of simple methods to isolate tens of milligrams of sea urchin sperm plasma membrane.

2. Materials

1. Seawater (FSW) or artificial seawater (ASW) passed through a 0.2-μm Millipore filter.
2. Benzamidine-HCl (Sigma, St. Louis, MO; cat. no. B-6506).
3. Protease inhibitor cocktail (Sigma, cat. no. P-8340), diluted 1:100 for use.
4. Artificial seawater: 454 m*M* NaCl, 9.7 m*M* KCl, 24.9 m*M* $MgCl_2$, 9.6 m*M* $CaCl_2$, 27.1 m*M* $MgSO_4$, 4.4 m*M* $NaHCO_3$.
5. pH 9.2 Seawater: 40 m*M* Tris-base, pH 9.2, 10 m*M* benzamidine-HCl in FSW or ASW, plus a 1:100 dilution of protease inhibitor cocktail, adjust to pH 9.2.
6. Nitrogen cavitation buffer: 480 m*M* NaCl, 10 m*M* $MgCl_2$, 10 m*M* KCl, 20 m*M* benzamidine-HCl, 1:100 dilution of protease inhibitor cocktail, 10 m*M* MES, pH 6.0.

7. KCl/MES seawater: 20 m*M* of 2-(4-morpholino)-ethane sulfonic acid (MES), pH 6.0, 40 m*M* KCl in FSW or ASW.
8. Ionophore A23187 (Calbiochem, La Jolla, CA, cat. no. 100105). Two milligrams of ionophore dissolved in 1 mL of dimethyl sulfoxide (DMSO) and stored at –20°C.
9. Ionophore nigericin (Sigma, cat. no. N-7143). Two milligrams of ionophore dissolved in 1 mL of DMSO and stored at –20°C.
10. Parr 4635 cell disruption bomb (Parr Instruments Co., Moline, IL).
11. Wheat germ agglutinin (WGA) conjugated to agarose beads (E-Y-Laboratories, San Mateo, CA; cat. no. A-21-01).
12. *N*-Acetyl-D-glucosamine (Sigma, cat. no. A-8625).
13. MES (Fisher Biotech, Fair Lawn, NJ; cat. no. BP300-100).
14. Sephadex-G25 equilibrated in FSW or ASW. (Any beaded molecular sieving resin will work.)

3. Methods

Unless specified, all procedures are on ice or 4°C.

3.1. Collecting Sperm

Adult sea urchins are injected intracoelomically with 0.5 M KCl (23°C). The creamy white semen comes out of the gonopores and is collected in 1.5-mL microcentrifuge tubes, which are capped and stored in ice. This is termed "dry" sperm and contains 4×10^{10} spermatozoa per milliliter. One microliter contains 100 µg of sperm protein. The dry sperm can be stored packed in ice for 5 d without noticeable changes in the plasma membrane proteins (*see* **Note 1**).

3.2. Removal of Coelomocytes

Coelomocytes are sea urchin immune cells that are in low abundance in "dry" sperm. It is critical to remove these cells from the sperm because they are rich in protease activity that will digest the plasma membrane proteins. One subpopulation of coelomocytes is red, so removal is monitored by looking for a red pellet following centrifugation.

1. Resuspend dry sperm in 50 vol FSW or ASW in conical 50-mL plastic conical centrifuge tubes. Stir the cell suspension by hand with a spatula to ensure uniform resuspension.
2. Centrifuge at 500*g* for 5 min.
3. Draw off the sperm suspension with a 10-mL pipet and discard the red pellet.
4. Repeat **steps 2** and **3** three times. This will remove all the coelomocytes (*see* **Note 2**).
5. Carefully pipet the sperm suspension into round-bottomed 40- or 50-mL tubes.
6. Sediment the sperm at 1500*g* for 20 min in a swinging-bucket rotor and resuspend the cell pellet in the appropriate buffer. Use a flat, plastic spatula to break up the pellet and resuspend the cells in the desired medium (*see* **Note 3**).

3.3. Isolating Sperm Heads and Flagella

It is possible to isolate a head and a flagellar fraction. Using a nonionic-detergent-solubilized head protein, adequate amounts of the receptor for egg jelly 3 (REJ3) were obtained to successfully coimmunoprecipitate the cleaved NH_2- and COOH-termini of the protein ***(16)***. Also, isolating heads is the first step in producing acrosome-reaction vesicles (described below). Sperm are resuspended in low pH and increased KCl concentration to prevent motility and the acrosome reaction. Once separated heads and flagella are obtained, their membranes can be isolated by the pH 9.2 method or nitrogen cavitation as described below.

1. Resuspend the 1500*g* sperm pellet from above in 50 volumes KCl/MES seawater in round-bottomed 50 mL tubes.
2. Sediment the sperm (1500*g* for 20 min) in a swinging-bucket rotor.
3. Discard the supernatant and resuspend the cells in 20 vol of KCl/MES buffer.
4. Shear flagella from heads with a tight-fitting glass ball homogenizer. Use phase-contrast microscopy to check continuously for the approximate percentage of sperm heads that lack flagella. Check with the microscope after every five passes with the glass ball. Some small pieces of flagella may remain associated with the heads (*see* **Note 4**).
5. Separate the sperm heads from flagella by differential centrifugation. A 10-min centrifugation at 800*g* in a swinging-bucket rotor will sediment most of the free heads. The supernatant is carefully removed with a pipet and placed in another tube and kept on ice. The head pellet is resuspended in the same volume of KCl/MES seawater and the 800*g* centrifugation is repeated until the head pellet is free of flagellar fragments. The first 800*g* supernatant is centrifuged repeatedly for 10 min at 800*g* until all the residual heads have been sedimented. Phase-contrast observation at approx 400× magnification will show that very few sperm heads are present among the broken flagella. Isolated flagella are finally sedimented by centrifugation at 5000*g* for 20 min. Silver-stained sodium dodecyl sulfate–polyacrylamide gel electrophoresis (SDS-PAGE) analysis shows the separation of heads from flagella can be very clean ***(1)***; *see* **Note 5**.

3.4. Isolating and Characterizing Plasma Membranes

3.4.1. Isolating Sealed, Right-Side-Out Vesicles (SMVs) by the pH 9.2 Method

The protocol is based on **ref. *21***. After isolation, the sperm membrane vesicle (SMV) pellet can be stored at –70°C. These vesicles have been used to identify many sperm membrane proteins, and the 40,000*g* and 200,000*g* supernatants have been used to isolate the receptor for egg jelly 1 (REJ1) in sufficient quantities for peptide sequencing ***(15)***.

1. Resuspend the coelomocyte-free, 1500*g* sperm pellet from above in 10 vol FSW or ASW in a beaker. Isolated sperm heads or flagella can also be used as the starting material.
2. While vigorously stirring the sperm suspension with a plastic spatula, slowly add an equal volume of pH 9.2 seawater. Place the final sperm suspension at 4°C for 5–18 h without agitation. Originally ***(21)*** the incubation was at 15°C, but now we use 4°C.
3. After 5–18 h, use a magnetic stir plate to vigorously stir the sperm suspension for 4 min.
4. Sediment the cell debris by two 6000*g* centrifugations for 30 min each. Carefully remove the first and second 6000*g* supernatants into clean centrifuge tubes with a 10-mL pipet.
5. Sediment the large SMVs by centrifugation at 40,000*g* for 1 h (*see* **Note 6**).
6. The 40,000*g* supernatant can be centrifuged at 200,000*g* for 1 h to obtain a pellet of membrane vesicles of smaller size. The 200,000*g* supernatant is an excellent starting material for the isolation of receptor for egg jelly 1 (REJ1) by wheat germ agglutinin affinity chromatography ***(15)***.
7. The 40,000*g* and 200,000*g* pellets of SMVs can be frozen as is or dissolved in detergents and then frozen. They have a characteristic composition of plasma membrane proteins ***(1,21)*** (*see* **Note 7**).

3.4.2. Characterizing SMVs

At pH 9.2, about 1 pH unit above that of normal seawater, the sperm plasma membranes bleb off of the cells as tightly sealed, right-side-out vesicles. If 2 m*M* carboxyfluorecein is present in the pH 9.2 buffer, the dye becomes trapped inside the SMVs. Free dye can be separated from trapped dye by resuspending the 40,000*g* SMV pellet in FSW or ASW and passing the vesicles though a 1 × 30-cm column of Sephadex-G25 (Pharmacia; any molecular sieving matrix can be used). The SMVs come out in the void volume, whereas the free dye runs in the included volume. Incubating these SMVs on ice for 22 h and then passing them back over the sieving column shows that very little carboxyfluorecein has leaked out during this time. This shows that impermeable compounds can be trapped inside SMVs if they are present as the vesicles bleb off of the cells and also that the SMVs are tightly sealed ***(21)***. Electron microscopy shows that SMVs are large, unilamellar vesicles that vary in diameter from 0.05 to 0.7 μm ***(21)***. That they are right-side-out in orientation is shown by the fact that all of the SMV protein binds to a wheat germ agglutinin affinity column ***(21)***.

The following characterization of SMVs used whole sperm cells as the starting material. The 40,000*g* pellet of SMVs can be resuspended in FSW or ASW and banded by ultracentrifugation in a sucrose gradient at a density of 1.17 g/cm^3. However, comparison of the 40,000*g* pellet with the sucrose-banded SMVs shows no difference in protein composition by silver-stained SDS-PAGE analysis. Therefore, no extra purity is gained by the ultracentrifuge gradient step usually employed in plasma membrane purification ***(21)***. There is no detectable DNA in the SMV pellet. Nuclei, mitochondria, and axonemes remain fairly intact during the procedure. There is little contamination by mitochondrial membrane, as shown by the 23-fold reduction in cytochrome oxidase in SMVs compared to whole sperm. Adenylyl cyclase is enriched sixfold in SMVs. There is a fourfold to eightfold increase in the specific radioactivity of SMVs isolated from sperm that were surface radiolabeled with ^{125}I. One-third of the total Commassie blue binding SMV protein is found in three proteins of 150 kDa [creatine kinase ***(10)***], 140 kDa [guanylyl cyclase ***(11)***], and 80 kDa ***(1,21)***. The carboxyfluorecein-containing SMVs bind species-specifically to unfertilized sea urchin eggs; this binding can be blocked by preincubation of the SMVs with soluble sea urchin egg jelly ***(21)***. If the SMVs are solubilized and either 3% cholate, 3% SDS, or 3% Triton X-100 on ice for 30 min and then centrifuged at 170,000*g* for 1 h, many of the SMV proteins sediment ***(21)***. This shows these proteins are either relatively insoluble or present in large aggregates that might be physiologically significant (*see* **Note 8**).

3.4.3. Isolating Sperm Plasma Membrane Vesicles of Both Orientations by Nitrogen Cavitation

This protocol ***(22)*** yields cavitated membrane vesicles (CMVs) that are a 50:50 mixture of inside-out and right-side-out vesicles. The inside-out vesicles can be used to study enzyme activities. For example, guanylyl cyclase was purified from sea urchin sperm by solubilizing CMVs and applying the protein to a concanavalin-A affinity column ***(22)***. Also, a protein kinase activity that phosphorylates only the guanylyl cyclase is present in CMVs ***(23)***.

1. Resuspend the 1500*g*, coelomocyte-free, sperm pellet from above in 10 vol of nitrogen cavitation buffer.
2. Subject sample to nitrogen cavitation at 400 psi for 10 min in a Parr bomb.
3. Collect the cavitate in 50-mL round-bottomed tubes and centrifuge for 10 min at 8000*g*.
4. Pipet off the 8000*g* supernatant and centrifuge it at 11,000*g* for 30 min.
5. Sediment the CMVs by ultracentrifugation of the 11,000*g* supernatant at 200,000*g* for 1 h. The CMVs have a protein composition similar to SMVs ***(22,23)*** (*see* **Note 9**).

3.4.4. Separating Right-Side-Out from Inside-Out CMVs

This is a simple, advantageous method to separate vesicles of different orientation. This can be used in experiments to study the directional movement of ions through the vesicle membrane, to load vesicles with substrates for enzymes or with fluorescent reporters, and to study the phosphorylation of regulatory proteins such as the guanylyl cyclase ***(11,12,22,23)***.

1. The CMV pellet is resuspended in FSW or ASW using a Pasteur pipet to produce a uniform suspension.
2. The suspension is centrifuged at 8000*g* for 10 min to sediment aggregated CMVs.
3. The supernatant is applied to a wheat germ agglutinin agarose column and the flowthrough volume passed back over the column a total of five times. The final flow through fraction contains the inside-out vesicles.
4. The WGA agarose is washed with 10 column volumes of ASW or FSW and the right-side-out CMVs eluted with the addition of ASW or FSW containing 100 m*M* *N*-acetylglucosamine to the column. The right-side-out vesicles bind the WGA agarose beads because of the sialic acid residues on the oligosaccharide chains on proteins facing to the external side of the sperm plasma membrane ***(21)*** (*see* **Note 10**).

3.4.5. Isolating Acrosome Reaction Vesicles

Acrosome reaction vesicles (ARVs) will be important for our further understanding of the acrosome reaction because they may contain many of the proteins involved in this most important event of animal reproduction. The acrosome reaction involves the fusion at many points of the plasma membrane with the acrosomal vesicle membrane. The two membranes fuse, forming small composite membrane vesicles that are shed from the cells. Shedding of ARVs can be a very "protein-selective" event, as shown by the fact that all the detectable syntaxin, VAMP ***(19)***, and SNAP-25 ***(20)***, are shed from sperm in the ARVs as a result of the acrosome reaction. ARVs have been incorporated into planar lipid bilayers and the activity of single ion channels detected ***(24)***.

1. Isolate sperm heads as described above.
2. Resuspend sperm heads in 20 vol of FSW or ASW.
3. While vigorously stirring with a spatula, add 1 vol of A23187 and 1 vol of nigericin in DMSO to 99 vol of sperm suspension and incubate for 5 min. The ionophores induce the acrosome reaction.
4. Remove cell debris by centrifuging at 10,000*g* for 30 min.
5. Sediment the ARVs by ultracentrifugation at 200,000*g* for 1 h (*see* **Note 11**).

4. Notes

1. It is important to keep the dry semen stored in crushed ice and not at 4°C or in the refrigerator.
2. It is very important to use conical tubes to sediment the coelomocytes as a tight red pellet. Do not pour off the sperm cell suspension. Pipet off the sperm cell suspension, leaving about 1 mL in the tube bottom over the red pellet.
3. If the sperm cell pellet does not immediately dissociate into single cells, put the tube on ice for 5 min and then stir again with the spatula until the cells are completely resuspended.

4. It is difficult to attain 100% removal of flagella from sperm heads. Closely monitor the damage done to the cells after every 10 stokes of the glass ball with the homogenizer. Use approx 400–1000× magnification (phase-contrast microscopy). Remove all excess water between the slide and cover slip. Scan rapidly, observing approx 100 cells for remnants of the flagella.
5. Silver-stained SDS-PAGE analysis of whole sperm compared to isolated heads and flagella is the best way to access contamination of isolated flagella by heads and isolated heads by flagella. The core histones of heads and the tubulin bands of flagella are excellent indicators of contamination of one fraction with the other.
6. The two 6000*g* pellets can be resuspended in pH 9.2 seawater and stirred vigorously for 5 min. A second batch of SMVs can then be isolated by repeating the two 6000*g* and the one 40,000*g* centrifugations.
7. The SMV proteins are more soluble if the SMVs are dissolved in nonionic detergent before freezing at –70°C.
8. In characterizing membrane proteins, it is important to report whether they sediment in 1% Triton X-100 after centrifugation at 100,000*g* for 1 h.
9. The degree of cell damage is directly proportional to the pressure used in the cavitation bomb. The optimal pressure level will differ with sperm of different species. The desired pressure can be empirically determined by microscopic observation of cavitates at different pressures.
10. Wheat germ agglutinin immobilized on agarose beads is very stable and will last for years if stored in a pH 7 buffer and 10 m*M* azide to inhibit microbial growth. WGA must never be exposed to reducing agents such as mercaptoethanol or dithiothreitol. Reducing agents irreversibly denature WGA.
11. The ARVs can be dissolved in a nonionic, dialyzable detergent such as octylglucoside and phospholipid and frozen for later formation of liposomes.

Acknowledgment

This research was supported by NIH grant no. HD12986 to VDV.

References

1. Vacquier, V. D. (1986) Handling, labeling, and fractionating sea urchin spermatozoa. *Methods Cell Biol.* **27,** 15–40.
2. Darszon, A., Beltran, C., Felix, R., et al. (2001) Ion transport in sperm signaling. *Dev. Biol.* **240,** 1–14.
3. Cross, N. L. (1983) Isolation and electrophoretic characterization of the plasma membrane of sea urchin sperm. *J. Cell Sci.* **59,** 13–25.
4. Tombes, R. M., Brokaw, C. J., and Shapiro, B. M. (1987) Creatine kinase-dependent energy transport in sea urchin spermatozoa. *Biophys. J.* **52,** 75–86.
5. Vacquier, V. D., Swanson, W. J., and Hellberg, M. E. (1995) What have we learned about sea urchin sperm bindin? *Dev. Growth Differ.* **37,** 1–10.
6. Hirohashi, N. and Vacquier, V. D. (2002) High molecular mass egg fucose sulfate polymer is required for opening both Ca^{2+} channels involved in triggering the sea urchin sperm acrosome reaction. *J. Biol. Chem.* **277,** 1182–1189.

7. Vilela-Silva, A.-C. E. S., Castro, M. O., Valente, A.-P., et al. (2002) Sulfated fucans from the egg jellies of the closely related sea urchins *Strongylocentrotus droebachiensis* and *S. pallidus* ensure species-specific fertilization. *J. Biol. Chem.* **277,** 379–387.
8. Hirohashi, N. and Vacquier, V. D. (2002) Egg sialoglycans increase intracellular pH and potentiate the acrosome reaction of sea urchin sperm. *J. Biol. Chem.* **277,** 8041–8047.
9. Hirohashi, N. and Vacquier, V. D. (2002) Egg fucose sulfate polymer, sialoglycan, and speract all trigger the sea urchin sperm acrosome reaction. *Biochem. Biophys. Res. Commun.* **296,** 833–839.
10. Wothe, D. D., Charbonneau, H., and Shapiro, B. M. (1990) The phosphocreatine shuttle of sea urchin sperm: flagellar creatine kinase resulted from a gene triplication. *Proc. Natl. Acad. Sci. USA* **87,** 5203–5207.
11. Garbers, D. L. (1989) Molecular basis of fertilization. *Annu. Rev. Biochem.* **58,** 719–742.
12. Dangott, L. J., Jordan, J. E., Bellet, R. A., et al. (1989) Cloning of the mRNA for the protein that crosslinks to the egg peptide speract. *Proc. Natl. Acad. Sci. USA* **86,** 2128–2132.
13. Mendoza, L. M., Nishioka, D., and Vacquier, V. D. (1993) A GPI-anchored sea urchin sperm membrane protein containing EGF domains is related to human uromodulin. *J. Cell Biol.* **121,** 1291–1297.
14. Gauss, R., Seifert, R., and Kaupp, U. B. (1998) Molecular identification of a hyperpolarization activated channel in sea urchin sperm. *Nature* **393,** 583–587.
15. Moy, G. W., Mendoza, L. M., Schulz, J. R., et al. (1996) The sea urchin sperm receptor for egg jelly is a modular protein with extensive homology to the human polycystic kidney disease protein, PKD1. *J. Cell Biol.* **133,** 809–817.
16. Mengerink, K. J., Moy, G. W., and Vacquier, V. D. (2002) suREJ3, a polycystin-1 protein, is cleaved at the GPS domain and localizes to the acrosomal region of sea urchin sperm. *J. Biol. Chem.* **277,** 943–948.
17. Su, Y.-H. and Vacquier, V. D. (2002) A flagellar K^+-dependent Na^+/Ca^{2+} exchanger keeps Ca^{2+} low in sea urchin spermatozoa. *Proc. Natl. Acad. Sci. USA* **99,** 6743–6748.
18. Mengerink, K. J. and Vacquier, V. D. (2002). An ATP-binding cassette transporter is a major glycoprotein of sea urchin sperm membranes. *J. Biol. Chem.* **277,** 40,729–40,734.
19. Schulz, J. R., Wessel, G. M., and Vacquier, V. D. (1997) The exocytotic regulatory proteins syntaxin and VAMP are shed from sea urchin sperm during the acrosome reaction. *Dev. Biol.* **191,** 80–87.
20. Schulz, J. R., Sasaki, J. D., and Vacquier, V. D. (1998). Increased association of synaptosome-associated protein of 25 kDa with syntaxin and vesicle-associated membrane protein following acrosomal exocytosis of sea urchin sperm. *J. Biol. Chem.* **273,** 24,355–24,359.
21. Podell, S. B., Moy, G. W., and Vacquier, V. D. (1984) Isolation and characterization of a plasma membrane fraction from sea urchin sperm exhibiting species specific recognition of the egg surface. *Biochim. Biophys. Acta* **778,** 25–37.

22. Ward, G. E., Garbers, D. L., and Vacquier, V. D. (1985) Effects of egg extracts on sperm guanylate cyclase. *Science* **227,** 768–770.
23. Ward, G. E., Moy, G. W., and Vacquier, V. D. (1986) Phosphorylation of membrane-bound guanylate cyclase of sea urchin spermatozoa. *J. Cell Biol.* **103,** 95–101.
24. Schulz, J. R. (1999) Sea urchin sperm acrosomal exocytosis: identification of acrosome reaction vesicle associated proteins. Ph.D. thesis, University of California, San Diego.

12

Detection of Centrosome Structure in Fertilized and Artificially Activated Sea Urchin Eggs Using Immunofluorescence Microscopy and Isolation of Centrosomes Followed by Structural Characterization with Field Emission Scanning Electron Microscopy

Heide Schatten and Amitabha Chakrabarti

1. Introduction

Sea urchin eggs have been used for over a century to study fertilization, cell division, cell differentiation, and embryo development. This system provides excellent and abundant material to investigate the basic mechanisms underlying germ cell functions and to explore tools that facilitate the studies of other less readily available germ cell systems. Immunofluorescence microscopy of microtubule organization performed in sea urchin eggs *(1,2)* led to numerous subsequent studies on the cytoskeleton in invertebrate *(3–8)* and mammalian *(9–12)* reproductive cell systems. The pioneering studies on centrosomes in sea urchin eggs were performed over 100 yr ago by Boveri *(13)* and have provided a wealth of information and insights that has led to a renaissance and an explosion in centrosome research during the past few years. By using iron hematoxylin as the primary staining technique in fertilized sea urchin eggs, Boveri showed elegantly that dominant centrosome material is contributed by sperm. He showed that sperm centrosomes reorganize after pronuclear fusion and separate to the mitotic poles to form the mitotic apparatus during cell division. Moreover, he showed that sea urchin eggs fertilized with more than one sperm contained supernumerary centrosomes, resulting in multipolar spindles that lead to unequal chromosome separation during cell division. These studies on sea urchin eggs gave thought to the fascinating idea that the formation of multiple centrosome organizations might be involved in the development and

From: *Methods in Molecular Biology, vol. 253: Germ Cell Protocols: Vol. 1 Sperm and Oocyte Analysis*
Edited by: H. Schatten © Humana Press Inc., Totowa, NJ

progression of malignant tumors *(14)*. Boveri's original ideas, derived from studies on sea urchin eggs, have been validated in numerous new studies in more recent years and extended the original findings to cell and molecular levels *(6,12,15–21)*. Taken together, the sea urchin germ cell system has yielded an enormous amount of information that has benefited other cell systems. In this chapter, we focus on methods for detection of centrosomes in fertilized *(12)* and artificially activated *(22)* sea urchin eggs by immunofluorescence microscopy and on the isolation of centrosomes from fertilized sea urchin eggs followed by structural characterization with immunofluorescence (IF) and field emission scanning electron microscopy (FESEM) *(23)*.

Centrosomes are the primary microtubule organizing centers that undergo molecular and structural reorganizations throughout the cell cycle. In germ cells, the centrosome is a key structure during fertilization that in most systems, except for some rodents, nucleates and organizes the sperm aster that moves the haploid genomes of male and female gametes in close proximity *(6,12)*. In most systems, centrosomes are tightly associated with the basal body in sperm and become decompacted in the egg's cytoplasm during fertilization. Centrosomes compact and decompact throughout the cell cycle to organize the dynamically changing microtubule formations of the sperm aster and the radial zygote stage aster. Centrosomes duplicate during S-phase *(24)* before separating and relocating to the mitotic poles where they organize the bipolar mitotic apparatus that separates the diploid genome during cell division. By organizing polarized microtubule configurations, centrosomes are also crucial for subsequent coordinated equal and unequal cell divisions.

The centrosome cycle is normally synchronized with the chromosome cycle in order to ensure equal genome separation during cell division *(25,26)*. At the onset of mitosis, centrosomes undergo significant molecular and structural reorganizations whereby phosphorylation of nuclear and centrosome proteins plays a key role in triggering the mitotic events *(27–30)*. Several factors play a role in centrosome functions that are typically also coordinated with the chromosome cycle and other cell-cycle-regulating events. Changes in pH and calcium have been shown to activate latent centrosome material in the unfertilized sea urchin egg *(22)*, which are changes normally induced by sperm *(6,7,31,32)*. Several factors are involved in the folding and unfolding of centrosome structure and include the reduction of disulfide bonds and the formation of sulfhydryl bonds *(33)*.

Centrosomes nucleate microtubules in a precisely determined organization in which the microtubule organizing protein γ-tubulin *(34–38)* plays a critical role. Centrosome abnormalities can be causes for infertility *(39)* and for diseases of various nature, including cancer *(16,18,20,40,41)*. Because centrosomes determine the polarity and orientation of microtubule assembly, they play a critical role in directing transport of cell organelles and maintaining cellular metabolism.

Although the molecular aspects of centrosomes are being well studied, few studies have focused on the structural behavior of centrosomes. Conventional transmission electron microscopy has not provided suitable methods to study centrosome structure ***(40,42–44)*** and direct methods to visualize centrosomes have not advanced as fast as molecular methods. Many technologies and methodologies that have been used to explore chromosome structure have not yet been applied to the study of centrosome structure. Immunofluorescence microscopy has allowed to some extent the conclusion that centrosomes contain a flexible structure that is able to compact and decompact ***(6,40,45)*** and it has been shown that the compaction and decompaction cycle can be impaired by using pharmacological agents such as chloral hydrate and diazepam ***(46)*** or the chaotropic agent formamide ***(40)***. By using low-voltage field emission scanning electron microscopy on isolated centrosomes, it was possible to gain knowledge on the nature of centrosome material in compacted and decompacted states ***(23)***. These studies have contributed new knowledge on the various shapes that centrosomes undergo to nucleate the diversity of microtubule configurations.

1.1. Immunofluorescence Microscopy

Immunofluorescence microscopy has first been employed on sea urchin eggs by Harris et al. ***(1)*** using cold MeOH as a fixative to determine microtubule organization after fertilization and during cell division. Bestor and Schatten ***(3)*** employed immunofluorescence microscopy to study microtubule organization during fertilization and artificial activation. Although these methods provided excellent information on microtubule organization, improved methods employed extraction protocols ***(5,47)*** to more clearly visualize the sea urchin embryo's cytoskeleton. With the finding that human autoimmune sera would stain centrosomes in mouse oocytes, Calarco-Gillam et al. ***(48)*** and Schatten et al. ***(6)*** used the autoimmune serum 5051 to stain centrosomes in fertilized sea urchin eggs ***(6,12)***. Monoclonal antibodies against isolated centrosomes from sea urchins allowed further analysis of specific centrosome proteins in sea urchin eggs ***(23)***.

1.2. Isolation of Centrosomes

The isolation of centrosomes from a variety of cell systems, including mammalian cells ***(49,50)***, *Drosophila* ***(51)***, *Spisula* ***(52)***, *Dictyostelium* ***(53)***, and sea urchin eggs ***(23)***, has resulted in new approaches to explore microtubule organization and the production of monoclonal antibodies to specific centrosome proteins. The isolation of centrosomes from sea urchin eggs resulted in the characterization of novel monoclonal antibodies, including one (4D2) that was specific for sea urchin centrosomes ***(23)***. Electrophoretic analysis of isolated centrosomes has revealed that centrosomes are composed of numerous low- and high-molecular-weight proteins, many of which have been characterized in recent years and specific functions have been ascribed to some of them ***(54–57)***. Taken together with data from other investigators ***(58,59)***, it can be speculated that centrosomes contain tissue- and organ-specific centrosome proteins. Further research may well result in the identification of specific centrosome proteins that can be targeted to correct centrosome abnormalities that are involved in centrosome impairment, including certain cases of infertility.

1.3. Characterization of Isolated Centrosomes by Low-Voltage Field Emission Scanning Electron Microscopy

Although in the past few years the molecular composition of centrosomes and factors involved in centrosome duplication have been explored by several investigators ***(24,29,60–66)***, the analysis of centrosome structure has lagged far behind the molecular analysis of centrosomes because appropriate methods have not yet been available. Only limited knowledge exists about the structure of centrosomes ***(23,41,67)***. This is in part related to the fact that centrosomes cannot be imaged well with conventional transmission electron microscopy (TEM), and TEM images of centrosomes only revealed undefined osmiophilic material that in most cases surrounds a pair of well-structured centrioles. The isolation of centrosomes combined with new and more powerful electron microscopy imaging methods such as FESEM ***(23)*** and new preparation methods ***(68)*** may provide new insights into the complex centrosome structure combining molecular biology and immunoelectron microscopy to correlate centrosome protein functions with centrosome structure.

2. Materials

2.1. Sea Urchin Gametes

The Californian sea urchins *Lytechinus pictus* or *Strongylocentrotus purpuratus* were obtained from Marinus, Inc. (Long Beach, CA, USA). *L. pictus* is typically fertile from April through October and *S. purpuratus* is typically fertile from October through April (*see* **Note 1**).

2.2. Artificial Seawater

Compounds	Concentration (m*M*)	g/L	Sigma cat. no.
NaCl	484.42	28.32	S7653
KCl	10.32	0.77	P9333
$MgCl_2 \cdot 6H_2O$	26.61	5.41	M2670

Compounds	Concentration (m*M*)	g/L	Sigma cat. no.
$MgSO_4·7H_2O$	115.70	7.13	M1880
$CaCl_2·2H_2O$	8.02	1.18	C3881
$NaHCO_3$	2.38	0.2	S6297

Adjust the pH to 8.3 and filter (0.22 µm). Store at 4°C.

2.3. Culture Medium (Calcium-Free Seawater)

Compounds	Concentration (m*M*)	g/L	Sigma cat. no.
NaCl	484.42	28.32	S7653
KCl	10.32	0.77	P9333
$MgCl_2·6H_2O$	26.61	5.41	M2670
$MgSO_4·7H_2O$	115.70	7.13	M1880
$NaHCO_3$	238.00	0.2	S6297
Tris	10.00	0.121	T6791
EGTA	25.00	1.17	E8145

2.4. Phosphate-Buffered Saline Buffer

Compounds	Concentration (m*M*)	g/L	Sigma cat. no.
NaCl	136.89	8.0	S7653
KCl	2.68	0.2	P9333
Na_2HPO_4	10.14	1.44	S7907
KH_2PO_4	1.76	0.24	P5379

Adjust the pH to 7.2. PBST = 1X PBS with 0.1% Tween-20 (Sigma, P1379)

2.5. Other Chemicals

Compounds	Concentration	Sigma cat. no.
KCl	500 m*M* (3.72 g/100 mL)	P9333
ATA (aminotriazole)	10 m*M* (0.0084 g/100 mL)	A8056
Poly-L-lysine	1 mg/mL dH2O	P1524
Paraformaldehyde	8 g/100 mL	P 6148
DAPI	4 µg/mL in 1X PBS	D9542
Mowiol 4-88	2.44 g/20 mL	From Calbiochem # 475904
Glycerol	6 g/20 mL	G6279
Methanol (absolute acetone-free)		M1775
NH_4Cl	10 m*M*	A 4514
Calcium–ionophore A23187	10 m*M* stock solution in dimethyl sulfoxide (DMSO)	I1647

2.6. Antibodies

1. Mouse monoclonal antibody to β-tubulin (E-7; Developmental Studies, Hybridoma Bank, IA).
2. Mouse monoclonal antibodies Ah6 ***(12,69)*** or 4D2 ***(23,46)*** to detect centrosomes.
3. Fluorescein isothiocyanate (FITC)- or rhodamine-conjugated goat anti-mouse secondary antibody (Zymed, South San Francisco, CA).
4. Human autoimmune serum against centrosome proteins (5051) ***(6,48)*** and SPJ ***(70,71)***.
5. FITC- or rhodamine-conjugated goat anti-human secondary antibody (Zymed).

3. Methods

3.1. Poly-L-Lysine-Coated Cover Glass Preparation

For immunofluorescence microscopy, clean #1 cover glass (Corning) in 100% ethanol and cover the entire surface of the cover glass with 1 mg/mL poly-L-lysine solution for 20 min, decant excess, air-dry, and store at 4°C. For FESEM, use #2 cover glass (Corning).

3.2. Paraformaldehyde Preparation

1. Add 8 g of paraformaldehyde (Sigma, cat. no. P6148) to 100 mL of water and heat to 60°C in a fume hood.
2. Add a few drops of 1 *N* NaOH to facilitate paraformaldehyde powder to completely dissolve in water.
3. Cool to room temperature; add 100 mL of 2X PBS.
4. Prepare fresh before use.

3.3. DAPI

Make up DAPI (Sigma, cat. no. D9542) in 1X PBS at a concentration of 4 μg/mL.

3.4. Mowiol Antifade

1. Add 2.4 g Mowiol 4-88 (Calbiochem, cat. no. 475904) to 6 g glycerol (Sigma, cat. no. G6279) and stir for 1 h.
2. Add 6 mL distilled water and stir for another 2 h.
3. Finally, add 12 mL of 0.2 *M* Tris-HCl, pH 8.5.
4. Incubate in a 50°C water bath with occasional stirring. Clarify the mixture by centrifugation at 5000*g* for 15 min.
5. Add 2.5% DABCO (Sigma, cat. no. D2522) to retard photobleaching and store at –20°C (*see* **Note 2**).

3.5. Gamete Collection and Test of Fertilizability

1. Wash entire sea urchin in artificial seawater (ASW) thoroughly (*see* **Note 3**).
2. Inject 0.5 *M* KCl in the soft intracoelomic part of the animal. Fully formed gametes are spawned through the five gonopores located at the opposite end of the animal within 2–3 min of KCl stimulation.

3. Collect sperm undiluted ("dry") into a small Petri dish placed on ice and check for sperm motility under a microscope (*see* **Note 4**).
4. Collect white-yellowish egg suspension at 4–6°C in a beaker containing ASW.
5. Wash eggs with ASW and perform test fertilization.
6. For test fertilization (*see* **Note 5**), dilute sperm (25 µL concentrated sperm in 3 mL ASW) and mix with a diluted (1%) (v/v) suspension of eggs. The sperm induces the elevation of the fertilization coat that is seen as a transparent membrane surrounding the egg. The fertilization coat serves as a marker for fertilization. Cultures exhibiting less than 90% fertilization rates should not be used for experiments.

3.6. Stripping of Fertilization Coats

1. Fertilize eggs in 10 m*M* aminotriazole (ATA) to prevent hardening of the fertilization coat *(72)*.
2. After 20 min postinsemination, pass fertilized egg suspension through 54 µm Nitex membrane (Small Parts, Inc., Miami Lakes, FL) and transfer denuded eggs to ASW at 14°C.

3.7. Activation Protocols with NH3 or Calcium Ionophore A23187

The unfertilized sea urchin egg contains latent centrosome material that can be activated to aggregate by either pH modulation using ammonium or by calcium modulation using the calcium ionophore A23187 *(22)*.

For treatment with ammonium:

1. Incubate eggs in 10 m*M* NH_4Cl for 30 min.
2. Wash three times in Millipore-filtered (0.22 µm) ASW.
3. Culture in either calcium-free or regular ASW.

For treatment with calcium ionophore:

1. Treat unfertilized eggs for 30 min with 10 µ*M* ionophore A23187 made from a 10 m*M* stock solution dissolved in DMSO in ASW containing 10 m*M* ATA.
2. Wash three times in Millipore-filtered (0.22 µm) ASW containing 10 m*M* ATA.
3. After 20 min following calcium ionophore activation, pass fertilized egg suspension through a 54-µm Nitex membrane (Small Parts, Inc., Miami Lakes, FL) and transfer denuded eggs to ASW at 14°C.
4. Culture in either calcium-free or regular ASW.

3.8. Fixation

1. For immunofluorescence microscopy, collect 100-µL aliquots at various timepoints throughout the first cell cycle and immobilize eggs by attaching to a poly-L-lysine-coated cover glass.
2. Fix eggs by immersion into –20°C prechilled 90% methanol/50 m*M* EGTA, pH 6.0, for 6 min. Alternatively, fix eggs in a 4% paraformaldehyde solution for 10 min at room temperature and wash twice with PBS. Permeabilize the fixed

eggs by incubating in –20°C prechilled 100% methanol for 5 min and rinse gently in PBS with four changes at 5 min each.

3.9. Application of Antibody

1. Rehydrate fixed samples in PBS for 10 min at room temperature.
2. Incubate samples in blocking buffer (5% milk in PBST) for 4 h at room temperature (*see* **Note 6**).
3. To detect β-tubulin or centrosomes, add mouse monoclonal anti-β-tubulin E-7 (1:10) or Ah6 (1:5) or mouse monoclonal anticentrosomal 4D2 (undiluted) primary antibody (*see* **Note 7**), respectively, for 2 h at room temperature in a humidified chamber (*see* **Note 8**). Wash three times in PBST for 5 min each. Apply the labeled secondary antibody (FITC- or rhodamine-linked goat anti-mouse) at a dilution of 1:50 for 2 h at room temperature in the humidified chamber. Wash three times with PBS for 5 min each.

3.10. DAPI Staining and Mounting

Add the DNA intercalating dye DAPI at a concentration of 4 μg/mL to the final rinse in PBS for 10 min. Place a small drop (approx 5 μL) of Mowiol on a glass slide and invert the coverslip (cells facing toward the glass slide) while slowly lowering the coverslip onto the glass slide containing the Mowiol solution. Carefully remove excess Mowiol. Seal off the edges of the coverslip with nail polish. Observe and photograph under the fluorescence microscope.

3.11. Isolation of Centrosomes

Sea urchin eggs provide an excellent system for centrosome isolation because they maintain almost 100% synchrony for the first and subsequent cell cycles. Large amounts of cells can be obtained for biochemical analysis or antibody production. The following isolation procedure was first published in 1996 ***(23)*** and followed the experiments originally suggested by Mazia et al. ***(73)*** based on microtubule disassembly at 0°C to isolate centrosomes without microtubule attachments.

1. Culture fertilized denuded eggs until nuclear envelope breakdown (NEB) occurs (*see* **Note 9**).
2. Let cells settle by gravity and resuspend in 2.5% calcium-free ASW at 4°C for 16–20 h with constant agitation.
3. Let cold-treated cells settle by gravity and rinse in 4°C buffered glycerol ***(74)***.
4. Lyse cells in 1 *M* glycerol medium (1 *M* glycerol, 10 m*M* piperazine-*N,N'*-bis(2-ethane-sulfonic acid [PIPES]; Sigma), 1 m*M* $MgSO_4$, 1 m*M* ethylene glycol-bis-(B-aminoethyl ether) *N,N,N',N'*-tetraacetic acid (EGTA; Sigma), pH 6.8, with 0.05% Triton X-100, 1 m*M* DL-dithiothreitol, 100 μg/mL soybean trypsin inhibitor, and 1 m*M* phenylmethylsulfonyl fluoride (PMSF) (*see* **Note 10**).
5. Incubate lysate on ice for 1 h.

6. Centrifuge lysate at 3000*g* for 30 min in a swinging-bucket rotor.
7. Resuspend in 2 *M* glycerol medium underlaid with 3 *M* glycerol medium.
8. Recentrifuge at 3000*g*; repeat two or three times (*see* **Note 11**).
9. Resuspend pellets in 2 *M* glycerol medium with 1 mg/mL heparin (Sigma, cat. no. 9339) and 50 µg/mL Dnase I (Sigma, cat. no. AMP-DI).
10. Incubate in water bath at room temperature for 20 min.
11. Add glycerol to 3 *M* and PMSF to 1 m*M*.
12. Centrifuge suspension at 3000*g* for 30 min in a swinging-bucket rotor.
13. Centrifuge supernatant at 55,000*g* for 30 min in a Sorvall SS-34 rotor.
14. Resuspend the final pellet in 50% glycerol, 10 m*M* PIPES, 1 m*M* $MgSO_4$, 1 m*M* EGTA, 1 m*M* PMSF, pH 6.8 (*see* **Note 12**).
15. Freeze at –80°C until use or use fresh.
16. The isolated centrosomes are then analyzed with scanning and transmission electron microscopy, with immunoelectron microscopy, and with Western blotting using either Ah-6 or 4D2 mouse monoclonal antibodies.

3.12. Imaging of Centrosomes with Field Emission Scanning Electron Microscopy

Cold-isolated centrosomes are imaged with low-voltage field emission scanning electron microscopy (LVFESEM) to visualize compacted centrosome structure.

Decompacted centrosome structures are analyzed from in vivo recoveries of cold-treated eggs. Eggs are incubated in the appropriate length of time in room-temperature buffered glycerol medium, then lysed and centrifuged as for nonrecovered cold-isolated cells.

1. Resuspend LVFESEM samples in 1 *M* glycerol medium without detergent and settle on silicon chips coated with Cell-Tak (Biopolymers, Inc., Farmington, CT), on ice.
2. Fix material in 2% glutaraldehyde in 1 *M* glycerol medium for 1 h, on ice.
3. Rinse samples in distilled water.
4. Dehydrate samples through an ethanol series at 30%, 50%, 70%, 90%, 95%, 4 × 100% for 10 min each.
5. Dry samples in a critical point dryer.
6. Coat samples held in a planetary jig apparatus with a layer of 1–2 nm platinum using ion-beam sputter coating.
7. View samples in a Hitachi S-900 LVFESEM or similar instrumentation.

4. Notes

1. The seasonal fertilizability of sea urchin gametes is of consideration for experimental purposes.
2. Other antifade agents can be used with equal results and include *p*-phenylenediamine (PPD), *N*-propylgallate (NPG), or other commercially available antifade agents from Sigma or Vector Laboratories.

3. Washing the entire sea urchin ensures that any contamination of sperm and eggs resulting from possible premature spawning during shipping is removed.
4. Sperm should be diluted and observed under a microscope; misshaped or immotile sperm should not be used for fertilization.
5. A test fertilization should always be performed, and gametes should not be used unless at least 90% eggs are fertilized.
6. To prevent nonspecific binding of antibody 3% BSA or 10% FBS can also be used with similar results. If desired, the blocking protein can be added to each antibody preparation and/or can be used to incubate the samples before addition of the antibody.
7. To determine the best primary antibody concentration prepare and test serial dilution.
8. A simple humidified chamber can be prepared by wetting two pieces of filter paper and inserting them into two chambers of a Petri dish that is divided into four parts (quadri-Petri dish). Place coverslip at the center of the quadri-Petridish, making certain that the wetted filter paper does not touch the coverslip). Cover with lid during incubation.
9. Remove small samples at frequent intervals and observe with light microscopy. NEB typically occurs at approx 70–75 min after insemination.
10. All manipulations and centrifugations are carried out at 4°C.
11. If necessary, repeat until pellets are white and supernatants are clear.
12. In some experiments, the final supernatant is applied to a discontinuous sucrose gradient (3 mL of 0.96 *M*, 2 mL of 1.25 *M*, 2 mL of 1.5 *M*, 2 mL of 1.75 *M*, 3 mL of 2.1 *M* in PIPES buffer) and centrifuged at 55,000*g* in a Beckman (Palo Alto, CA) SW 28.1 rotor for 1 h at 4°C.

References

1. Harris, P., Osborn, M., and Weber, K. (1980) A spiral array of microtubules in the fertilized sea urchin egg cortex examined by indirect immunofluorescence and electron microscopy. *Exp. Cell Res.* **126,** 227–236.
2. Harris, P. J. (1986) Cytology and immunocytochemistry. *Methods Cell Biol.* **27,** 243–262.
3. Bestor, T. H. and Schatten, G. (1981) Anti-tubulin immunofluorescence microscopy of microtubules present during the pronuclear movements of sea urchin fertilization. *Dev. Biol.* **88,** 80–91.
4. Hertzler, P. L. and Clark, W. H., Jr. (1993) The late events of fertilisation in the penaeoidean shrimp Sicyonia ingentis. *Zygote* **4,** 287–296.
5. Salmon, E. D. (1982) Mitotic spindles isolated from sea urchin eggs with EGTA lysis buffers. *Methods Cell Biol.* **25,** 69–105.
6. Schatten, H., Schatten, G., Mazia, D., et al. (1986) Behavior of centrosomes during fertilization and cell division in mouse oocytes and sea urchin eggs. *Proc. Natl. Acad. Sci. USA* **83,** 105–109.
7. Sluder, G., Miller, F. J., Lewis, K., et al. (1989) Centrosome inheritance in starfish zygotes: selective loss of the maternal centrosome after fertilization. *Dev. Biol.* **131,** 567–579.

8. Staiber W. (1994) Immunofluorescence study of spindle microtubule arrangements during differential gonial mitosis of Acricotopus lucidus (Diptera, Chironomidae). *Cell Struct. Funct.* **19,** 97–101.
9. Gard, D. L. (1991) Organization, nucleation, and acetylation of microtubules in *Xenopus laevis* oocytes: a study by confocal immunofluorescence microscopy. *Dev. Biol.* **143,** 346–362.
10. Lee, J., Miyano, T., and Moor, R. M. (2000) Spindle formation and dynamics of gamma-tubulin and nuclear mitotic apparatus protein distribution during meiosis in pig and mouse oocytes. *Biol. Reprod.* **5,** 1184–1192.
11. Meng, L. and Wolf, D. P. (1997) Sperm-induced oocyte activation in the rhesus monkey: nuclear and cytoplasmic changes following intracytoplasmic sperm injection. *Hum. Reprod.* **5,** 1062–1068.
12. Schatten, H., Walter, M., Biessmann, H., et al. (1987) Centrosome detection in sea urchin eggs with a monoclonal antibody against *Drosophila* intermediate filament proteins: characterization of stages of the division cycle of centrosomes. *Proc. Natl. Acad. Sci. USA* **84,** 8488–8492.
13. Boveri, T. (1901) Zellen–Studien IV: Ueber die Natur der Centrosomen Jena. *Zeitschr. Naturwiss* **35,** 1–220.
14. Boveri, T. (1914) *Zur Frage der Entstehung maligner Tumoren*, G. Fisher, Jena, Germany.
15. Bornens, M. (2002) Centrosome composition and microtubule anchoring mechanisms. *Curr. Opin. Cell Biol.* **14,** 25–34.
16. Brinkley, B. R. and Goepfert, T. M. (1998) Supernumerary centrosomes and cancer: Boveri's hypothesis resurrected. *Cell Motil. Cytoskel.* **41(4),** 281–288.
17. Doxsey, S. (2001) Re-evaluating centrosome function. *Nat. Rev. Mol. Cell Biol.* **9,** 688–698.
18. Lingle, W. L., Lutz ,W. H., Ingle, J. N., et al. (1998) Centrosome hypertrophy in human breast tumors: Implications for genomic stability and cell polarity. *Proc. Natl. Acad. Sci. USA* **95,** 2950–2955.
19. Nigg, E. A. (2002) Centrosome aberrations: cause or consequence of cancer progression? *Nat. Rev. Cancer* **11,** 815–825.
20. Pihan, G., Purohit, A., Knecht, H., et al. (1998) Centrosomes and cancer. *Cancer Res.* **58,** 3974–3985.
21. Rieder, C. L., Faruki, S., and Khodjakov, A. (2001) The centrosome in vertebrates: more than a microtubule-organizing center. *Trends Cell Biol.* **10,** 413–419.
22. Schatten, H., Walter, M., Biessmann, H., et al. (1992) Activation of maternal centrosomes in unfertilized sea urchin eggs. *Cell Motil. Cytoskel* **23,** 61–70.
23. Thompson-Coffe, C., Coffe, G., Schatten, H., et al. (1996) Cold-treated centrosomes from mitotic sea urchin eggs. Production of an anticentrosomal antibody, and novel ultrastructural imaging. *Cell Motil. Cytoskel.* **33,** 197–207.
24. Balczon, R. (1996) The centrosome in animal cells and its functional homologs in plant and yeast cells. *Int. Rev. Cytol.* **169,** 25–82.
25. Mazia, D. (1987) The chromosome cycle and the centrosome cycle in the mitotic cycle. *Int. Rev. Cytol.* **100,** 49–92.

26. Meraldi, P. and Nigg E. A. (2002) The centrosome cycle. *FEBS Lett.* **521,** 9–13.
27. Vandre, D., Davis, F., Rao, P., et al. (1984) Phosphoproteins are components of mitotic microtubule organizing centers. *Proc. Natl. Acad. Sci. USA* **81,** 4439–4443.
28. Saredi, A., Howard, L., and Compton, D. A. (1997) Phosphorylation regulates the assembly of NuMA in a mammalian mitotic extract. *J. Cell Sci.* **110,** 1287–1297.
29. Hinchcliffe, E. H., Li, C., Thompson, E. A., et al. (1999) Requirement of Cdk2-cyclin E activity for repeated centrosome reproduction in *Xenopus* egg extracts. *Science* **283,** 851–854.
30. Stearns, T. and Kirschner, M. (1994) In vitro reconstitution of centrosome assembly and function: the central role of γ-tubulin. *Cell* **76,** 623–637.
31. Vacquier, V. D. (1975) The isolation of the intact cortical granule from sea urchin eggs: calcium ions trigger granule discharge. *Dev. Biol.* **43,** 62–74.
32. Epel, D. (1977) The program of fertilization. *Sci. Am.* **237,** 128–138.
33. Schatten, H. (1994) Dithiothreitol prevents membrane fusion but not centrosome or microtubule organization during the first cell cycles in sea urchins. *Cell Motil. Cytoskel.* **27,** 59–68.
34. Horio, T., Uzawa, S., Jung, M. K., et al. (1991) The fission yeast gamma-tubulin is essential for mitosis and is localized at microtubule organizing centers. *J. Cell Sci.* **99,** 693–700.
35. Joshi, H. C., Palacios, M. J., McNamara, L., et al. (1992) γ-Tubulin is a centrosomal protein required for cell cycle-dependent microtubule nucleation. *Nature* **356,** 80–83.
36. Moritz, M., Brownfeld, M. B., Sedat, J. W., et al. (1995) Microtubule nucleation by gamma-tubulin-containing rings in the centrosome. *Nature* **378,** 638–640.
37. Oakley, C. D. and Oakley, B. R. (1989) Identification of γ-tubulin, a new member of the tubulin superfamily encoded by mipA gene of *Aspergillus nidulans. Nature* **338,** 662–664.
38. Zheng, Y., Wong, M. L., Alberts, B., et al. (1995) Nucleation of microtubule assembly by a gamma-tubulin–ring complex. *Nature* **378,** 578–583.
39. Simerly, C., Wu, G. J., Zoran, S., et al. (1995) The paternal inheritance of the centrosome, the cell's microtubule-organizing center, in humans, and the implications for infertility. *Nature Med.* **1,** 47–52.
40. Schatten, H., Wiedemeier, A. M., Taylor, M., et al. (2000) Centrosome–centriole abnormalities are markers for abnormal cell divisions and cancer in the transgenic adenocarcinoma mouse prostate (TRAMP) model. *Biol. Cell* **92,** 331–340.
41. Schatten, H., Hueser, C. N., and Chakrabarti, A. (2000) From fertilization to cancer: the role of centrosomes in the union and separation of genomic material. *Microsc. Res. Tech.* **49,** 420–427.
42. Paweletz, N., Mazia, D., and Finze, E.-M. (1984) The centrosome cycle in the mitotic cycle of sea urchin eggs. *Exp. Cell Res.* **152,** 47–65.
43. Sathananthan, A. H. (1997) Ultrastructure of the human egg. *Hum. Cell* **10,** 21–38.
44. Zeligs, J. D. and Wollman, S. H. (1979) Mitosis in rat thyroid epithelial cells in vivo. II. Centrioles and pericentriolar material. *J. Ultrastruct. Res.* **66,** 97–108.
45. Mazia, D. (1984) Centrosomes and mitotic poles. *Exp. Cell. Res.* **153,** 1–15.

46. Schatten, H. and Chakrabarti, A. (1998) Centrosome structure and function is altered by chloral hydrate and diazepam during the first reproductive cell cycles in sea urchin eggs. *Eur. J. Cell Biol.* **75,** 9–20.
47. Balczon, R. and Schatten, G. (1983) Microtubule containing detergent-extracted cytoskeletons in sea urchin eggs from fertilization through cell division. *Cell Motil. Cytoskel.* **3,** 213–226.
48. Calarco-Gillam, P. C., Siebert, M. C., Hubble, R., et al. (1983) Centrosome development in early mouse embryos as defined by an autoantibody against pericentriolar material. *Cell* **35,** 621–629.
49. Mitchison, T. J. and Kirschner, M. W. (1986) Isolation of mammalian centrosomes. *Methods Enzymol.* **134,** 261–268.
50. Blomberg-Wirschell, M. and Doxsey, S. J. (1998) Rapid isolation of centrosomes. *Methods Enzymol.* **298,** 228–238.
51. Moritz, M. and Alberts, B. M. (1999) Isolation of centrosomes from *Drosophila* embryos. *Methods Cell Biol.* **61,** 1–12.
52. Palazzo, R. E. and Vogel, J. M. (1999) Isolation of centrosomes from *Spisula solidissima* oocytes. *Methods Cell Biol.* **61,** 35–56.
53. Graf, R. (2001) Isolation of centrosomes from Dictyostelium. *Methods Cell Biol.* **67,** 337–357.
54. Gergely, F. (2002) Centrosomal TACCtics. *BioEssays* **24,** 915–925.
55. Mack, G. M., Ou, Y., and Rattner, J. B. (2000) Integrating centrosome structure with protein composition and function in animal cells. *Microsc. Res. Tech.* **49(5),** 409–419.
56. Zeng, C. (2000) NuMA: a nuclear protein involved in mitotic centrosome function. *Microsc. Res. Tech.* **49,** 467–477.
57. Zinovkina, L. A. and Nadezhdina, E. S. (1996) Centrosomal proteins. *Biokhimiia* **61,** 1347–1365.
58. Joswig, G. and Petzelt, C. (1990) The centrosomal cycle: visualization in PtK cells by a monoclonal antibody to centrosomal 32kd protein. *Cell Motil. Cytoskel.* **15,** 181–192.
59. Joswig, G., Petzelt, C., and Werner, D. (1991) Murine cDNAs coding for the centrosomal antigen centrosomin A. *J. Cell Sci.* **98(Pt. 1),** 37–43.
60. Balczon, R., Varden, C. E., and Schroer, T. A. (1999) Role for microtubules in centrosome doubling in Chinese hamster ovary cells. *Cell Motil. Cytoskel.* **42,** 60–72.
61. Daunderer, C. and Graf, R. O. (2002) Molecular analysis of the cytosolic *Dictyostelium* gamma-tubulin complex. *Eur. J. Cell Biol.* **81,** 175–184.
62. Helfant, A. H. (2002) Composition of the spindle pole body of Saccharomyces cerevisiae and the proteins involved in its duplication. *Curr. Genet.* **40,** 291–310.
63. Kalt, A. and Schliwa, M. (1996) A novel structural component of the Dictyostelium centrosome. *J. Cell Sci.* **109,** 3103–3112.
64. Lacey, K. R., Jackson, P. K., and Stearns, T. (1999) Cyclin-dependent kinase control of centrosome duplication. *Proc. Natl. Acad. Sci. USA* **96,** 2817–2822.

65. Lange, B. M., Bachi, A., Wilm, M., et al. (2000) Hsp90 is a core centrosomal component and is required at different stages of the centrosome cycle in *Drosophila* and vertebrates. *EMBO J.* **19,** 1252–1262.
66. Ou, Y. Y., Mack, G. J., Zhang, M., et al. (2002) CEP110 and ninein are located in a specific domain of the centrosome associated with centrosome maturation. *J. Cell Sci.* **115,** 1825–1835.
67. Moritz, M., Braunfeld, M. B., Fung, J. C., et al. (1995) Three-dimensional structural characterization of centrosomes from early *Drosophila* embryos. *J. Cell Biol.* **30,** 1149–1159.
68. Schatten, H. and Ris, H. (2002) Unconventional specimen preparation techniques using high resolution low voltage field emission scanning electron microscopy to study cell motility, host cell invasion, and internal cell structures in *Toxoplasma gondii. Microsc. Microanal.* **8,** 94–103.
69. Falkner, F. G., Saumweber, H., and Biessmann, H. (1981) Two *Drosophila melanogaster* proteins related to intermediate filament proteins of invertebrate cells. *J. Cell Biol.* **91,** 175–183.
70. Balczon, R. and West, K. (1991) The identification of mammalian centrosomal antigens using human autoimmune anticentrosome antisera. *Cell Motil. Cytoskel.* **20,** 121–135.
71. Rattner, J. B. (1992) Ultrastructure of centrosome domains and identification of their protein components, in *The Centrosome* (Kalnins, V. I., ed.), Academic, San Diego, CA, pp. 45–68.
72. Foerder, C. and Shapiro, B. H. (1977) Release of ovoperoxidase from sea urchin eggs hardens the fertilization membrane with tyrosine crosslinker. *Proc. Natl. Acad. Sci. USA* **74,** 4214–4218.
73. Mazia, D., Schatten, H., Coffe, G., et al. (1987) Aggregation of the mitotic centrosomes into a single spherical centrosome by cold treatment in sea urchin eggs. *J. Cell Biol.* **105,** 206a.
74. Suprenant, K. A. (1986) Tubulin-containing structures. *Methods Cell Biol.* **27,** 189–215.

13

Microscopic Techniques for Studying Sperm–Oocyte Interaction During Fertilization and Early Embryonic Development

Mark W. Tengowski

1. Introduction

The union between sperm and oocyte is species-specific, and the courses of events that follow are some of the most dramatic surrounding fertilization and early development. How does the highly motile, yet very small, sperm from one species recognize the immotile, very large oocyte from its own species? How does the sperm penetration and subsequent membrane fusion take place? What happens to this successful sperm inside the ooplasm prior to decondensation and pronuclear fusion? What is the best microscopy method to capture these events? Questions such as these have been asked ever since humans have questioned their own existence on the cellular level. Only recently have molecular answers to these questions been found. The use of the electron microscope has increased our appreciation of these events because of its outstanding resolution. However, specimen preparation captures events in a static fashion. New optical techniques offer promise in the investigation of these events in real time.

There are multiple levels of interaction between sperm and oocyte that are interposed between the initial contact and the initiation of the cortical reaction. The idea that sperm must penetrate an oocyte to stimulate any further development is the cornerstone of this process. Sperm must traverse a long, tortuous distance through a sometimes hostile mucosal environment. In the oviduct, a sperm contacts an oocyte after passing through a layer of cumulus cells and binds to the zona pellucida, the oocyte's initial extracellular matrix barrier. It is here that capacitated sperm undergo an exocytic event we now call the

From: *Methods in Molecular Biology, vol. 253: Germ Cell Protocols: Vol. 1 Sperm and Oocyte Analysis*
Edited by: H. Schatten © Humana Press Inc., Totowa, NJ

acrosome reaction and begins to penetrate the glycoprotein coat surrounding the oocyte. Interaction of the sperm with the oolemma surface activates the oocyte, removing it from an arrested state of meiosis and preventing future sperm from entering via the cortical granule release and subsequent zona pellucida changes. Now is the time when centrosome reconstitution and diploid genome formation begin, and the fertilized oocyte now begins its long road through cleavage, development, parturition, and repetition of the cycle in the next generation. This is an old basic idea of how fertilization occurs; yet, the entire molecular sequence of events for all species has not been completely unraveled.

It has been reported that of the approx 280 million sperm that are deposited in the human female reproductive tract, only about 200 are found at the region of the fertilization site in the Fallopian tube *(1)*, a loss in sperm numbers of over six orders of magnitude. In cattle, pigs, and rabbits, it has been shown that a considerable fraction of the spermatozoa ejaculated into the female reproductive tract remains in the isthmus of the fallopian tube until ovulation occurs; it is at this time that they resume their motility and reach the fertilization site in the ampulla within minutes *(2)*. Natural mating or assisted reproductive technologies such as artificial insemination result in the deposition of millions of sperm cells into the mammalian female reproductive tract. Does this apparent loss of sperm cells signify a sorting of healthy sperm from this large pool or does it suggest that factors released from the oocyte follicular fluid at ovulation have a chemoattractant effect *(3)*? The ability of the small motile sperm to find the large, relatively immotile oocyte could be considered to be the first step in the sperm–ovum interaction.

1.1. Sperm En Route to the Zona Pellucida

When freshly ejaculated sperm are deposited into the female reproductive tract, they are not capable of fertilizing the oocyte right away *(4,5)*. They must undergo changes that ready the sperm for successful fertilization—a process known as sperm capacitation. The capacitation events are reversible; some of these capacitated properties are decreased negative net charge *(6)*, disappearance of tetracycline fluorescence *(7)*, removal of surface-bound seminal plasma components *(8)*, lectin-binding sites *(9)*, specific sperm antibody-binding sites *(10)*, and calcium-dependent zona pellucida binding and penetration *(11)*.

Braden et al. *(12)* first described the concept that a penetrating sperm alters the zona pellucida, rendering the oocyte impenetrable by other sperm. This "zona reaction" does create a decreased likelihood that more than one sperm will gain entry into the oocyte, thus preserving the diploid chromosome complement. However, the molecular mechanism by which the sperm gains entry or is excluded remains a mystery. One component of this process is the description of cortical granule products altering the zona pellucida, thereby resulting in the zona reaction ***(13)***. The interaction between sperm and oocyte at the level of the zona pellucida has been studied with many different techniques in an effort to understand how this results in sperm penetration and zona reaction.

The ultrastructural studies of mammalian sperm membranes suggest that two basic head designs exist: The rodent sperm head is large, with a hook containing the acrosomal contents; the other morphology commonly seen in human and other domestic species is the paddle-shaped head with the acrosomal cap covering the most anterior segment. When capacitated sperm are exposed to intact zona pellucida fragments, the vast majority of sperm can be seen bound to the external, latticelike multilayered matrix of the zona pellucida ***(14)***, suggesting that sperm binding to the zona pellucida is specific and limited to the zona exterior. In order for sperm to fertilize oocytes in vivo, they would have to penetrate a layer of cumulus cells. It has been shown that the cumulus cells reduce the number of sperm surrounding the oocyte twofold to fourfold and that sperm that pass through the cumulus cells do so with their acrosome intact, suggesting that the acrosome reaction occurs at the level of the zona pellucida ***(15)***. Shur and Hall ***(16)*** showed that surface β1,4-galactosyltransferases on uncapacitated sperm were loaded with polylactosaminyl substrates and that sperm capacitation in vitro involves the requisite release of these glycoside substrates from the surface, thereby exposing the β1,4-galactosyltransferase. The species-specificity of fertilization must lie at the level of the zona pellucida because removal of the extracellular glycoprotein usually eliminates that barrier allowing cross-species fertilization.

Once an oocyte is activated by a sperm or artificially with calcium ionophore A23187, modifications in the zona pellucida glycoprotein (ZP2) can also occur. Bleil et al. ***(17)*** showed that the ZP2 of embryos, but not unfertilized oocytes, undergoes modifications. Once oocyte activation occurs, this normally 120-kDa protein is modified to a 90-kDa proteolytic fragment, as seen under reducing conditions. It would appear that intramolecular bonds hold the glycoprotein together and modifications of this structural arrangement may have an affect on further sperm binding and penetration. No matter what the chronology is at this level, once an oocyte has been fertilized and development is underway, the sperm receptor ligands are inactivated via the cortical granule enzymes.

1.2. Sperm at the Level of the Oolemma

Sperm–oolemma fusion occurs only after the acrosome-reacted sperm has completely penetrated the zona pellucida. While the sperm head is within the perivitelline space, it comes into contact with numerous oolemma microvilli and its tail may still be outside the zona pellucida. The angle at which the sperm penetrates the zona pellucida appears to be unimportant because fusion can occur at either a right or oblique angle, with the head eventually turning to a flat position against the oolemma surface ***(18)***.

A sperm that would be found in the perivitelline space displays distinct membrane architecture, meaning that the zona interactions have altered its appearance. At this point, there are distinct membrane domains that are visible on the sperm surface. The critically important domain is the equatorial domain, a part of the acrosomal membrane that is not removed during the acrosome reaction. Electron microscopy studies suggest that this is the initial site of oolemma fusion ***(19)*** followed shortly by the postacrosomal region ***(20)***.

The best studied interaction between sperm and oocyte at the level of the zona pellucida was performed using the mouse model. Evans et al. ***(21)***, using the mouse, furthered our understanding at the level of the sperm–oolemma interaction, presenting molecular sequence data and function blocking data based on this new information. By this time in the literature, the PH-30β-subunit, which contains the disintegrin containing the RGD peptide, has been renamed fertilin. Testing the hypothesis that the disintegrin of fertilin was the ligand for a yet unidentified integrin on the surface of the oolemma, mouse fertilin divalent cation-dependence and sequence were determined. Sperm–oocyte fusion studies investigating the importance of calcium, magnesium, or manganese in the culture media suggested that calcium is the most important divalent cation mediating sperm binding and fusion. It was also found that 2.5 m*M* calcium in the medium yielded the highest percentage of sperm binding and fusion. In contrast to the Bronson and Fusi report described earlier ***(22)***, an RGD-containing peptide construct in the mouse caused only a mild (30–58%) decrease in fertilization. This could be explained by the mouse fertilin sequence data. It was shown that mouse fertilin was 55% identical and 70% homologous to guinea pig fertilin. In the mouse disintegrin domain, it was shown that it contained a different tripeptide, QDE instead of RGD, in its cell recognition region. Peptides constructed containing the QDE sequence were now found to be more efficient in reducing sperm binding and fusion by 73% and 69%, respectively. These data in the guinea pig, mouse, and human sperm studying the disintegrin domain of fertilin provide strong evidence that sperm interact with an integrin expressed on the oolemma surface. The only aspect missing was the identification of the oocyte integrin.

The discovery that mouse oocytes express integrins shortly followed the work describing the disintegrin domain. Almeida et al. ***(23)*** utilized polymerase chain reaction (PCR) analysis of oocyte mRNA, immunolocalization, peptide construct and antibody function blocking trials, and integrin-expressing cell-culture-binding assays to support their hypothesis that mouse oocytes do, indeed, express integrins. Of the 14 α and 8 β integrin subunits described to date, the mouse oocyte was shown to express 3 α-subunits (α5, α6, and αv) and 3 β-subunits (β1, β3, and β5). The use of polyclonal and monoclonal antibodies in immunofluorescence localization studies suggested that the α6 and αv integrin subunits localized to the oolemma and did not localize to the area of the smooth-surfaced cortical granule-free membrane region, which has been shown to be the location of the meiotic spindle. Coupling the known pairings of other integrin subunits with subunit expression in the oocyte, it was deduced that α6β1, a laminin receptor, and αvβ3, a fibronectin/vitronectin receptor, were present on the oolemma surface. Because RGD-containing peptides have been shown to bind to αvβ3 ***(24,25)*** but not α6β1 ***(26)***, a peptide inhibition assay was performed with the results, suggesting that RDG peptides did not prevent sperm binding and further suggesting that the integrin α6β1 is the surface integrin. An α6 integrin function blocking monoclonal antibody was used in binding and fusion assays and suggested that it is the α6β1 integrin that is the best candidate for mediating sperm binding. To test the concept that this integrin receptor α6β1 mediated sperm binding, a culture cell system assaying sperm binding was used. F9 mouse embryonal carcinoma cells express α6β1 and other integrins on their surface; an F9 β1 knockout cell line was also used as a control. Sperm bound avidly to F9 cells, but displayed less than 25% binding affinity for the F9 cell line devoid of the β1-subunit. These data, taken with the fertilin data, begin to define a molecular model in the mouse at the level of the oolemma: sperm expressing fertilin with its disintegrin domain and species-specific tripeptide sequence interact with an oolemma-expressed integrin to mediate sperm binding and fusion. The most likely candidate for the oolemma integrin in the mouse is the laminin receptor, α6β1.

The importance of integrins on the surface acting as sperm receptors is much more than just being the complement to a disintegrin. Integrins may play an important function in cell signaling and cytoskeletal reorganization. Integrins, as already described, are comprised of α and β single pass membrane subunits with short cytoplasmic tails (β4 an exception); the extracellular domains are noncovalently linked. The α-chain is the subunit that confers a divalent cation-binding region; the β-chain contains the ligand-binding domain ***(27)***. Integrin-mediated signaling pathways identified in other systems include the regulation of a Na^+/H^+ antiporter ***(28)***, calcium influx ***(28–30)***, stimulation of inositol lipid synthesis ***(31)***, protein tyrosine phosphorylation of a group of 100- to 130-kDa cytoplasmic proteins ***(32–35)***, and increased activities associated with cyclin A and cdc2 ***(46)***. It is possible that the influx of sodium and calcium associated with integrin stimulation is involved with the fast and slow block to polyspermy, respectively, each mediating cytoplasmic alkalinization and cortical granule exocytosis critical for properly resuming meiosis and development.

1.3. Using Technology to Study Fertilization

One important aspect of the fertilization process is the union of male and female gametes. Mature sperm deposited in the female reproductive tract undergo several changes in membrane and cytoskeletal reorganization before the oocyte penetration event. Studies have been performed evaluating these membrane interactions with light and transmission electron microscopy; however, scanning electron micrographs that capture the surface events with detailed resolution are scarce. It is now possible to study these events with the advent of a unique high-resolution low-voltage scanning electron microscope (LVSEM) optimized for biological sample viewing. Conventional chemical fixation and ultrathin platinum metal coatings can be used to produce samples with unparalleled resolution. The overall goal of this project is to couple the outstanding resolution of the LVSEM with the power of fluorescence light microscopy to investigate the surface ultrastructure and cytoskeletal changes in ejaculated bull sperm prior to oolemma penetration. The role of sperm–oolemma recognition and the cytoskeleton in the incorporation process, including studies detailing postfusion membrane dynamics, will hopefully provide a molecular mechanism for sperm–oocyte fusion in the bovine model. The information gained from these experiments will complement what is currently known about the membrane events of gamete fusion from studies utilizing light and transmission electron microscopy techniques. The development of a molecular model of gamete fusion may provide insight into various areas of fertility research.

A mature sperm appears to be morphologically inactive; however, the ability of the sperm to reorganize its cytoskeleton during capacitation and acrosome reaction predict its success in completing its mission to fertilize the oocyte. Of these cytoskeletal components, actin is believed to be a major component ***(36)***. In boar sperm, it has been shown that filamentous actin reorganization occurs at the equatorial segment during capacitation and that inhibiting this with the cytoskeleton-altering drug cytochalasin D, actin localization is below detectable levels and physiologically important, because this treatment disrupted the number and ability of sperm to penetrate porcine oocytes ***(37)***. These results indicate that the cytoskeleton plays an important role in the sequential events of sperm penetration and fertilization.

Sperm that locate the zona pellucida and are acrosome-intact bind and undergo the acrosome reaction, an exocytic event vital to sperm penetration via zona pellucida digestion. In addition, it has been shown that the glycolipids present in the boar sperm membrane leaflets undergo a mobile exchange from storage in the apical ridge subdomain, ultimately concentrating in the equatorial segment, and are enhanced by the addition of the desulfating enzyme arylsulfatase A ***(38)***. These results suggest the importance of membrane lipid alterations prior to plasma membrane and extracellular matrix interactions favoring acrosome protection and fusigenic potential.

It has been shown that sperm interact with the zona pellucida via the glycosylated ZP3 protein ***(39,40)***. Two reports in the journal *Science* offered some insight into the specific nature of the sperm–zona pellucida interaction, suggesting that specificity may be multifactoral. Bookbinder et al. ***(41)*** presented evidence suggesting that mouse and hamster postmeiotic sperm contain a peripheral membrane-associated protein (sp56). These species of sperm specifically recognize the mouse zona pellucida, whereas guinea pig and human sperm, which lack the sp56-like protein transcripts, fail to bind to mouse oocytes. Burks et al. ***(42)*** presented a model of signal transduction based on tyrosine phosphorylation that could result from the interaction of a human sperm zona receptor kinase (similar to the mouse 95-kDa ZP3 receptor) interacting with the zona pellucida glycoprotein ZP3. The interaction with ZP3 can only occur with acrosome-intact sperm. Once ZP3 binding induces the acrosome reaction, secondary interactions may occur between ZP2 and sperm ZP2 receptors located on the inner acrosomal membrane with the ultimate penetration of the zona matrix ***(19)***. This specific interaction in bound sperm could initiate a signal transduction cascade resulting in the acrosome reaction, zona penetration, and cytoskeletal reorganization prior to sperm–oolemma fusion.

A sperm that would be found in the perivitelline space displays distinct membrane architecture, meaning that the zona interactions have altered its appearance. At this point, there are distinct membrane domains that are visible on the sperm surface. The critically important domain is the equatorial domain, a part of the acrosomal membrane that is not removed during the acrosome reaction. Electron microscopy studies suggest that this is the initial site of oolemma fusion ***(19)*** followed shortly by the postacrosomal region ***(20)***, which is shown to be heavily concentrated with the PH-30 protein ***(43)***. Found in guinea pig sperm, this fusigenic disintegrin peptide most likely interacts with an integrin expressed on the oolemma. Studies utilizing high-resolution scanning electron microscopy (SEM) may answer questions at this point because there are some questions as to what is the fate of the respective membranes during fusion and what function does the cytoskeleton play in gamete incorporation.

The objective of this chapter is to provide a technical methods guide for using gametes and advanced imaging technologies for the study of fertilization. Current methods used in studying gamete interactions employ the use of video microscopy, epifluorescence microscopy, or transmission electron microscopy (TEM). Although not discounting the advances made in the study of reproduction using these techniques, studies evaluating the membrane and cytoskeleton of the gametes during preparation and fusion at the level of surface resolution that can be obtained with the LVSEM and advanced optical techniques are generally lacking.

The LVSEM resolution is far superior to that obtained by a conventional SEM (cSEM) operating at a similar voltage, and this technique provides that aspect of depth which does not accompany transmission electron micrographs of ultrathin sections or the low resolution of light microscopy. In rendering a biological sample conductive, it is a common practice to "coat" a sample with a metal, such as gold or platinum. This metal also serves as a source of secondary electrons, adding to the overall signal for SEM. This metal also provides a path to ground, and LVSEM with its lower accelerating voltage requires less coating to protect the surface from beam damage. Examples of oocytes prepared with both techniques illustrate the value in LVSEM in producing superior images compared with cSEM. However, if access to LVSEM instrumentation is limiting, then methods contained within will enable you to preserve the architecture at its finest. The values in performing these experiments are in furthering our understanding of the basic cell and molecular biology of gametes. The information gained from these studies will not only answer questions in the area of gamete interactions but may also have clinical significance in estimating or predicting the reproductive efficiency of a particular bull if established parameters can be described using LVSEM as a measuring tool in the area of capacitation and acrosome reaction. The ability of LVSEM techniques to detect membrane damage as a result of sperm preparation and/or cryopreservation could lead to advances in cryopreparative techniques, increasing the viability and fertilizability of these cryopreserved sperm. Recent works with light microscopy and immunocytochemistry have advanced our knowledge of gamete interactions during fertilization. Advances in electron and optical microscopy instrumentation now make it possible to study the changes in membrane and cytoskeletal morphology. The methods contained here illustrate the preparative techniques that should enable those developmental biologists interested in studying the differences in surface membrane and cytoskeletal reorganization during capacitation, zona pellucida interactions, acrosome reaction, and gamete incorporation through first mitosis in boving oocytes.

2. Materials

2.1. Oocyte Culture Supply Checklist

1. Abattoir-derived bovine ovaries.
2. 18-gauge needles (pink hub).

3. 10 mL syringes.
4. 20-μL unopette tips.
5. 100-μL glass tuberculin syringe (or mouth aspirator to pick up cells).
6. Syringe filters (0.22 μm).
7. 100-mL Bottles (sterile).
8. 50-mL Conical centrifuge tubes.
9. 10-mL Test tubes (sterile).
10. Mineral oil, embryo-tested.
11. 35, 60, and 100-mm Plastic plates (sterile).
12. Dissecting scope.
13. 5% CO_2 incubator at 39°C for bovine culture.

2.2. Media

1. Maturation medium: 9 mL TC199 medium (Gibco), 1 mL heat-inactivated fetal bovine serum (FBS), 50 μg follicle stimulating hormone (FSH)-p (*see* **Note 1**), 10 μL gentamicin (25-mg/mL stock solution stored at 4°C); filter-sterilize this combination of reagents, add 10 μL estrogen (*see* **Note 2**), from 1-mg/mL estrogen stock in 95% ethanol stored at –20°C.
2. Wash medium/TL–HEPES stock, bring to 500 mL using double-distilled water (ddH_2O), good for 2 wk in a refrigerator: 3.3311 g sodium chloride, 0.1193 g potassium chloride, 0.0841 g sodium bicarbonate, 0.0240 g sodium monophosphate monobasic anhydrous, 0.9341 mL DL-lactic acid (60% syrup), 0.0299 g penicillin, 0.0050 g phenol red (optional), 0.1470 g calcium chloride dihydride, 0.0508 g magnesium chloride hexahydrate.
3. 100 mL TALP–HEPES wash medium: 100 mL TL–HEPES stock, 0.0220 g sodium pyruvate, 0.3000 g bovine serum albumin (BSA) fraction V, filter-sterilize (*see* **Note 3**).
4. Fertilization medium/TL stock: Bring to 500 mL using ddH_2O, good for 2 wk in a refrigerator: 3.3311 g sodium chloride, 0.1193 g potassium chloride, 1.0502 g sodium bicarbonate, 0.0240 g sodium monophosphate monobasic anhydrous, 0.9341 mL DL-lactic acid (60% syrup), 0.0299 g penicillin, 0.0050 g phenol red (optional), 0.1470 g calcium chloride dihydride, 0.0508 g magnesium chloride hexahydrate.
5. 10 mL Fertilization medium: 10 μL of 20 m*M* pyruvate stock (0.022 g sodium pyruvate in 10 mL TL stock), 10 μL gentamicin (25-mg/mL stock solution stored at 4°C), 60 mg BSA fatty acid-free fraction; bring volume to 10 mL with TL stock; filter-sterilize.
6. Sperm TL: 2.2922 g Sodium chloride, 0.1155 g potassium chloride, 1.0502 g sodium bicarbonate, 0.0174 g sodium monophosphate monobasic anhydrous, 2.0178 mL DL-lactic acid (60% syrup), 0.0299 g penicillin, 0.0050 g phenol red (optional), 0.1470 g calcium chloride dihydride, 0.0406 g magnesium chloride hexahydrate.
7. 10 mL of Sperm TL medium: 500 μL of 20 m*M* pyruvate stock (0.022 g sodium pyruvate in 10 mL TL stock), 60 mg BSA fraction V; bring volume to 10 mL with sperm TL stock; filter-sterilize.

8. PH stock components: 3.00 mg penicillamine in 10 mL of saline (2 m*M* stock), 1.09 mg hypotaurine in 10 mL of saline (1 m*M* stock).
9. Epinephrine diluent: 165 mg Na lactate syrup (60%), 50 mg Na metabisulfite, 50 mL ddH_2O, acidify to pH 4.0 with HCl, 1.83 mg epinephrine in 40-mL epinephrine diluent (250-μ*M* stock).
10. PHE cocktail stock: 5 mL Stock penicillamine, 5 mL stock hypotaurine, 2 mL stock epinephrine, 8 mL of 0.9% saline; aliquot PHE cocktail in 100 μL volumes (*see* **Note 4**).
11. Sodium heparin stock: 10 mg Sodium heparin in 10 mL of 0.9% saline; aliquot in 100-μL volumes.

3. Methods

3.1. Gamete Collection and Maturation

1. Prepare culture plates with 10–15 50-μL drops of maturation medium in sterile 60-mm culture plates.
2. Slowly pipet 10 mL of embryo-tested mineral oil onto an area of the plate not occupied by a medium drop, being careful not to disturb the medium drops; oil should completely cover the drops.
3. Replace plate cover and transfer to a 39°C humidified 5% CO_2 incubator to equilibrate for at least 1 h.
4. Ovaries should be washed with 0.9% normal saline and kept at 37°C during collection.
5. Aspirate 5- to 10-mm follicles using a 10 mL syringe with an 18-gauge needle.
6. Generate vacuum when the needle is inserted into the follicle by drawing the plunger.
7. Transfer follicular fluid to a 50-mL conical tube in a rack in the water bath.
8. Allow cells to settle while continuing to aspirate (*see* **Note 5**).
9. Prepare a 100-mL culture plate containing 10 mL TALP–HEPES wash medium.
10. Aspirate the pellet from the bottom of the follicular fluid conical tube.
11. Transfer cellular material to the 100-mL culture plate containing 10 mL TALP–HEPES wash medium (*see* **Note 6**).
12. Prepare a second transfer plate by pipetting approx 2.5 mL TALP–HEPES wash medium in a separate 35-mm culture plate.
13. Search plate for bovine oocytes with cumulus cells. These are approx 200 μm in diameter and easily differentiated under the dissecting scope.
14. Harvest the oocytes with at least three layers of cumulus cells, avoiding those without cumulus cells and those with expanded cumulus cells; this is best accomplished using a 20-μL Unopette tip attached to a 100-μL glass tuberculin syringe using light aspiration pressure.
15. Transfer oocytes to 35-mm culture plate; wash the clutch of oocytes through two more wash TALP–HEPES plates.
16. Collect approx 10 oocytes from wash plate.
17. Transfer to one of the drops in the equilibrated 60-mm plate containing the maturation medium.

18. Return to the incubator and culture for 18–20 h.
19. Cumulus cells will expand.
20. Oocytes will be ready for manipulation.

3.2. Sperm Preparation

1. Add 1 mL sperm in extender to 9 mL wash medium.
2. Place in centrifuge for 10 min at 700*g*.
3. Aspirate wash medium, leaving soft sperm pellet.
4. Resuspend pellet in new 10-mL volume of wash medium.
5. Repeat centrifugation step.
6. Aspirate wash medium, leaving soft sperm pellet.
7. Dilute pellet 1:1 with fertilization medium.
8. Count sperm using a hemocytometer.
9. Dilute sperm pellet to a final concentration of 20×10^6/mL, using fertilization medium.
10. Acrosome reaction in sperm: Dilute sperm suspension 1:1 with 200 μ*M* dilauroylphosphatidic acid (PC-12); incubate for 10 min at 39°C (*see* **Note 7**).

3.3. In Vitro Fertilization

3.3.1. Zona Pellucida Intact

1. In a similar manner to the maturation drops, create several plates of drops using fertilization medium.
2. Cover drops with oil.
3. Equilibrate in a 39°C humidified 5% CO_2 incubator to equilibrate for at least 1 h.
4. Transfer cumulus-expanded oocytes to a 35-mm plate with wash medium.
5. Place 10 mature oocytes in each fertilization drop.
6. Add 2 μL heparin stock.
7. Add 2 μL PHE stock.
8. Add 2 μL sperm dilution.
9. Return fertilization drop plate to incubator for desired time.

3.3.2. Zona Pellucida Removed

1. In a similar manner to the maturation drops, create several plates of drops using fertilization medium.
2. Cover drops with oil.
3. Equilibrate in a 39°C humidified 5% CO_2 incubator to equilibrate for at least 1 h.
4. Transfer cumulus-expanded oocytes to a 35-mm plate with wash medium.
5. Collect clutch of oocytes in as little volume wash medium as possible.
6. Transfer oocytes to a 1.5-mL microfuge tube.
7. Pulse oocytes on a desktop vortex shaker for up to 60 s; the physical action will remove the cumulus cells.
8. Add 1 mL wash to the microfuge tube.
9. Transfer entire contents to a new 100-mL plate containing wash medium.

Table 1
Effect of Cytochalasin B (30 μ*M*) Treatment on Sperm Entry into In Vitro Matured Bovine Oocytes Fertilized In Vitro with Ejaculated Bull Sperm

	Fertilized	Unfertilized	% Fertilized
Vehicle-treated oocytes			
Vehicle-treated sperm	106	16	86.8%
Cytochalasin B-treated sperm	153	21	87.9%
Cytochalasin B-treated oocytes			
Vehicle-treated sperm	26	92	22.0%
Cytochalasin B-treated sperm	26	122	17.6%
Chi-square test = 262.3, df = 3, $p < 0.001$			

Note: These data suggest that the actin cytoskeleton of the oocyte was affected by the presence of cytochalasin B, independent of sperm treatment. In cytochalasin-treated oocytes, there was a significant inhibition of fertilization. Cytochalasin B (30 μ*M* final) or an appropriate amount of DMSO acting as a vehicle control was added to sperm for a 1-h incubation prior to in vitro fertilization. Sperm were washed free of cytochalasin B or vehicle and diluted to 20 million sperm/mL. Then, 50,000 sperm were added to each fertilization drop containing 10 oocytes in media containing cytochalasin B (30 μ*M* final) or an appropriate amount of DMSO acting as a vehicle control. Fertilization was assessed 18 h after sperm addition using immunocytochemical and Nomarski DIC microscopy techniques. An oocyte was considered positive if there was at least one sperm head and tail within the oocyte. The cytochalasin B oocyte treatment prevented the expulsion of the second polar body. Oocytes were scored as unfertilized when not penetrated by sperm and were still at metaphase arrest of second meiosis or they contained two pronuclei, neither possessing a sperm tail (parthenogenotes). Data represents number of oocytes and percent fertilized. A chi-square test statistic was used to indicate significance.

10. Search plate for cumulus-free oocytes.
11. Transfer to fresh wash medium in a 35-mm plate.
12. Remove zona pellucida using 1 mg/mL pronase in wash medium.
13. Oocytes will deform just as the zona pellucida dissolves.
14. Transfer oocytes to fresh wash medium to recover for 30 min.
15. Transfer 10 oocytes to each fertilization drop.
16. Add 2 μL heparin stock.
17. Add 2 μL PHE stock.
18. Add 4 μL sperm acrosome-reacted sperm dilution.
19. Return fertilization drop plate to incubator for desired time.

3.4. Oocyte Treatments

1. For cytochalasin B treatment, add 2 μL of 750 μ*M* cytochalasin B stock to each fertilization medium drop; this produces a final concentration of 30 μ*M* cytochalasin B (**Table 1**).

Table 2
Effect of Nocodazole (20 μ*M*) Treatment on Sperm Entry into In Vitro Matured Bovine Oocytes Fertilized In Vitro with Ejaculated Bull Sperm

	Fertilized	Unfertilized	% Fertilized
Vehicle-treated oocytes			
Vehicle-treated sperm	104	19	84.6%
Nocodazole-treated sperm	109	13	89.3%
Nocodazole-treated oocytes			
Vehicle-treated sperm	107	24	81.7%
Nocodazole-treated sperm	91	14	86.7%

Chi-square test = 3.19, df = 3, $0.25 < p < 0.5$

Note: These data suggest that treatment of sperm or oocytes with the microtubule inhibitor nocodazole has no significant effect on inhibiting sperm entry. Nocodazole (20 μ*M* final) or an appropriate amount of DMSO acting as a vehicle control was added to sperm for a 1-h incubation prior to in vitro fertilization. Sperm were washed free of nocodazole or vehicle and diluted to 20 million sperm/mL. Then, 50,000 sperm were added to each fertilization drop containing 10 oocytes. After 8 h of sperm exposure, nocodazole (20 μ*M* final) or an appropriate amount of DMSO acting as a vehicle control was added to each fertilization drop. The 8-h experimental time-point has been shown to allow the sperm to activate the oocyte and cell cycle and disrupt sperm aster formation. Fertilization was assessed at 18 h after sperm addition using immunocytochemical and Nomarski DIC microscopy techniques. An oocyte was considered positive if there was at least one sperm head and tail within the oocyte. Oocytes were scored as unfertilized when not penetrated by sperm and were still at metaphase arrest of second meiosis or they contained two pronuclei, neither possessing a sperm tail (parthenogenotes). Data represents number of oocytes and percent fertilized. A chi-square test statistic is used to indicate significance.

2. Add oocytes to these treatment drops for 1 h at 39°C in 5% CO_2 under mineral oil.
3. For nocodazole treatment, oocytes are incubated in fertilization medium for 8 h in the presence of sperm and capacitation factors (**Table 2**).
4. Add 2 μL of 750 μ*M* nocodazole to fertilization drops; this produces a final concentration of 20 μ*M* nocodazole.
5. Continue incubation for 10 h at 39°C in 5% CO_2 under mineral oil.
6. For cytochalsin B and nocodazole, add 2 μL of 750 μ*M* cytochalasin B stock and 2 μL of 750 μ*M* nocodazole to fertilization drops at coinciding times as individual treatments.
7. Washed sperm are incubated in fertilization medium containing 30 μ*M* cytochalasin B or 20 μ*M* nocodazole for 1 h at 39°C in 5% CO_2.
8. Incubations with fertilization media containing domethyl sulfoxide (DMSO) are under the same conditions and serve as controls.
9. Prior to the fertilization step, sperm are washed free of drug treatments and diluted.

10. For cytochalasin E treatment, recovered zona pellucida-free oocytes are incubated in fertilization medium containing 30 μ*M* cytochalasin E for 1 h at 39°C in 5% CO_2 under mineral oil.
11. Acrosome-reacted sperm, 10 μ*M* hypotaurine, 20 μ*M* penicillamine, 1 μ*M* epinephrine, and 2 μg/mL heparin are added to the fertilization drop containing cytochalasin E.
12. Oocytes are removed at the appropriate time point for microscopy preparation.
13. Sperm swimming in the drops at the 240-min time-point suggest that cytochalasin E does not alter sperm motility, serving as an important control.

3.5. Fixation Protocols

3.5.1. Preparation

1. Using a no. 2 cover glass and a diamond pencil, score a coverslip to create several 8-mm × 20-mm rectangles (or other size to accommodate specimen holder).
2. Rinse the stubs in acetone to remove any fingerprints.
3. Clean and buff glass stub with lens paper.
4. Store in a desiccator until needed.
5. Coat coverslip stub surface with 20 μL poly-L-lysine solution (1 mg/mL).
6. Allow to dry on a warming plate
7. Cumulus cells are removed from the oocytes by vortexing cumulus-expanded mature oocytes for 1 min in a microfuge tube
8. Transfer oocyte and cumulus medium to a wash plate (*see* **Note 8**).
9. Separate oocytes from cumulus cells (*see* **Note 9**) in the wash plate.
10. Add 5 mL zero protein/zero cation medium (*see* **Note 10**) to a separate agar-bottom plate.
11. Using this wash medium stock that is devoid of protein and divalent cations (zero protein/zero cation), transfer oocytes to this new plate.
12. Prepare a six-well culture plate with 2.5 mL zero protein/zero cation medium in each well.
13. Place one clean coverslip stub in each well
14. Transfer oocytes/zygote (zona-intact or zona-free) from agar plate to wells with a chip, slowly releasing the gametes from the pipet.
15. Let gravity settle cells onto the coated chip for adhesion; do not try to reposition settled gametes.

3.5.2. Low-Voltage Scanning Electron Microscopy

1. Gametes bound to stubs are transferred to primary fixative in six-well plates for 1 h at room temperature.
2. Cells are washed in 0.15 *M* cacodylate buffer (pH 7.4).
3. When indicated, a 1% osmium tetroxide postfixation for 2 h at room temperature is added.
4. Cells are washed in 0.15 *M* cacodylate buffer (pH 7.4).
5. Cells are dehydrated in increasing concentrations of ethanol.

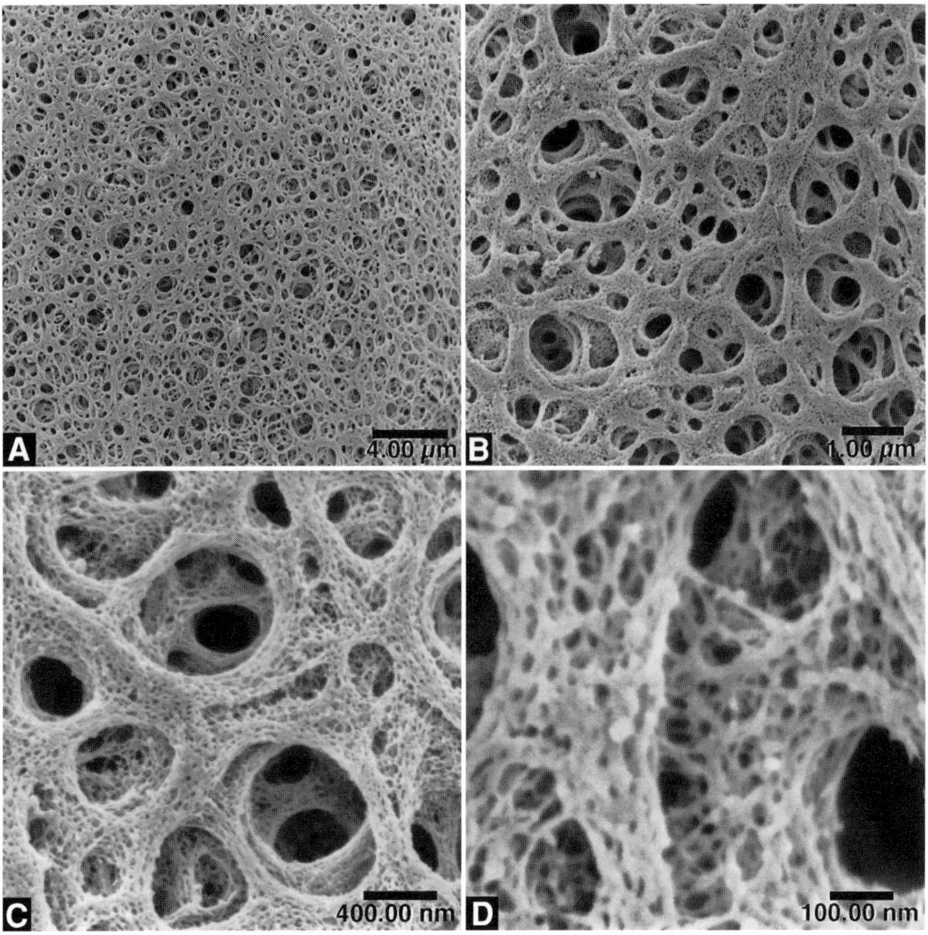

Fig. 1. Bovine oocyte chemically fixed with glutaraldehyde and formaldehyde without postfixation. Sample prepared for LVSEM with platinum coating. Increasing magnification images of the zona pellucida show the large holes created by the multilayer matrix. Fine fibrillar features can be discerned in the zona pellucida fibers at high magnification. Magnification: **(A)** 3000×, **(B)** 10,000×, **(C)** 30,000×, **(D)** 100,000×. (From **ref. *45***; reprinted with the permission of the Microscopy Society of America.)

6. Primary fixative (1% glutaraldehyde + 1% formaldehyde overnight at room temperature [*see* **Fig. 1**]): 0.5 mL glutaraldehyde (50%, electron microscope [EM] grade), 0.5 mL formaldehyde (37% methanol-free, EM grade), 49 mL of 0.15 *M* cacodylate buffer (pH 7.4).
7. Secondary fixation (1% osmium tetroxide [*see* **Fig. 2**]): 1 g osmium tetroxide crystal, 100 mL ddH_2O.

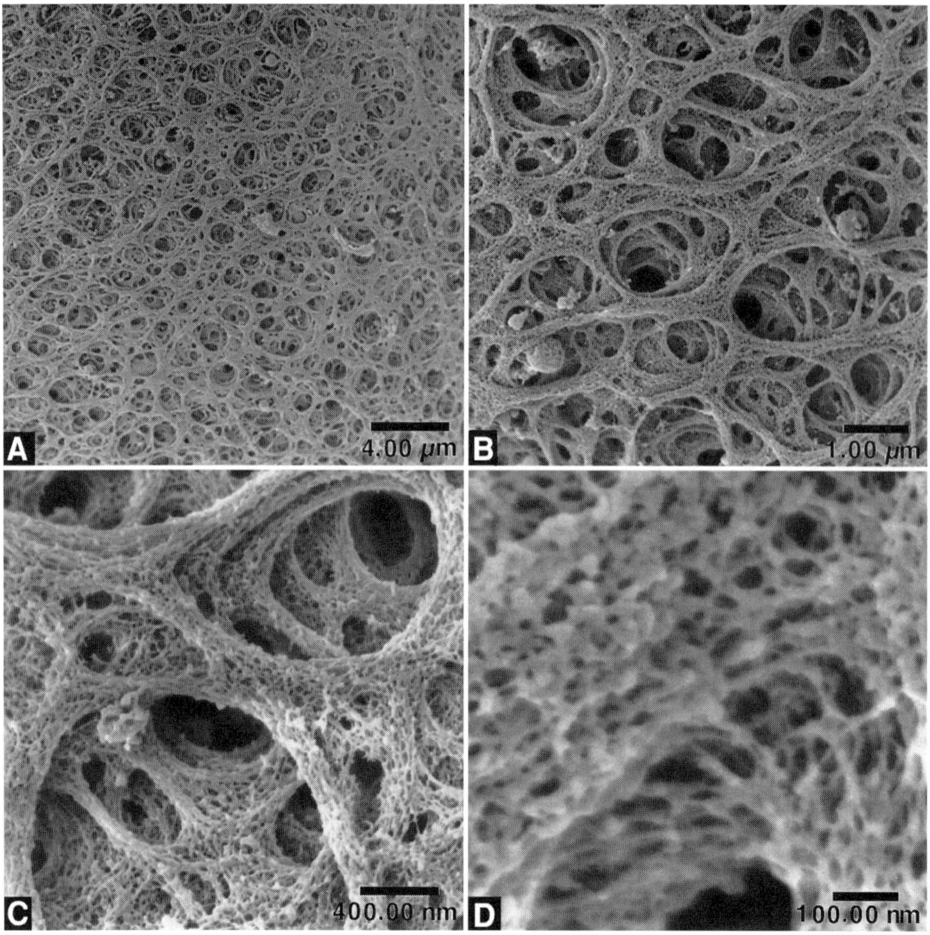

Fig. 2. Bovine oocyte chemically fixed with glutaraldehyde and formaldehyde and postfixed with thiocarbohydrazide/osmium tetroxide. Sample prepared for LVSEM with platinum coating. Increasing magnification images of the zona pellucida show the large holes created by the multilayer matrix. Fine fibrillar features can be discerned in the zona pellucida fibers at high magnification. Osmium postfixation does not alter resolution. Magnification: **(A)** 3000×, **(B)** 10,000×, **(C)** 30,000×, **(D)** 100,000×. (From **ref.** ***45***; reprinted with the permission of the Microscopy Society of America.)

8. Conductive staining (T/O/T/O method [*see* **Fig. 3**]): (A) Prefixation solution (45 min at room temperature): 0.010 g saponin, 0.500 g ruthenium red, 50 mL of 0.15 *M* cacodylate buffer (pH 7.4). (B) Fixation (overnight fixation at room temperature): 0.5 mL glutaraldehyde (50%, EM grade), 0.5 mL formaldehyde (37% methanol-free, EM grade), 0.010 g saponin, 0.500 g ruthenium red,

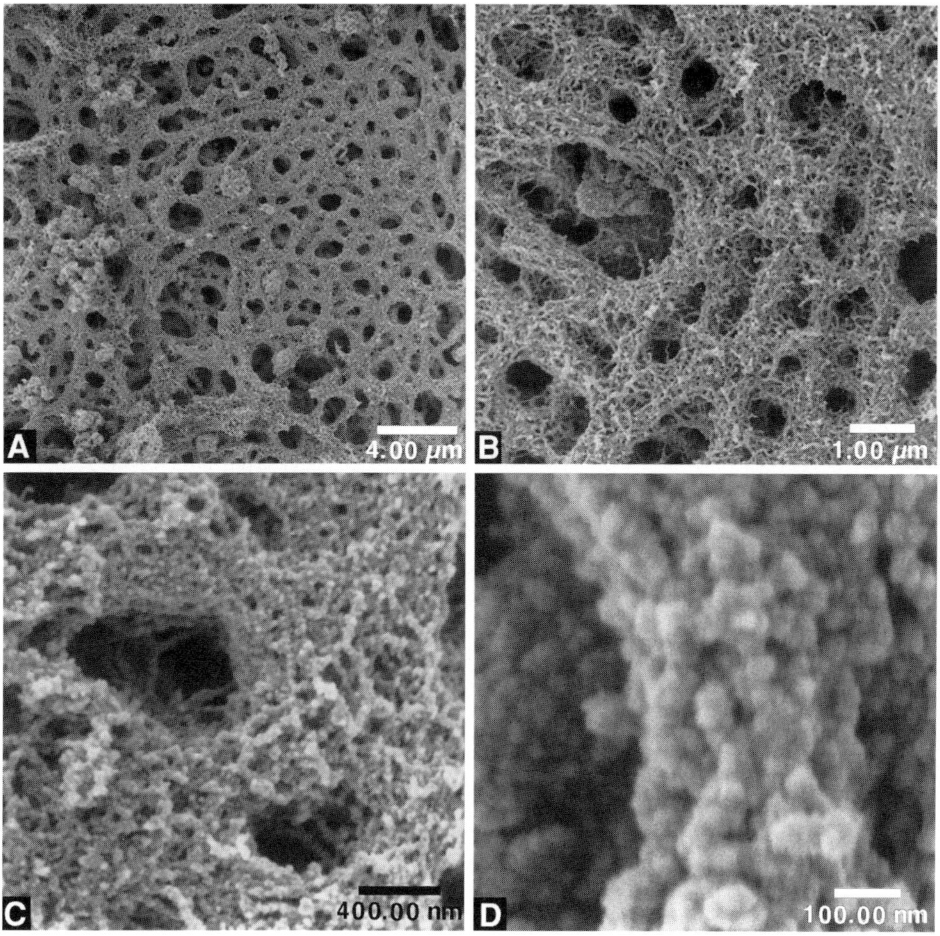

Fig. 3. Bovine oocyte prepared with the T /O /T /O conductive staining protocol and viewed uncoated in the LVSEM. Increasing magnification images of the zona pellucida show the large holes in the matrix; however, the fine detail seen in **Figs. 1–3** is lost when viewed at high magnification. Magnification: **(A)** 3000×, **(B)** 10,000×, **(C)** 30,000×, **(D)** 100,000×. (From **ref.** ***45***; reprinted with the permission of the Microscopy Society of America.)

49 mL 0.15 *M* cacodylate buffer (pH 7.4). (C) Wash (1 h at room temperature) in 20 mL of 0.15 *M* cacodylate buffer (pH 7.4). (D) Postfixation (2 h at room temperature): 0.5 g osmium tetroxide crystal, 0.010 g saponin, 0.375 g ruthenium red, 50 mL of 0.15 *M* cacodylate buffer (pH 7.4).

9. Thiocarbohydrazide (1% solution): 0.5 g thiocarbohydrazide, 50 mL of 0.15 *M* cacodylate buffer (pH 7.4).

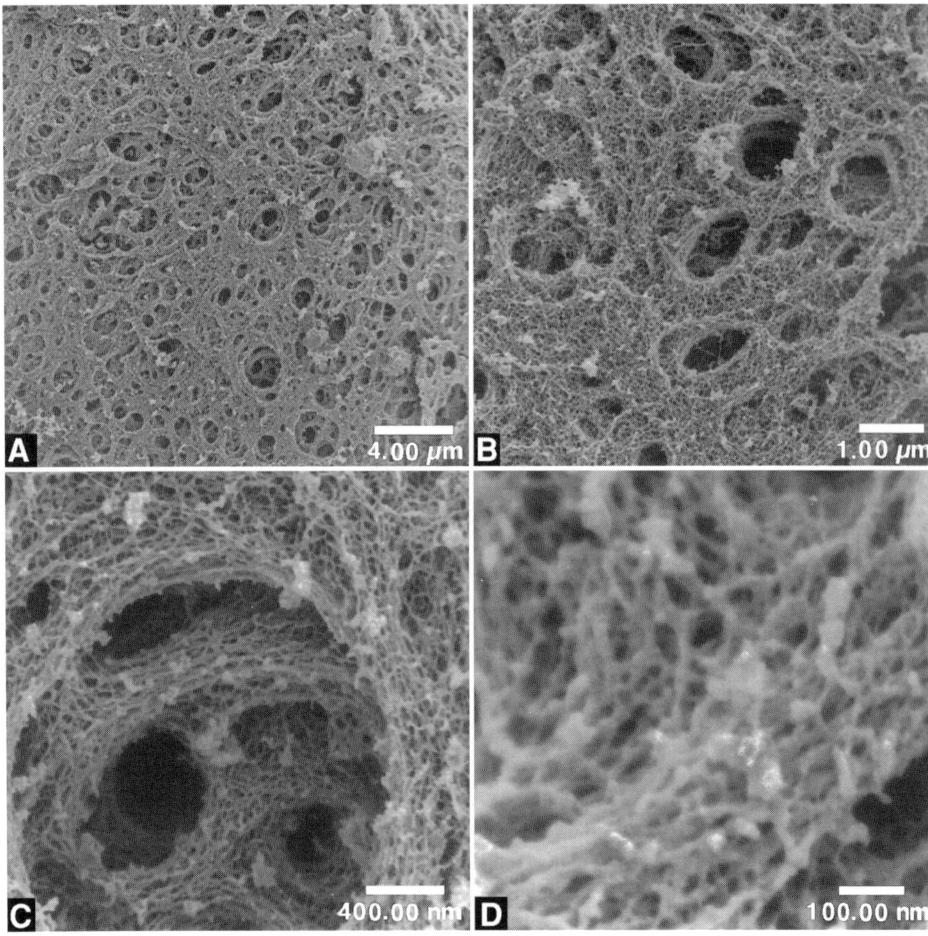

Fig. 4. Bovine oocyte cryopreserved by high-pressure freezing and freeze-substitution. Sample prepared for LVSEM with platinum coating. Increasing magnification images of the zona pellucida show the detail of the large holes created by the multilayer matrix. Detail in the zona pellucida matrix is conserved with this preparative technique. Detail in the zona pellucida fibers at high magnification is similar to that of the aldehyde-fixed samples. Magnification: **(A)** 3000×, **(B)** 10,000×, **(C)** 30,000×, **(D)** 100,000×. (From **ref. *45***; reprinted with the permission of the Microscopy Society of America.)

10. T/O/T/O conductive staining: 20 min 1% thiocarbohydrazide, 60 min 1% osmium tetroxide, 20 min 1% thiocarbohydrazide, 60 min 1% osmium tetroxide.
11. Ethanol dehydration (*see* **Note 11**): 30% ethanol, 10 min; 50% ethanol, 10 min; 70% ethanol, 10 min; 80% ethanol, 10 min; 90% ethanol, 10 min; 90% ethanol,

10 min; 95% ethanol, 10 min; 95% ethanol, 10 min; 100% ethanol, 10 min; 100% ethanol, 10 min; transfer to critical-point dryer, 100% ethanol, 10 min; critical-point dry samples in liquid CO_2.

12. Cryoimmobilization (**Fig. 4**): Oocytes are loaded into planchettes with yeast paste and 1-hexadecene, high pressure (up to 2.1 kbar pressure) during a liquid-nitrogen jet freezing is applied over 500 ms (Balzers), frozen oocytes transferred under liquid nitrogen to cryovials containing a 1% osmium tetroxide solution in molecular-sieve-dried acetone for freeze-substitution at –80°C for 3 d (Balzers); oocytes are warmed to room temperature, washed with fresh molecular-sieve-dried acetone, and adhered to 8-mm × 20-mm silicon wafer chips, chips are transferred to the critical-point dryer, store dried samples in a desiccator until viewing (*see* **Note 12**).

3.6. Microscopy

1. LVSEM and cSEM: Fixed and dried samples are stored in a desiccator until viewing (**Figs. 1–6**).
2. Platinum is argon ion-beam sputter coated onto the sample, which is held in a planetary jig apparatus, producing an ultrathin platinum layer approx 1–2 nm in thickness.
3. Samples are transferred to the specimen holder for insertion into the Hitachi S-900 field emission LVSEM, with an accelerating voltage of 1.5 keV equipped with secondary and backscatter detectors.
4. Control aldehyde-fixed samples processed for viewing in a Hitachi S-570 (cSEM) are sputter coated with 30 nm gold and viewed at either 10 keV or 30 keV accelerating voltage.
5. LVSEM images are captured using Gatan's Digital Micrograph 2.5 and archived on optical disk.
6. cSEM images were recorded on Type 55 Instant Polaroid film.
7. Fluorescence microscopy: Oocytes attached to coverslips are washed and permeabilized in 1% Triton X-100 in phosphate-buffered saline (PBS) for 40 min.
8. Nonspecific binding of antibodies is prohibited with the incubation of coverslips in 1% Triton X-100 and 3% nonfat dry milk in PBS for 40 min.
9. Antitubulin immunostaining is performed using a mouse monoclonal antibody that recognizes formaldehyde-fixed β-tubulin (E7, 1:10 dilution).
10. Primary antibody incubation for each coverslip is carried out in 1% Triton X-100 and 3% nonfat dry milk in PBS for 40 min.
11. Coverslips are washed free of excess antibody.
12. A secondary antibody (goat anti-mouse IgG, 1:400 dilution) conjugated to Oregon Green 514 is used as the fluorophore.
13. Actin is visualized via a phalloidin-conjugate staining. Phalloidin is conjugated to the fluorochrome Texas Red (1:40 dilution, 5 units per coverslip; Molecular Probes).
14. DNA is visualized via with the addition of DAPI (5 μg/mL) added briefly to the next to last rinse.

15. Coverslips are mounted with VectaShield antifade medium.
16. Coverslips are sealed with nail polish, refrigerated until viewing, and stored flat.
17. When viewing, do your best to keep slides in the dark with a box or cover.
18. Multiple-photon-excitation microscopy: Zona-free oocytes are adhered to coverslips in special coverslip-bottom culture wells, optimized for use in an inverted microscope system.
19. Ejaculated sperm are washed, incubated with the vital DNA dye, Hoechst 33342 (5 μg/mL), and acrosome reacted with PC-12.
20. A dilute amount of sperm is added to the culture-dish fertilization drop containing several oocytes.
21. Sperm are visualized for forward motility to confer viability.
22. Each oocyte is imaged in an attempt to capture sperm entry in an active fashion.
23. In these images, a Z-series is captured; each image was generated by averaging five fields over 15 s of exposure.
24. Time-course series are taken every 2 min for the duration of the experiment.
25. Z-Series are adjusted as needed to accommodate increasing sperm head size over time.
26. Concurrent transmitted light images aid localization of fluorescence.

3.7. Fluorescence Microscopy

1. Prepare agar plate with poly-L-lysine coverslips, pipet 5 mL zero protein/zero cation medium to each well of a six-well culture plate (*see* **Note 13**).
2. For fixation, add 500 μL of 10% formaldehyde to each well for 40 min at room temperature.
3. Fixed zygotes are permeabilized with a 1% Triton X-100, 0.1 *M* PBS solution overnight at room temperature.
4. Aspirate the permeabilizing solution and add 3 mL of 0.1 *M* PBS containing 150 m*M* glycine for 30 min at room temperature to reduce the remaining free aldehydes and reduce autofluorescence.
5. Block nonspecific binding of primary and secondary antibodies by incubating fixed, permeabilized cells with 1% Triton X-100, 0.1 *M* PBS containing 3% nonfat dry milk; use this diluent for subsequent wash steps and antibody incubations.
6. Addition of primary and secondary antibodies: Dilute E-7 supernatant (20 μL E-7 to 180 μL of 1% Triton X-100, 0.1 *M* PBS containing 3% nonfat dry milk) on a sheet of parafilm.
7. Pipet a 100-μL drop for each coverslip; aspirate blocking medium from the culture well containing coverslip.
8. Transfer coverslip to the raised center of a 100-mm four-well culture plate for support; pipet 150 mL primary antibody to each coverslip (one per plate).
9. Incubate on a warming plate for 45 min.
10. Prepare secondary antibody (goat anti-mouse Oregon Green 514) at this time using a 1:400 dilution using 1% Triton X-100, 0.1 *M* PBS containing 3% nonfat dry milk.

11. Wash each coverslip through three 5 mL of 1% Triton X-100, 0.1 *M* PBS containing 3% nonfat dry milk beakers.
12. Blot slip, return to the four-well plates, and add 150 μL of secondary antibody dilution to each coverslip.
13. Incubate on a warming plate for 45 min.
14. Actin staining: phalloidin conjugated to Texas Red is diluted 1:40 using 1% Triton X-100, 0.1 *M* PBS containing 3% nonfat dry milk.
15. Remove secondary antibody and wash each coverslip through three 5 mL of 1% Triton X-100, 0.1 *M* PBS containing 3% nonfat dry milk beakers.
16. Blot slip, return the coverslip to the four-well plates, and dispense 150 μL Texas Red–phalloidin to each coverslip.
17. Incubate on a warming plate for 45 min.
18. DNA staining: Prepare a 100-μg/mL DAPI stock solution.
19. Remove the Texas Red–phalloidin with aspiration.
20. Wash each coverslip through two 5 mL of 1% Triton X-100, 0.1 *M* PBS containing 3% nonfat dry milk beakers.
21. Blot slip, return to the four-well plates, and add 100 μL of 1% Triton X-100, 0.1 *M* PBS containing 3% nonfat dry milk to each coverslip.
22. Pipet 5 μL of the DAPI stock to each coverslip.
23. Let stand 5 min.
24. Wash each coverslip through two 5 mL 1% Triton X-100, 0.1 *M* PBS containing 3% nonfat dry milk beakers.
25. Mounting coverslip to glass slide: Prepare glass slide with appropriate label.
26. Break small glass spacers using a diamond pencil (i.e., 2 mm × 18 mm).
27. Pipet 20 μL antifade medium to edge of slide.
28. Place two spacers on either side of the antifade medium.
29. Remove coverslip from final wash beaker, hold perpendicular to the bench on an absorbent towel (i.e., Kim-wipe), and slowly lower coverslip on to glass slide, letting the meniscus of the antifade medium travel with the cover glass so as to not create any bubbles, yet keeping the spacers from crushing the cells.
30. Seal the edges of the coverslip with nail polish.
31. Store flat in the refrigerator, keeping them in the dark.
32. It is best to view on the fluorescence system within 48 h, as aqueous-mounted cell fluorescence quality does degrade over time (*see* **Note 14**).

4. Notes

1. FSH-p is normally sold in 50-IU vials, with each lot having a specific protein content (i.e., 28 mg). Based on this number (e.g., dilute 28 mg FSP-p in 37.333 mL of 0.9% saline), you can create 100-μL aliquots containing 750 μg FSH-p, which you lyophilize and then store at –20°C. Reconstitute each aliquot with 150 μL TC199, adding 100 μL to the maturation medium, yielding 50 μg FSH-p.
2. Maturation medium requires estrogen, but filter-sterilizing the solution destroys the activity in some way, so it is vital that you filter *first* and then add the estrogen from a stock solution.

3. Scale this recipe to meet your needs. Discard any unused wash medium at the end of the day.
4. Epinephrine is easily oxidized; make this stock just prior to making PHE cocktail, keep vials of PHE out of the light, freeze and store stocks and aliquots, thaw out of direct light, add 2 µL per 50-µL drop of fertilization medium.
5. Maintain ovaries in a beaker with a little 0.9% saline, placed in a water bath; follicular fluid tubes can be placed in a test tube rack in the same circulating 37°C water bath until collection is complete. Avoid the large, cyctic follicles, as the quality of the oocyte within these cysts is poor. The pellet that forms at the bottom of the conical tube is a mixture of cells, you will search this pellet for cumulus cell–oocyte complexes.
6. To keep track of where you have searched under the dissection scope; draw a few parallel lines approx 1 cm apart along the underside of your 100-mL Petri dish before you begin. This helps organize your scan, but does make it hard to see oocytes lying on the line. Red Sharpie pens help reduce this, although a really fine blue line is a close second.
7. This is an effective way to acrosome react sperm, but you need to work fast once you add PC-12. The liposome disrupts the sperm's outer membrane. Acrosome reaction can be confirmed with phase-contrast microscopy. This procedure produces >85% acrosome-reaction-positive sperm. Sperm do not live long following this treatment. Time this accordingly with the readiness of the zona-free oocytes for fertilization.
8. If desired, remove zona pellucida using 1% pronase in wash medium; the zona pellucida will deform and dissolve quickly; pipet oocytes into fresh wash medium, and recover oocytes for 30 min on a warming plate.
9. Normal fertilization requires the presence of expanded cumulus cells, so be sure to include this step in your experimental design. Avoid mature oocytes without expanded cumulus cells and those with expanded and focal clumpy cumulus cells.
10. Make a TL–HEPES solution without calcium, magnesium, and BSA fraction V; supplement only with pyruvate. Using an agar-bottomed plate helps. Add 100 mg agar to a 10-mL volume of water, bring to a boil over a flame, and transfer dissolved agar solution to a 60-mm culture plate, coating the plate; allow to cool. Transfer 5 mL of the zero protein/zero cation TL–HEPES to the agar-coated culture plate. Oocytes in this medium become very sticky, so it is best to work in small clutches. Alternatively, the plate is large enough to accommodate several cover-glass stubs—place oocytes in the center of the plate with stubs arranged on the periphery, dividing oocytes among the stubs. Control of the oocytes is best obtained with a pulled 20 µL Unopette tip, with an approx 150-µm opening, this tip also is useful in removing zona pellucida fragments. Allow oocytes to settle via gravity onto the poly-L-lysine stubs ***(44)***, do not attempt to move them once they make contact with the coverslip—any attempt to move once attached will destroy the gamete. Gametes bound to stubs can be transferred to fixative, or in the case of six-well plates, allow for the addition of aldehyde stock to final concentration.

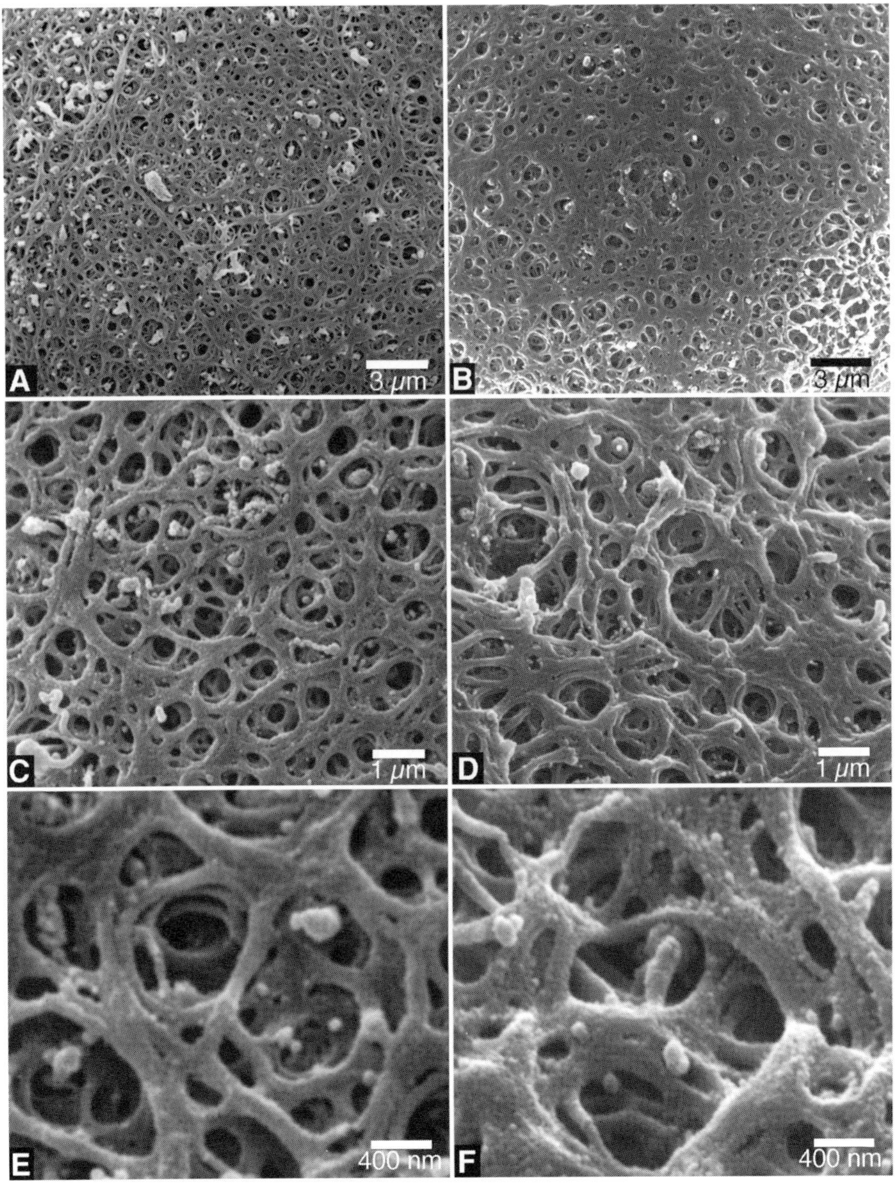

Fig. 5. Bovine oocyte chemically fixed with glutaraldehyde and formaldehyde without postfixation. Sample prepared for cSEM with gold coating. Increasing magnification images of the zona pellucida show the large holes, but lack the detail of the samples prepared for LVSEM (*see* **Fig. 1**). Increased accelerating voltage decreases image contrast. Magnification: **(A,B)** 3000×, **(C,D)** 10,000×, **(E,F)** 30,000×; accelerating voltage 10 keV **(A,C,E)**, 30 keV **(B,D,F)**. (From **ref. *45***; reprinted with the permission of the Microscopy Society of America.)

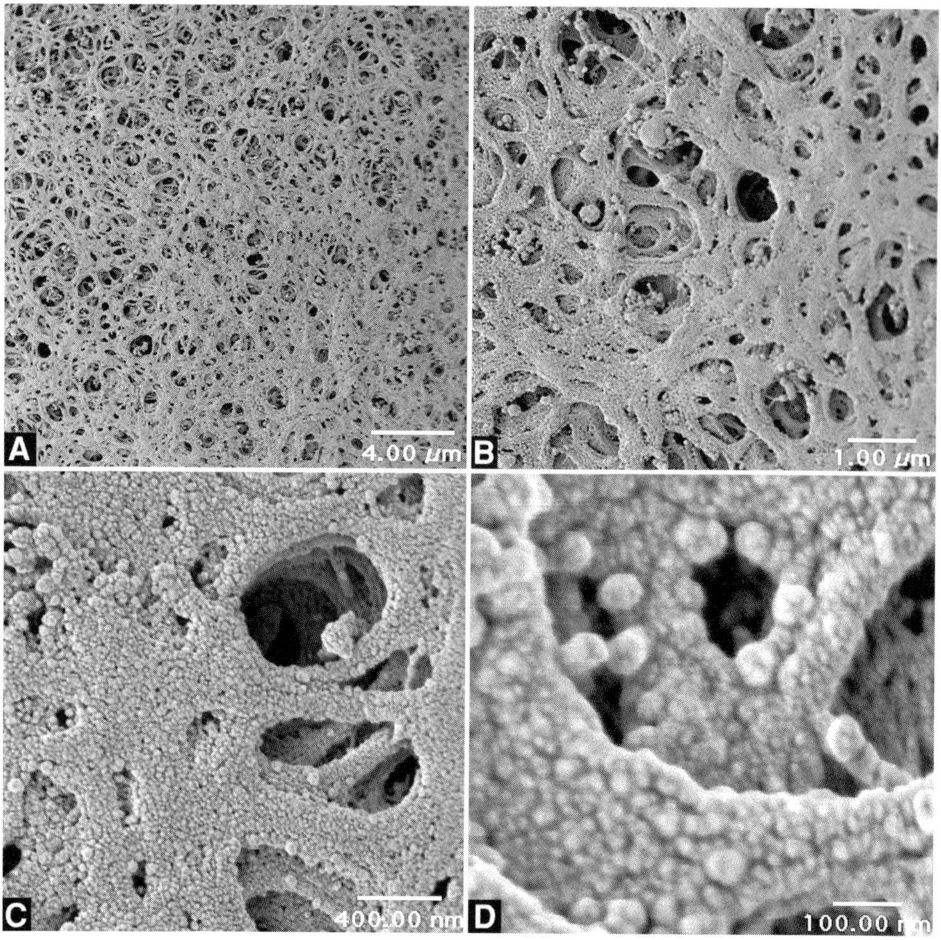

Fig. 6. Bovine oocyte chemically fixed with glutaraldehyde and formaldehyde without postfixation. Sample prepared for cSEM with gold coating and viewed with the LVSEM. Increasing magnification images of the zona pellucida show that the gold coating needed for imaging in the cSEM obscures the detail that can be seen at high magnification of samples prepared for and viewed in the LVSEM. Magnification: **(A)** 3000×, **(B)** 10,000×, **(C)** 30,000×, **(D)** 100,000×. (From **ref. *45***; reprinted with the permission of the Microscopy Society of America.)

11. Avoid exposing cells to air by slowly adding the next alcohol dilution to the top of the sample; aspirate the bottom layer of liquid (heavier, more dense) using a faucet vacuum and glass pipet. This is unavoidable during the transfer to the critical-point dryer, so work quickly during this step.
12. The LVSEM samples are coated with 1–2 nm platinum using evaporation techniques in an argon plasma. This renders them conductive and produces an excel-

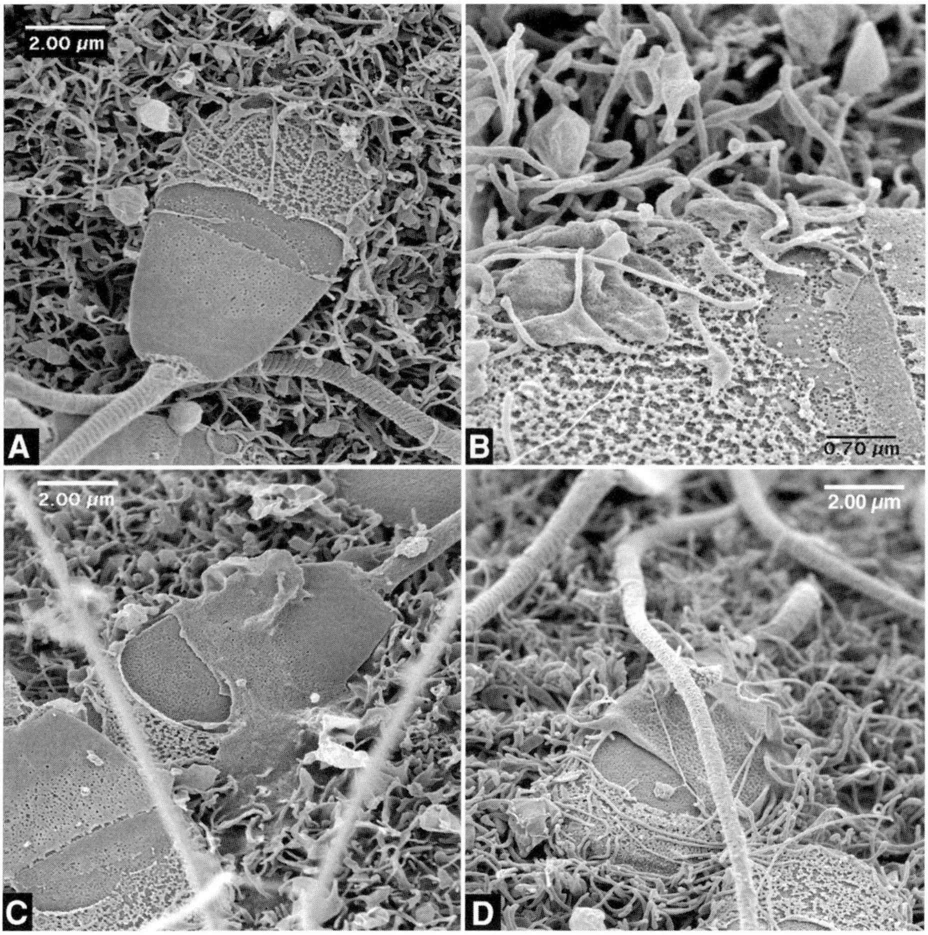

Fig. 7. Acrosome-reacted bull sperm on the surface of zona pellucida-free bovine oocytes. **(A)** Microvilli can be seen uniformly distributed over the oocyte surface. After a 10-min exposure to sperm, microvilli from the oocyte can be seen making contact with the sperm over the anterior head in the vesiculated outer acrosomal membrane and equatorial segment membrane areas. Several spherical structures are apparent around the sperm head, most likely the products of cortical granule release. **(B,C)** Microvilli on the oolemma surface are in contact with the sperm head after 20 min of sperm exposure. It is at this time-point that the earliest images of membrane fusion can be seen in the LVSEM. This fusion event appears to begin primarily over the area of the sperm head corresponding to the equatorial segment. **(D)** At 30 min after sperm addition, the area over the equatorial segment appears to be completely fused with the oolemma membrane, yet the oocyte microvilli are seen interacting with the acrosomal and postequatorial domains. A portion of the sperm membrane is still exposed to the media.

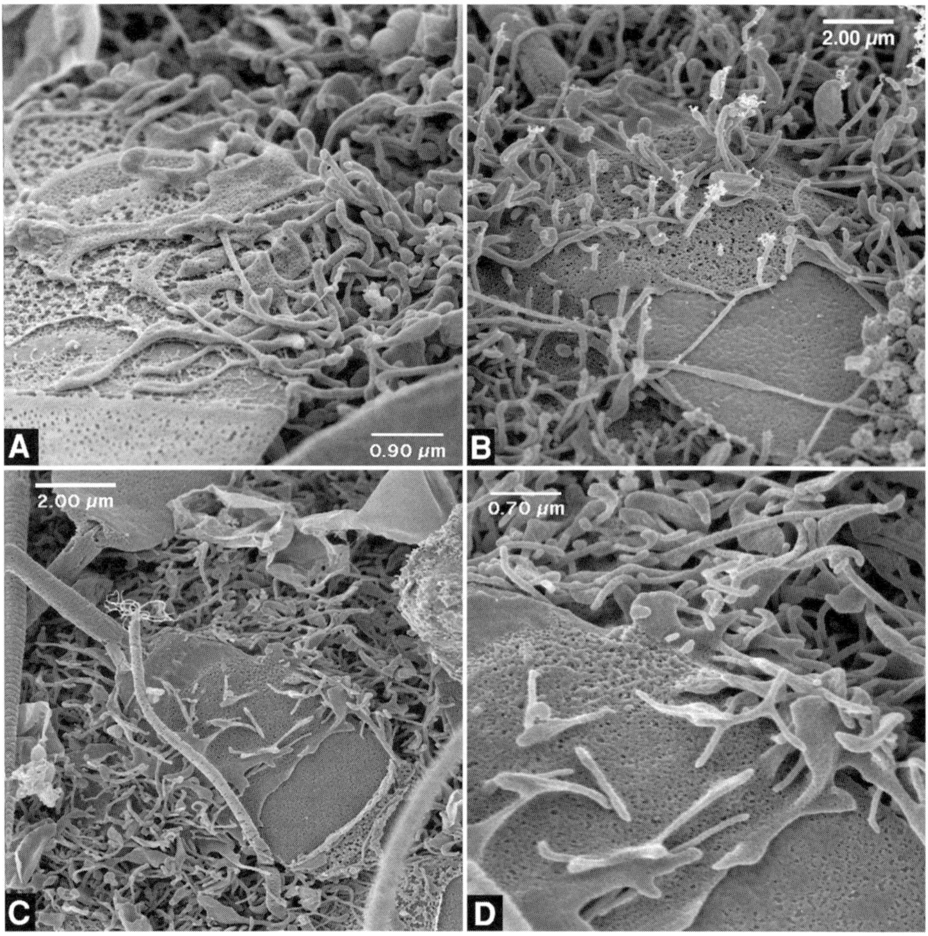

Fig. 8. Acrosome-reacted bull sperm in various stages of oolemma fusion 60 min after sperm addition. **(A)** Microvilli from the oocyte can still be seen interacting with the sperm equatorial segment and head. Processes are extended to the equatorial region in many sperm. **(B,C)** The appearance of microvilli over the equatorial and postequatorial region of an incorporating bull sperm. The sperm tail and implantation fossa are not incorporated at this time-point; the inner acrosomal membrane can still be discerned. **(D)** Higher magnification of **(C)** showing detail of the microvilli over the equatorial region.

lent source of secondary electrons. Conductive stained samples are not coated with heavy metal. Oocytes processed for conventional SEM (*see* **Fig. 5**) are coated with 30 nm gold, using plasma-sputtering techniques (*see* **Fig. 5**, revealed by LVSEM in **Fig. 6**).

13. Experimental design will drive the time-points for observation (**Figs. 7–11**). In all cases, processing steps will be the same. For zona pellucida-intact fertiliza-

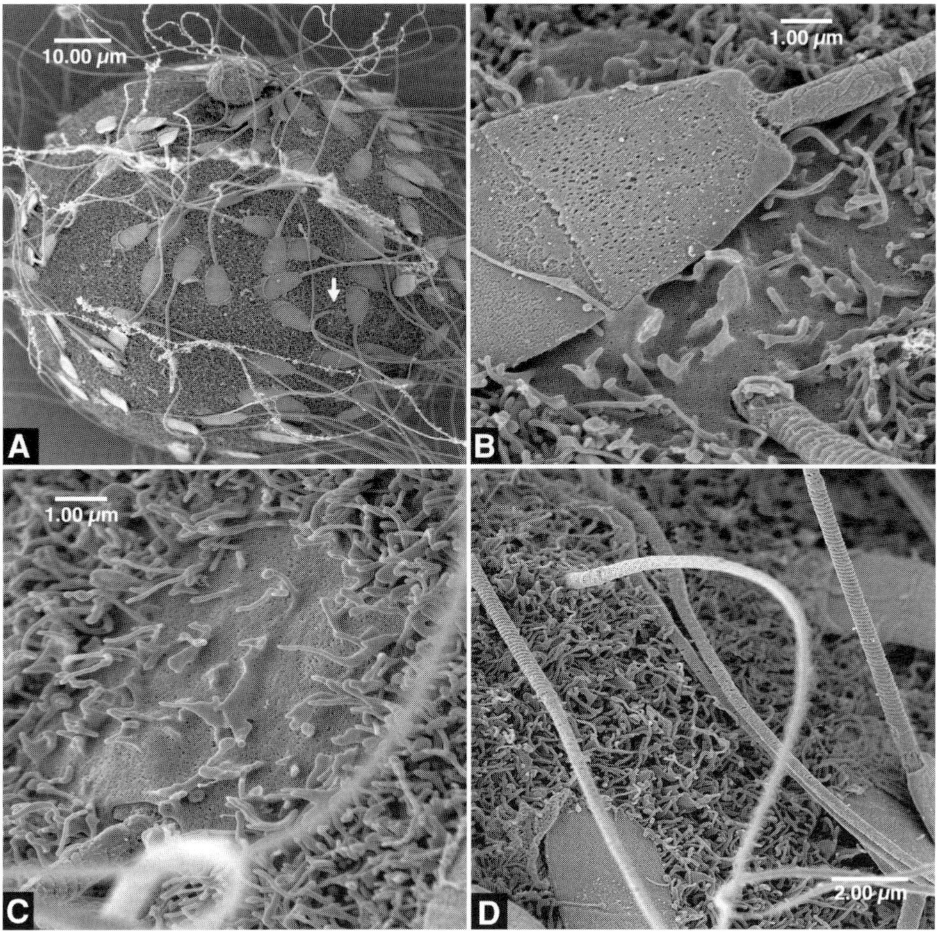

Fig. 9. One hundred twenty min after the addition of acrosome-reacted bull sperm to zona pellucida-free bovine oocytes. **(A)** The appearance of a polar body at the top of one oocyte suggests that the penetrating sperm, indicated with a small arrow, activated the oocyte. Many other sperm are bound to the surface at various stages of incorporation; many do not get captured at these short time-points in states of fusion. **(B,C)** At this time-point, the complete sperm head is under the oolemma, and the tail can be seen entering at one point of contact. The sperm head is completely under the oolemma membrane and microvilli can be seen over the area of the sperm head. The implantation fossa is in the process of being incorporated into the ooplasm and the mitochondrial sheath can still be seen, suggesting that the head is still very close to the oolemma membrane. **(D)** Tail incorporation occurs at only one point in the bovine system; it is possible to follow the tail from the incorporation site to the distal end without encountering an alternative entry site.

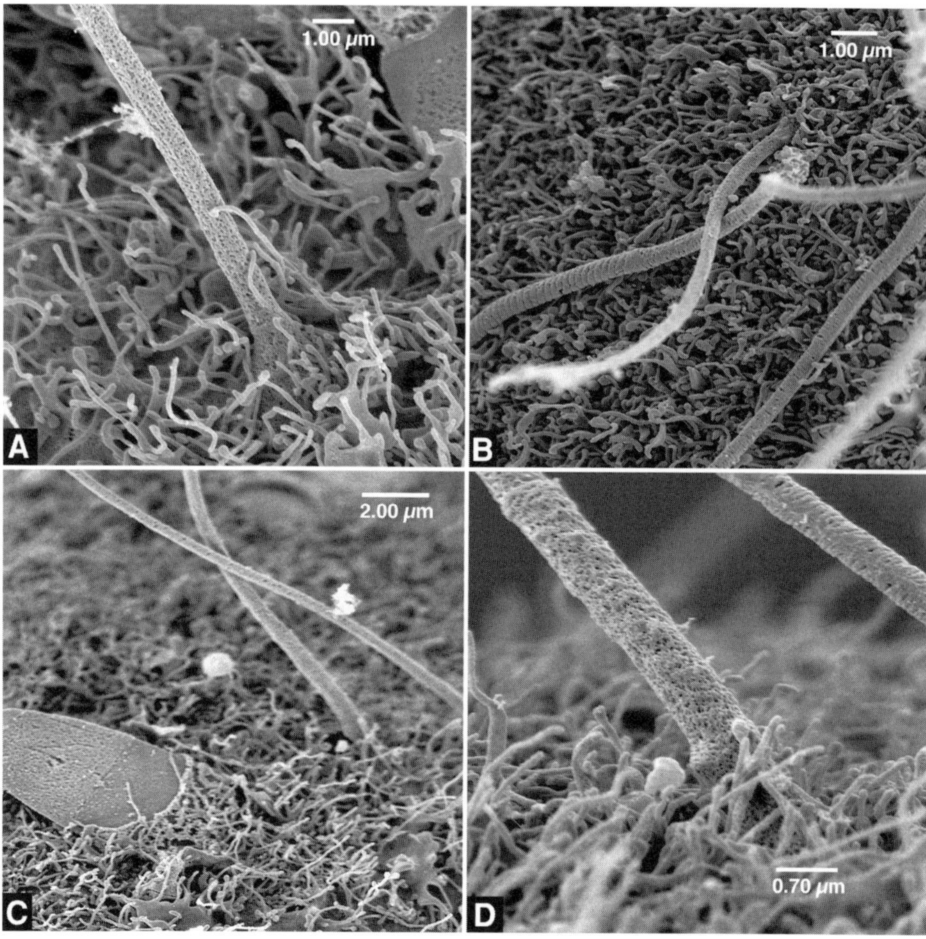

Fig. 10. Fertilized oocytes 240 min after the addition of acrosome-reacted bull sperm to zona pellucida-free bovine oocytes. **(A–D)** Sperm tails can be seen entering various oocytes, each with only one point of tail incorporation. There does not appear to be a large or prominent fertilization cone in this species.

tion, you will need to remove the cumulus cells and zona pellucida using the 1% pronase wash medium, recover cells in fresh fertilization medium without PHE, heparin, or sperm. Because this is an optical technique, use a no. 1.5 cover glass to match optics. Work in a six-well culture plate for best results; this also facilitates future fixation and processing. In this example, the antibody will recognize fixed epitopes; alternatively, you could use an acid–alcohol fixation or ice-cold methanol; follow the directions for fixation provided by your antibody supplier (**Figs. 12–14**).

14. Think of this as a hanging drop, where the adherent cells are suspended in antifade medium. These cells can be imaged using epifluorescence, confocal, or multiple-

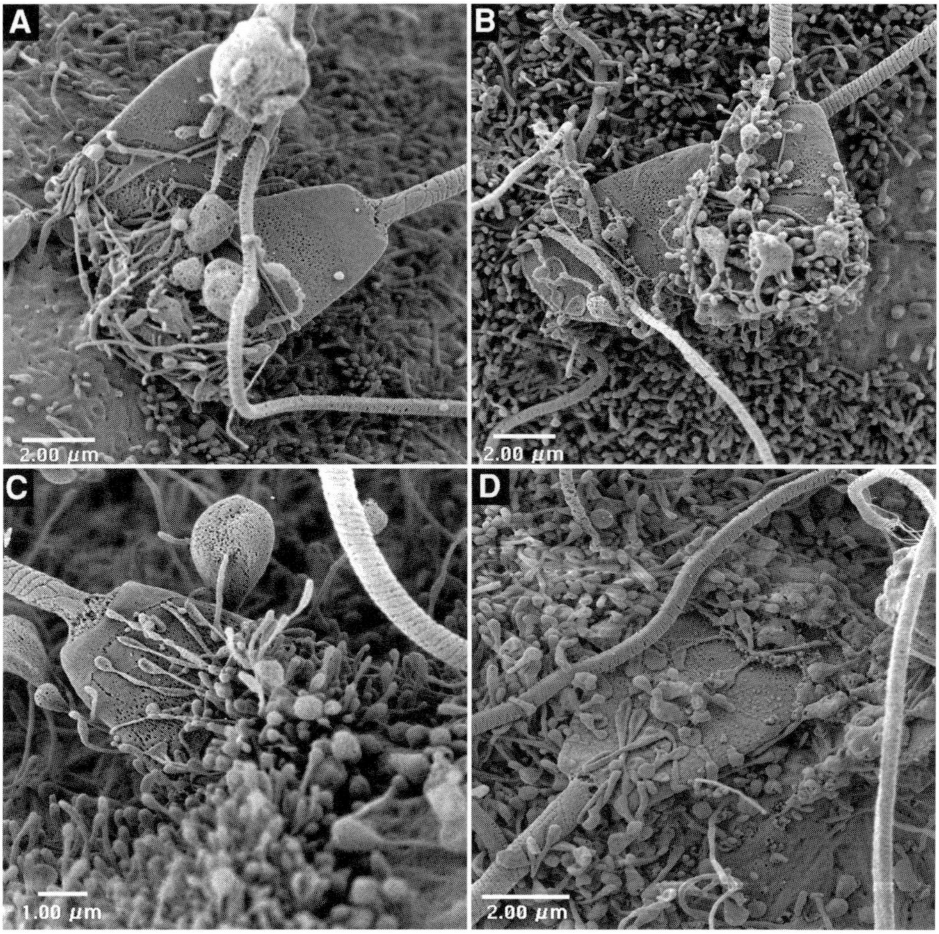

Fig. 11. Cytochalasin-treated zona pellucida-free oocytes exposed to acrosome-reacted bull sperm display markedly different incorporation. Cytochalasin treatment alters the morphology of the oocyte microvilli, producing areas devoid of microvilli and microvilli with short, flat, clublike processes. After either 30 min **(A)**, 60 min **(B)**, 120 min **(C)**, or 240 min **(D)** of sperm exposure with the oocytes in the presence of cytochalasin, the remaining oocyte microvilli are seen contacting the sperm head in various locations, yet the posterior sperm head, implantation fossa, and sperm tail with mitochondria remain outside the ooplasm, suggesting that oocyte microfilament assembly is important for successful sperm entry in the bovine system.

photon excitation techniques. In this chapter, we used a monoclonal supernatant to the E-7 cell line, which specifically recognizes formaldehyde-fixed β-tubulin subunits, a rhodamine–phalloidin conjugate, and DAPI. Follow the manufacturer's directions if different from this method.

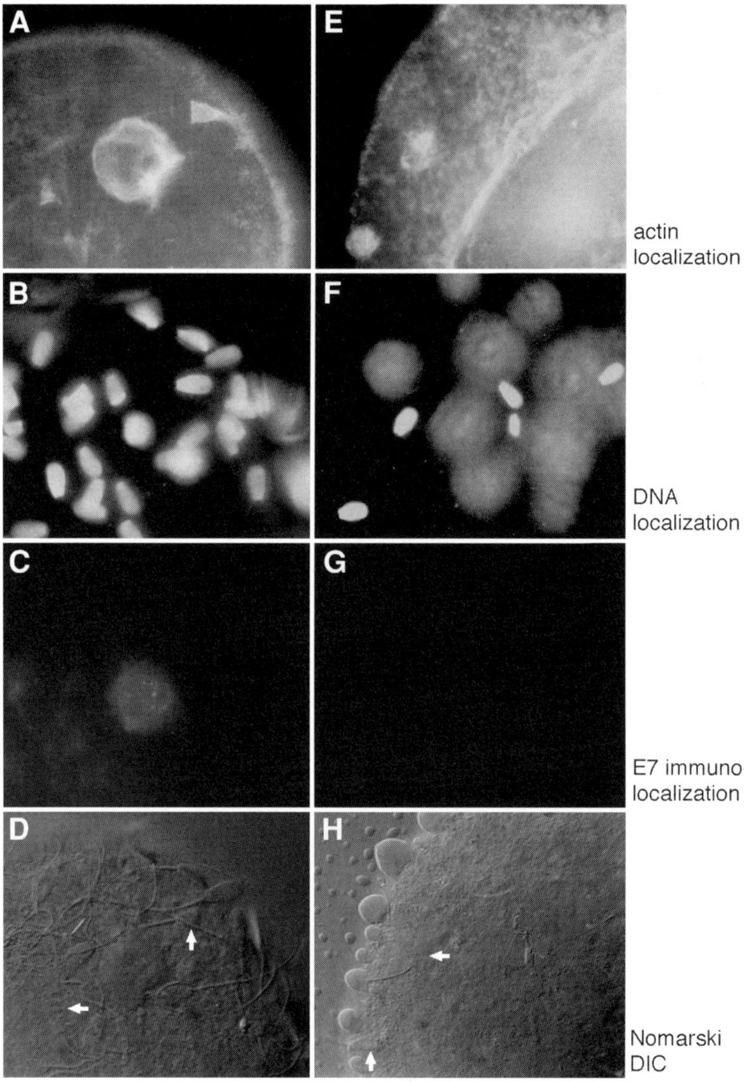

Fig. 12. Immunocytochemistry of ejaculated bull sperm incorporation into zona pellucida-free bovine oocytes. Images **A–D** (120 min) and **E–H** (240 min) display different stages of incorporation post-sperm addition. Actin localization (**A** and **E**) as indicated by Texas Red–phalloidin conjugate shows a localization at the sperm head, oolemma, and site of polar body extrusion (**A**). Multiple sperm heads can be seen bound to the surface of the oocytes as detected by DAPI staining (**B** and **F**). At this time-point of sperm incorporation, there was no significant microtubule immunolocalization (meiotic spindle was visible in **C**). Nomarski microscopy confirms the presence of numerous sperm bound to the surface of the oocytes; incorporating sperm are indicated with arrows.

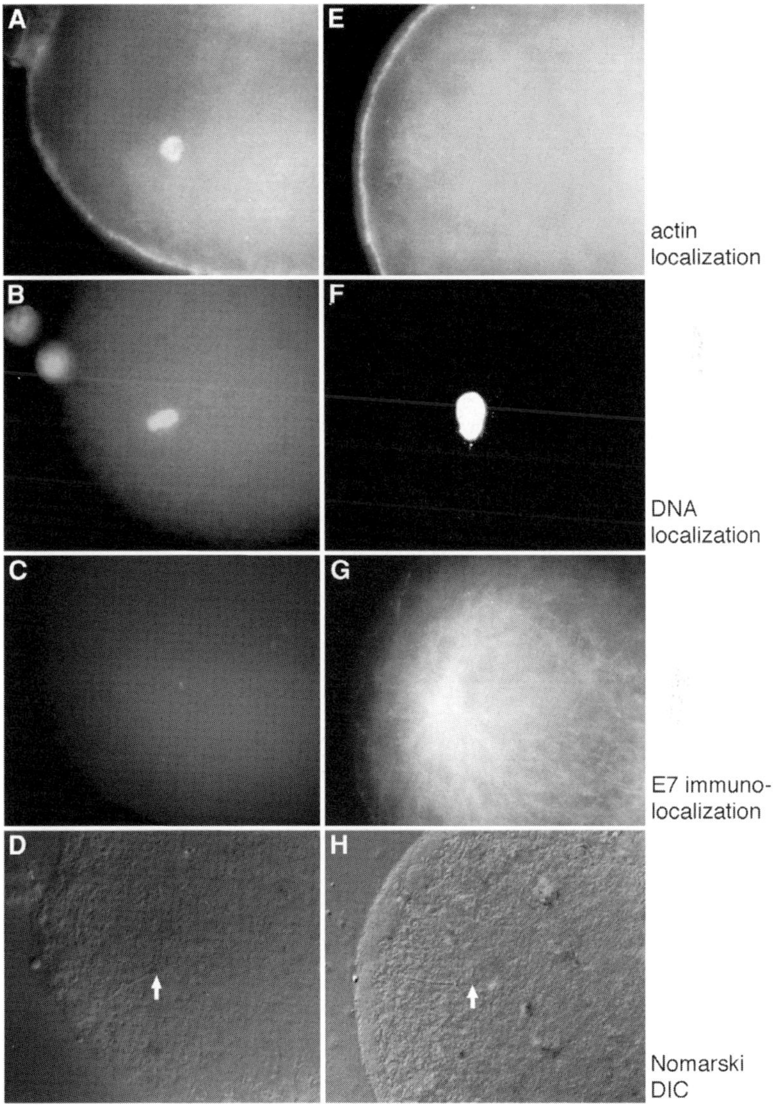

Fig. 13. Immunocytochemistry of ejaculated bull sperm incorporation into zona pellucida-intact bovine oocytes. The images **A–D** were 8 h post-sperm addition and images **E–H** were 12 h post-sperm addition. Actin localization (**A** and **E**) as indicated by Texas Red–phalloidin conjugate suggests that actin localization around the sperm head may be transient, whereas the oolemma microvilli have a consistent localization pattern throughout. Sperm heads can be localized with DAPI staining (**B** and **F**) as well as the extrusion of the polar body (**B**). A microtubule sperm aster can be immunolocalized in **G**, but not in **C**. Nomarski microscopy confirms the presence of a single sperm bound to the surface of the oocyte; incorporating sperm is indicated with an arrow.

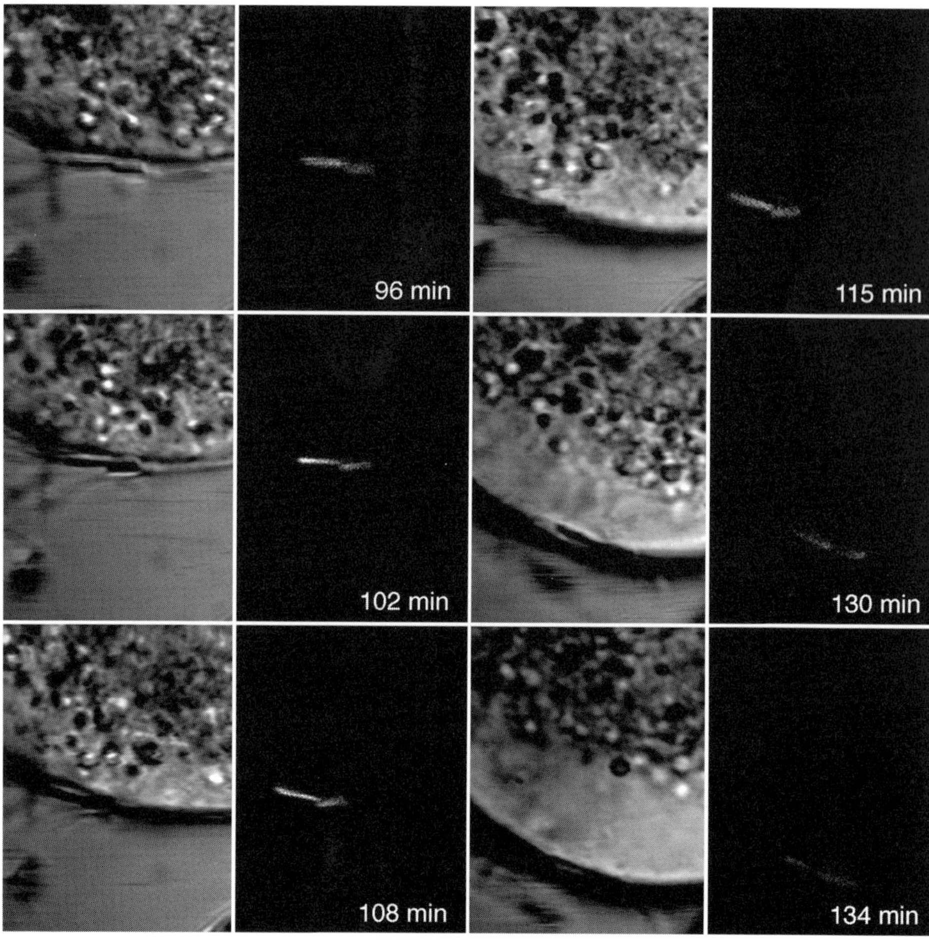

Fig. 14. Multiple photon excitation fluorescence imaging of live bovine zona pellucida-free fertilization. Ejaculated bull sperm acrosome-reacted with PC-12 treatment and labeled with Hoechst 33342 (5 μg/mL) were exposed to zona pellucida-free oocytes and observed with transmitted light and three-photon optical sectioning fluorescence microscopy. The time course of events suggests that sperm head decondensation can be observed with Z-series collections and that cortical rearrangement occur with sperm incorporation. The exact molecules involved in these processes are not identified, but one of them may be actin. The use of this microscopy technique offers researchers the opportunity to perform observations in real time, provided appropriate excitable probes exist.

Acknowledgments

The author wishes to thank Dr. Gerald Schatten for his mentorship during the creation of the experimental data, and Colleen A. Lavin and Heidi R. Barnhill for their technical expertise with the high-pressure cryo-immobilization/freeze-substitution and conventional scanning electron microscopy techniques, respectively. The Integrated Microscopy Resource located on the University of Wisconsin–Madison campus is funded by NIH grant no. RR570, and this investigation was sponsored by research grants from the USDA and NIH. Without these sources of support, this research could not have been possible.

References

1. Harper, M. J. K. (1982) Sperm and egg transport, in *Reproduction in Mammals* (Austin, C. R. and Short, R. V., eds.), Cambridge University Press, Cambridge, pp. 102–127.
2. Hunter, R. H. (1987) Human fertilization in vivo, with special reference to progression, storage and release of competent spermatozoa. *Hum. Reprod.* **2,** 329–332.
3. Ralt, D., Goldenberg, M., Fetterolf, P., et al. (1991) Sperm attraction to a follicular factor(s) correlates with human egg fertilizability. *Proc. Natl. Acad. Sci. USA* **88,** 2840–2844.
4. Austin, C. R. (1951) Observations on the penetration of the sperm into the mammalian egg. *Aust. J. Biol. Sci.* **4,** 581–596.
5. Chang, M. C. (1951) Fertilizing capacity of spermatozoa deposited in the Fallopian tubes. *Nature* **168,** 697–698.
6. Vaidya, R. A., Glass, R. H., Dandekar, P., et al. (1971) Decrease in the electrophoretic mobility of rabbit spermatozoa following intra-uterine incubation. *J. Reprod. Fertil.* **24,** 299–301.
7. Ericsson, R. J. (1967) A fluorometric method for measurement of sperm capacitation. *Proc. Soc. Exp. Biol. Med.* **125,** 1115–1118.
8. Oliphant, G. and Brackett, B. G. (1973) Capacitation of mouse spermatozoa in media with elevated ionic strength and reversible decapacitation with epididymal extracts. *Fertil. Steril.* **24,** 948–955.
9. Gordon, M., Dandekar, P. V., and Bartoszewicz, W. (1975) The surface coat of epididymal, ejaculated and capacitated sperm. *J. Ultrastruct. Res.* **50,** 199–207.
10. Koehler, J. K. (1976) Changes in antigenic site distribution on rabbit spermatozoa after incubation in "capacitating" media. *Biol. Reprod.* **15,** 444–456.
11. Saling, P. M., Storey, B. T., and Wolf, D. P. (1978) Calcium-dependent binding of mouse epididymal spermatozoa to the zona pellucida. *Dev. Biol.* **65,** 515–525.
12. Braden, A. W. H., Austin, C. R., and David, A. H. (1954) The reaction of the zona pellucida to sperm penetration. *Aust. J. Biol. Sci.* **7,** 391–409.
13. Austin, C. R. and Braden, A. W. H. (1956) Early reactions of the rodent egg to spermatozoa penetration. *J. Exp. Biol.* **33,** 358–365.

14. Swenson, C. E. and Dunbar, B. S. (1982) Specificity of sperm-zona interaction. *J. Exp. Zool.* **219,** 97–104.
15. Storey, B. T., Lee, M. A., Muller, C., et al. (1984) Binding of mouse spermatozoa to the zonae pellucidae of mouse eggs in cumulus: evidence that the acrosomes remain substantially intact. *Biol. Reprod.* **31,** 1119–1128.
16. Shur, B. D. and Hall, N. G. (1982) Sperm surface galactosyltransferase activities during in vitro capacitation. *J. Cell Biol.* **95,** 567–573.
17. Bleil, J. D., Beall, C. F., and Wassarman, P. M. (1981) Mammalian sperm–egg interaction: fertilization of mouse eggs triggers modification of the major zona pellucida glycoprotein, ZP2. *Dev. Biol.* **86,** 189–197.
18. Gaddum-Rosse, P. (1985) Mammalian gamete interactions: what can be gained from observations on living eggs? *Am. J. Anat.* **174,** 347–356.
19. Yanagimachi, R. (1988) Mammalian fertilization, in *The Physiology of Reproduction* (Knobil, E. and Neill, J. D., eds.), Raven, New York, pp. 135–185.
20. Oura, C. and Toshimori, K. (1990) Ultrastructural studies on the fertilization of mammalian gametes. *Int. Rev. Cytol.* **122,** 105–151.
21. Evans, J. P., Schultz, R. M., and Kopf, G. S. (1995) Mouse sperm–egg plasma membrane interactions: analysis of roles of egg integrins and the mouse sperm homologue of PH-30 (fertilin) beta. *J. Cell Sci.* **108,** 3267–3278.
22. Bronson, R. A. and Fusi, F. (1990) Evidence that an Arg-Gly-Asp adhesion sequence plays a role in mammalian fertilization. *Biol. Reprod.* **43,** 1019–1025.
23. Almeida, E. A., Huovila, A. P., Sutherland, A. E., et al. (1995) Mouse egg integrin alpha 6 beta 1 functions as a sperm receptor. *Cell* **81,** 1095–1104.
24. Cheresh, D. A., Smith, J. W., Cooper, H. M., et al. (1989) A novel vitronectin receptor integrin (alpha v beta x) is responsible for distinct adhesive properties of carcinoma cells. *Cell* **57,** 59–69.
25. Pytela, R., Pierschbacher, M. D., and Ruoslahti, E. (1985) A 125/115-kDa cell surface receptor specific for vitronectin interacts with the arginine-glycine-aspartic acid adhesion sequence derived from fibronectin. *Proc. Natl. Acad. Sci. USA* **82,** 5766–5770.
26. Mercurio, A. M. and Shaw, L. M. (1991) Laminin binding proteins. *BioEssays* **13,** 469–473.
27. Hynes, R. O. (1992) Integrins: versatility, modulation, and signaling in cell adhesion. *Cell* **69,** 11–25.
28. Schwartz, M. A., Ingber, D. E., Lawrence, M., et al. (1991) Multiple integrins share the ability to induce elevation of intracellular pH. *Exp. Cell Res.* **195,** 533–535.
29. Pardi, R., Bender, J. R., Dettori, C., et al. (1989) Heterogeneous distribution and transmembrane signaling properties of lymphocyte function-associated antigen (LFA-1) in human lymphocyte subsets. *J. Immunol.* **143,** 3157–3166.
30. Miyauchi, A., Alvarez, J., Greenfield, E. M., et al. (1991) Recognition of osteopontin and related peptides by an alpha v beta 3 integrin stimulates immediate cell signals in osteoclasts. *J. Biol. Chem.* **266,** 20,369–20,374.

31. McNamee, H. P., Ingber, D. E., and Schwartz, M. A. (1993) Adhesion to fibronectin stimulates inositol lipid synthesis and enhances PDGF-induced inositol lipid breakdown. *J. Cell Biol.* **121,** 673–678.
32. Burridge, K., Turner, C. E., and Romer, L. H. (1992) Tyrosine phosphorylation of paxillin and pp125FAK accompanies cell adhesion to extracellular matrix: a role in cytoskeletal assembly. *J. Cell Biol.* **119,** 893–903.
33. Defilippi, P., Bozzo, C., Volpe, G., et al. (1994) Integrin-mediated signal transduction in human endothelial cells: analysis of tyrosine phosphorylation events. *Cell Adhes. Commun.* **2,** 75–86.
34. Guan, J. L., Trevithick, J. E., and Hynes, R. O. (1991) Fibronectin/integrin interaction induces tyrosine phosphorylation of a 120-kDa protein. *Cell Regul.* **2,** 951–964.
35. Kornberg, L., Earp, H. S., Parsons, J. T., et al. (1992) Cell adhesion or integrin clustering increases phosphorylation of a focal adhesion-associated tyrosine kinase. *J. Biol. Chem.* **267,** 23,439–23,442.
36. Virtanen, I., Badley, R. A., Paasivuo, R., et al. (1984) Distinct cytoskeletal domains revealed in sperm cells. *J. Cell Biol.* **99,** 1083–1091.
37. Castellani-Ceresa, L., Mattioli, M., Radaelli, G., et al. (1993) Actin polymerization in boar spermatozoa: fertilization is reduced with use of cytochalasin D. *Mol. Reprod. Dev.* **36,** 203–211.
38. Gadella, B. M., Gadella, T. W., Jr., Colenbrander, B., et al. (1994) Visualization and quantification of glycolipid polarity dynamics in the plasma membrane of the mammalian spermatozoon. *J. Cell Sci.* **107,** 2151–2163.
39. Bleil, J. D. and Wassarman, P. M. (1980) Structure and function of the zona pellucida: identification and characterization of the proteins of the mouse oocyte's zona pellucida. *Dev. Biol.* **76,** 185–202.
40. Bleil, J. D. and Wassarman, P. M. (1983) Sperm–egg interactions in the mouse: sequence of events and induction of the acrosome reaction by a zona pellucida glycoprotein. *Dev. Biol.* **95,** 317–324.
41. Bookbinder, L. H., Cheng, A., and Bleil, J. D. (1995) Tissue- and species-specific expression of sp56, a mouse sperm fertilization protein. *Science* **269,** 86–89.
42. Burks, D. J., Carballada, R., Moore, H. D., et al. (1995) Interaction of a tyrosine kinase from human sperm with the zona pellucida at fertilization. *Science* **269,** 83–86.
43. Primakoff, P., Hyatt, H., and Tredick-Kline, J. (1987) Identification and purification of a sperm surface protein with a potential role in sperm-egg membrane fusion. *J. Cell Biol.* **104,** 141–149.
44. Mazia, D., Schatten, G., and Sale, W. (1975) Adhesion of cells to surfaces coated with polylysine. Applications to electron microscopy. *J. Cell Biol.* **66,** 198–200.
45. Tengowski, M. W. and Schatten, G. (1997) LVSEM of mammalian fertilization in vitro: the preparation of bovine oocytes for low-voltage scanning electron microscopy. *Microsc. Microanal.* **3,** 193–202.
46. Symington, B. E. (1992) Fibronectin receptor modulates cyclin-dependent kinase activity. *J. Biol. Chem.* **267,** 25,744–25,747.

14

Immunocytochemical Staining of the Metaphase II Spindle in Mammalian Oocytes

Steven F. Mullen and John K. Critser

1. Introduction

Investigation into microtubule function in mature mammalian oocytes in relation to cryopreservation is one area of study for the present authors. An increasing number of couples are seeking the assistance of physicians for reproductive dysfunction, and cryopreservation of gametes and embryos plays a role in this type of therapy (http://www.cdc.gov/nccdphp/drh/ART99/index99.htm). To some individuals the cryopreservation of human embryos raises ethical concerns *(1)*. Unfortunately, the success of oocyte cryopreservation is low. The metaphase II (MII) spindle in mature oocytes has been shown to be sensitive to the nonphysiologic conditions imposed upon cells during cryopreservation, and it is believed that the disruption of the MII spindle is a principal cause of the loss of developmental potential in frozen–thawed human oocytes *(2–4)*.

Using fluorescence immunochemistry with tubulin-specific antibodies allows one to visualize microtubular structures in cells. This technique has been applied in an effort to understand normal and abnormal microtubule structure and function in many of the reports cited in this chapter. Although we use this technique exclusively on oocytes, it is applicable to embryos and other cell types. In conjunction with confocal microscopy, this technique allows high-resolution assessment of microtubules, which may provide a means of understanding their function and, in many cases, malfunction.

1.1. Microtubules

Microtubules are one of the three principal components of the cytoskeleton. They can be very static structures, such as those found in cilia or flagella, or

From: *Methods in Molecular Biology, vol. 253: Germ Cell Protocols: Vol. 1 Sperm and Oocyte Analysis*
Edited by: H. Schatten © Humana Press Inc., Totowa, NJ

dynamic structures, such as those found in the cytoplasm. They have several functions, most involved with either movement of the cell or components therein. It is the structure of microtubules that confers their ability to facilitate movement and this ability allows such important processes as protein and organelle translocation and DNA segregation *(5)*.

Microtubules are assemblies of subunits of α- and β-tubulin proteins. These subunits associate via noncovalent interactions to form tubelike structures with a diameter of approx 24 nm. The αβ dimers can associate and disassociate rapidly, causing the growth and shrinkage of microtubules, a process termed "dynamic instability" *(6)*. This dynamic behavior allows microtubules to "search" the three-dimensional space inside of a cell for specific targets. As an example, this phenomenon has been visualized in living newt lung cells as a mechanism to capture kinetochores to the developing mitotic spindle *(7)*.

Movements of proteins, vesicles, and entire organelles within a cell are facilitated by microtubules in conjunction with microtubule motor proteins, of which kinesins and dyneins are two major families *(8)*. Microtubules have an inherent polarity, with both (+) and (–) ends dictated by the orientation of the tubulin dimers. Generally, kinesins are (+)-end-directed motors, moving toward the (+) end of the microtubule, whereas dyneins translocate cargo toward the (–) end. There are, however, exceptions to this rule. Movements that rely upon microtubule motor proteins include the beating of cilia and flagella (dyneins) *(9)*, positioning of the endoplasmic reticulum and Golgi apparatus and movement of vesicles between *(10)*, and the migration of chromosomes during mitosis and meiosis *(11)*. It is this last example that is of particular importance in developing oocytes.

1.2. Microtubules and the Spindle

The separation of chromosomes during mitosis by the mitotic spindle is one of the earliest studied and most fascinating events occurring in cell biology *(12)*. During the M-phase of the cell cycle, a highly choreographed series of events occurs, eventually culminating in cytokinesis *(13)*. Microtubules play an integral part in all of these events, and changes in microtubule dynamics and location occur during this period.

In mammalian cells, prior to the dismantling of the nuclear membrane in preparation for mitosis, the centrosome duplicates, and each daughter centrosome becomes located at opposite sides of the nucleus. Microtubules and kinesin motor proteins facilitate this movement, and along with the microtubule-organizing centers (centrosomes), they develop into the mitotic spindle *(14)*. Unlike G-phase microtubules, M-phase microtubules are restricted in location to the meiotic and mitotic spindles. Not only is the location of these microtubules modified, but their dynamics also change rapidly. The rates of polymerization and depolymerization in spindle microtubules increases markedly compared to their G-phase counterparts *(15)*. The average lifetime of a mitotic microtubule is less than 1 min, compared to approx 10 min for G-phase microtubules. Because of this dynamic nature, the spindle is continuously changing in size and shape during mitosis.

Upon nuclear envelope breakdown, the kinetochore microtubules of the spindle probe the cytoplasm and "search" for the chromosomes *(7)*. It is believed that either the (+) end of the microtubule achieves a direct hit on the kinetochore or it passes the kinetochore and a lateral interaction occurs whereby the microtubule is captured and the chromosome is moved to the (+) end via kinesin motors. Either way, the kinetochore eventually caps the (+) end and stabilizes it, preventing further depolymerization. A microtubule from the opposite pole eventually attaches in a similar manner to the kinetochore of the sister chromatid and the sister pair migrates to the midpoint of the spindle. Once all of the chromatids have become aligned between the spindle poles, the proteins holding the sister chromatids together are cleaved and the chromatids migrate toward the spindle poles. After all of the chromosomes have segregated to the respective spindle poles, cytokinesis commences and two daughter cells are formed that contain exact copies of the genome along with one compacted centrosome. Cytoplasmic microtubules are again nucleated from this centrosome, and the cell cycle begins anew.

1.3. Microtubules During Mammalian Oogenesis and Fertilization

In the mouse, meiosis is initiated as early as embryonic day (E) 12 in the oogonia, transforming the cells into oocytes. Shortly after birth, all of the oocytes have reached the dictyate stage of meiotic prophase, where they will remain arrested until stimulated to resume meiosis at the time of ovulation *(16)*. This development of oocytes is similar in all vertebrate species, and in those with a long reproductive span, such as humans, the oocytes can remain arrested at prophase I for decades.

Oocytes are recruited to resume growth and development during the life of the organism. They proceed through the stages of meiosis during maturation, and the pattern of microtubule activity resembles that described for somatic cells in the M-phase. Investigations of the microtubule-organizing center (MTOC) position and microtubule morphology in oocytes from mouse, bovine, porcine, and human have been reported and show very similar patterns during the stages of oogenesis *(17–20)*.

At the time when oocytes are arrested at the first meiosis, only very short microtubules are present, and they are nucleated from the centrosome in mouse oocytes. Upon breakdown of the germinal vesicle and resumption of meiosis, a dramatic change in the arrangement of the centrosomes occurs, with one remaining

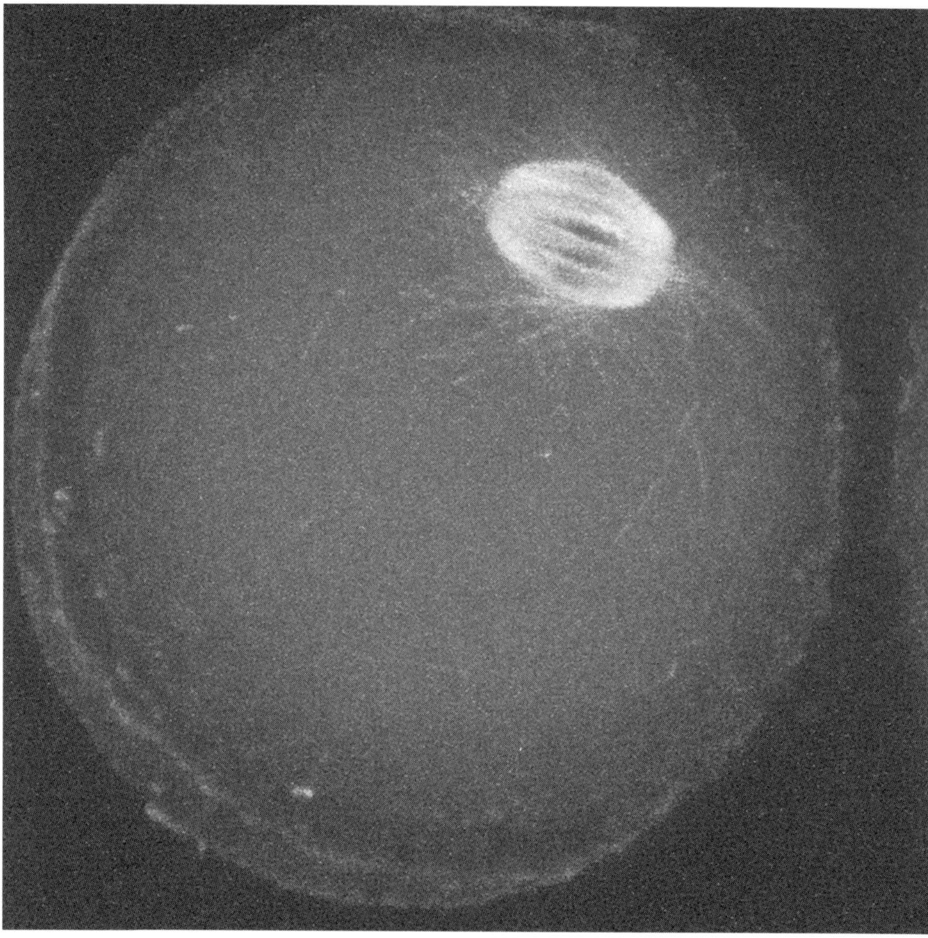

Fig. 1. The microtubule pattern seen in a mature mouse oocyte. Microtubules are located exclusively in the MII spindle at this stage of the cell cycle.

near the chromatin and another migrating into the cytoplasm. In addition, microtubules begin to nucleate around the chromosomes. In many mammalian species, the centrosome lacks the pair of centrioles in the maturing oocyte, the centrioles having been lost during oocyte maturation. It is not until further embryonic development that centrioles reappear ***(21)***. A bipolar spindle forms during this time as the number of microtubules increases. In addition, although remnants of the microtubule organizing centers in the cell periphery still exist, they do not nucleate microtubules. Thus, upon the onset of metaphase I, microtubules are only located within the spindle.

In the anaphase I stage, the pairs of sister chromatids are separated by the microtubule-based spindle and half of the DNA is extruded from the cell in the first polar body. The cell then enters metaphase II (MII) and another spindle complex develops. As in metaphase I, microtubules are only localized in the spindle of MII oocytes (*see* **Fig. 1**). At this stage, the oocyte is arrested for a second time, with activation usually commencing upon interaction with a spermatozoon during fertilization.

Microtubules play a predominant role in the events that occur during fertilization *(22)*. Initially, activation triggers the resumption of meiosis and results in the final meiotic division. The remaining sister chromatids are separated by the spindle and half are extruded into the second polar body. This results in the creation of a haploid set of maternal chromosomes to complement the haploid set of paternal chromosomes introduced by the spermatozoon. The sperm is dismantled during the early stages of fertilization, with the chromatin decondensing and the male pronucleus developing. A pronucleus also develops around the female chromosomes at this time.

Upon activation in mice, cytoplasmic microtubules again form, radiating from the peripheral MTOC, and these microtubules eventually transport the male pronucleus from the cortex to the center of the egg cytoplasm. During this transport, the female pronucleus is also captured by the microtubules and both pronuclei migrate toward the same location. At the time when the pronuclei are in close apposition, a sheath of microtubules forms around each. After pronuclear dissolution and intermingling of the chromosomes from the two pronuclei, a process called syngamy, the chromosomes and microtubules develop into a spindle and the first embryonic cleavage division commences. A similar process occurs in fertilization from other mammalian species with an interesting exception. The focus of microtubule organization is the centrosome that develops from the centriole contributed by the spermatozoon. An MTOC develops from this organelle and nucleates microtubules, which eventually cause the migration of the pronuclei and syngamy as in the mouse.

1.4. Pathologies Associated with Microtubule Dysfunction in the Oocyte

Given the importance of microtubules for gamete interaction and subsequent embryonic development, it is not surprising that microtubule dysfunction in oocytes and zygotes has serious repercussions. Incorrect chromosome segregation during meiosis and improper union of the two haploid sets of chromosomes can lead to developmental failure of the embryo *(23,24)*.

There have been several reports on the correlations between male infertility and sperm centrosome failure. Navara et al. *(25)* investigated the correlation between fertility and centrosome organization and sperm aster morphology in bulls and found a strong correlation between those factors. The bull with the highest fertility demonstrated superior aster formation and morphology in comparison to bulls with lower fertility. In another study investigating 211 presumed failed-to-fertilize human oocytes ***(26)***, many of the oocytes had been penetrated by one or more sperm but had arrested development at very early stages. Although not rigorously tested, it was thought that some causes of this arrest were defects in the sperm centrosome. In a similar study ***(27)***, failed-fertilized and pronuclear-arrested oocytes derived from clinical cases where the males had severe oligoteratozoospermia were examined for microtubule patterns. Abnormal patterns were seen, and the conclusions were that centrosome dysfunction may have contributed to the developmental arrest.

These studies show that male factor incidence of infertility can result from defects in microtubule organization and function, an even more common problem with fertility related to microtubule function in humans results from maternal aneuploidy.

The estimated level of aneuploidy in humans, given that it is nearly always lethal, is quite surprising. Although it is estimated that only 0.3% of newborns are aneuploid ***(28)***, it is estimated that greater than 20% of human conceptions result in aneuploid embryos ***(34,35)***. It is believed that most aneuploidies are maternal in origin, and most of those arise from missegregation during the first meiotic division ***(29)***.

Although the precise mechanism or mechanisms that result in aneuploidy are unclear, a correlation between maternal age and aneuploid frequency has been known for decades ***(30,31)***. The increase in incidence of trisomy in pregnancies in older women is striking, with the numbers of recognized pregnancies being trisomic rising from approx 2% in women 25 yr old to approx 35% in women over 40 yr old. The trend in delaying childbearing in developed countries is well documented ***(32)***, as is the change in microtubule patterns in aged mammalian oocytes ***(17,33)***. A better understanding of the causes of microtubule alterations in aging oocytes could assist in maintaining the fertility of those women who desire to have children yet choose to delay childbearing.

2. Materials

1. 35-mm Petri dishes.
2. Oocyte transfer pipets.
3. Methanol with 5 m*M* EGTA, prechilled in a –20°C freezer.
4. Thick glass slides with a concave well (e.g., Fisher cat. no. 12-565A).
5. Immunochemistry blocking buffer: Phosphate-buffered saline (PBS), 0.2□□% (v/v) Triton X-100, 3 m*M* sodium azide, 0.1% saponin (mass/v), 130 m*M* glycine, 2% bovine serum albumin (BSA) (mass/v), and 5% horse or donkey serum (v/v) (*see* **Note 1**).

6. Microscope slides.
7. Microscope slide coverslips #1.5 (*see* **Note 2**).
8. Antitubulin primary antibody, preferably clone DM1-A (*see* **Note 3**).
9. Secondary antibody labeled with fluorophore of choice. We use a donkey–anti-mouse antibody with a Texas Red® conjugate (Jackson Immunoresearch) (*see* **Note 4**).
10. Sytox Green DNA stain (Molecular Probes, cat. no. S7020) (*see* **Note 5**).
11. ProLong antifade reagent (Molecular Probes, cat. no. P7418) (*see* **Note 6**).
12. Waxy material for mounting coverslips to slides, to prevent the coverslips from squishing the cells upon mounting: 50% petroleum jelly and 50% paraffin wax (*see* **Note 7**).

3. Methods

1. All treatments described here are performed at room temperature unless otherwise noted. However, because microtubules are cold labile, ensure that the cells are maintained at physiologic temperature prior to fixation.
2. Fix the cells by transferring them into cold methanol (–20°C) for 6 min; transfer as little liquid with the cells as possible (*see* **Notes 8** and **9**).
3. Evacuate the transfer pipet of all solution (*see* **Note 9**).
4. After fixing the cells, transfer them to approx 2 mL of blocking buffer in a 35-mm Petri dish and hold them at 4°C (*see* **Note 10**).
5. Dilute the primary antibody in blocking buffer at a ratio of 1 μL of antibody solution to 9 mL of blocking buffer. Keep the solution on ice until you are ready to use it. Mix it thoroughly by carefully inverting the tube at least three times, and transfer the cells to a 35-mm dish containing 1.5–2 mL of the diluted primary antibody.
6. Maintain the cells in this solution for at least 1 h at room temperature. If desired, the cells may be held in this solution overnight at 4°C (*see* **Note 11**).
7. Transfer the cells to a 35-mm dish containing 1.5–2 mL of blocking buffer as a wash. Transfer the cells with as little volume as necessary. They should be held in this solution for at least 1 h. They may be washed for a few hours or overnight if desired (*see* **Note 12**).
8. Dilute 5 μL of the secondary antibody in 7.5 mL of the blocking buffer (*see* **Note 4**).
9. Transfer the cells to a 35-mm dish containing 1.5–2 mL of the secondary antibody diluted as described in **step 7**. Hold the cells in this solution for at least 1 h or overnight at 4°C (*see* **Note 11**).
10. Transfer the cells to a 35-mm dish containing 1.5–2 mL of blocking buffer as a wash. They should be held in this solution for at least 1 h (*see* **Note 13**).
11. Get one each of the two types of tubes from the ProLong antifade reagent box (*see* **Note 6**). The liquid in the clear tube needs to be thawed. This can be done at room temperature or in a water bath. *Carefully* (see below) transfer 200 μL of the clear solution into the amber tube. There is a crystalline substance in the bottom of the amber tube. Upon transfer, use the pipet tip to suspend the crystals in the liquid (with the action of a mortar and pestle). Once the crystals are suspended,

let the solution sit for at least 15 min to allow the crystals to dissolve. If DNA counterstaining is desired, to the remainder of the liquid in the clear tube, add 0.25 µL of the Sytox Green stock solution (*see* **Note 5**). After 15 min, draw the solution in the amber tube into a clear pipet tip to make sure that all of the crystals have dissolved. Return the 200 µL of the solution from the amber tube to the clear tube. Mix thoroughly but *carefully*! This solution is very viscous and foams easily. If excessive foaming occurs, centrifuge the tube at maximum rpm in a benchtop microcentrifuge. Keep the tubes capped as often as possible.

12. Draw up 20 µL of antifade reagent into a pipet and set the pipet down on the bench next to you. Using a dissecting microscope, transfer the cells to a microscope slide with as little liquid as possible. Suck up as much of the liquid as you can without removing the cells. Grab the pipet containing the antifade reagent and watch the remaining solution on the slide evaporate. When it has almost completely evaporated (this will facilitate the cells sticking to the glass), quickly but carefully add the antifade reagent to the cells. Adding more just makes a mess; adding less will create air bubbles in the solution as it dries. Once the cells are covered with the drop, add a small amount of the waxy substance to the corners of a coverslip and carefully place the coverslip onto the slide, waxy side down, avoiding the creation of air bubbles in the antifade reagent. Carefully press down on the corners of the coverslip with a pipet tip just until you see the oocytes being squished. The coverslip will rebound slightly and the distance between the coverslip and oocytes will be minimal (*see* **Note 14**).
13. Give the slides at least 1 h to dry prior to looking at them under the fluorescence microscope. Although it may not be necessary, we keep the slides under aluminum foil during the drying process. Seal the coverslips with clear nail polish after the drying time. We recommend imaging the cells within 2 d to maximize the signal-to-noise ratio.

4. Notes

1. Serum from a species closely related to the host of the secondary antibody is important during the blocking step. We use horse serum because we are using a donkey antibody. Donkey serum is available from Jackson Immunoresearch, and either should work well for this application. The blocking buffer we use was derived empirically. Because of the presence of sodium azide, this material is considered toxic and should be handled and disposed of appropriately. This buffer should be stored at 4°C. Periodic filtration through a 0.45-µm filter can be performed if precipitates form over time. We make 1-L volumes if we will be using all of it within approx 30 d. If longer storage is expected, we make up a 1-L volume of the solution without the BSA and serum and then make a 200-mL volume of working solution by adding BSA and serum to 200 mL of the stock solution.
2. Most high-power microscope objectives are designed to work with microscope coverslips with a thickness of 0.17 mm, which is equivalent to #1.5. Check the objectives on your microscope to verify this before mounting the cells and use appropriate coverslips. Although this sounds trivial, it is important for

obtaining good images. We use 18-mm square coverslips. If larger coverslips are used, use a proportionately larger volume of antifade reagent when mounting the cells.

3. We use a primary antibody from Sigma (St. Louis, MO, USA; cat. no. T9026) without further purification. We also aliquot our stock solution in 5-µL volumes and store these in a –80°C freezer. Monoclonal antibody clone DM1-A has proven to be an excellent antibody for general use in labeling tubulin structures. The antibody dilutions discussed in this chapter are only relevant to the antibody sources we cite. Other sources of antibodies may work just as well, but the dilutions necessary to achieve a good signal-to-noise ratio will differ and will need to be determined empirically. If a different antibody is used, ensure that a secondary antibody is chosen that will bind to it. (In other words, if the primary antibody is not from a mouse, do not choose a secondary antibody that binds to a mouse antibody.)
4. The secondary antibody comes in the form of a lyophilized powder and we use the manufacturer's recommendations for resuspension. Add 400 µL of quality, room-temperature water (DI or Milli-Q) to the powder and gently mix by pipetting. After about 10 min, draw the solution into a clear pipet to ensure complete dissolution. We either maintain this solution in a refrigerator if we expect it to be used within 2 mo or we freeze aliquots diluted 50% in glycerol. To do this, add 400 µL of glycerol to the stock solution and aliquot 10-µL volumes of this dilution into 0.5-mL tubes and store the aliquots at –80°C. To use this solution, thaw one tube and add the entire contents to 7.5 mL of blocking buffer. Add 1.5–2 mL of this solution to a 35-mm Petri dish for the final working solution. We use glycerol with these antibodies for long-term storage but not for the primary antibody, only because that is what the manufacturer recommends. We believe that either way would work for both antibodies, but assume that the manufacturers have recommended the methods that work best for their products.
5. Sytox Green from Molecular Probes is a superb DNA fluorophore. We aliquot the original product in 1-µL volumes and dilute 0.25 µL of this in 1 mL of the antifade reagent. We recommend its use if it can be obtained.
6. The Prolong antifade reagent has proven to be an exceptional antifade reagent. We have received positive comments from multiple core facilities on its photostable properties. The kit comes with multiple amber tubes, each containing a crystalline precipitate, and two bottles of a transparent solution. We thaw these bottles and aliquot this solution in 1-mL volumes into transparent 1.5-mL microcentrifuge tubes. In this way, a tube can be thawed for each procedure instead of having to repeatedly thaw and freeze the entire solution in the bottle. We use transparent tubes because it is easier to tell when the solution has been thoroughly mixed. The final solution is very viscous and foams easily, so handle it carefully. Be sure to mix the solution thoroughly prior to its use. We recommend using this reagent if it can be obtained.
7. The "waxy material" is a homemade concoction of half petroleum jelly ("Vaseline") and half paraffin wax. To make the waxy material, take a 500-mL beaker and put

it on a hot plate. Add approximately an equal volume of paraffin wax (judge by the size of the blocks) and petroleum jelly. The wax and jelly can be purchased at any large grocery store. When the wax has fully melted, stir the two together to mix well and draw the solution into 30-mL syringes (without needles). The material will solidify as it cools. Like many things in life, the precise ratio of the two materials is not critical. As long as it is solid but flows when pressure is put on the syringe plunger, it will be fine.

8. We find that methanol fixation provides a quick, safe, and easy alternative to fixation in formalin. In addition, it seems to provide better access to intracellular antigens during immunocytochemistry. If a few oocytes are being fixed, we recommend using thick glass slides with a concave well. Adding about four drops of methanol to the well and transferring as little medium with the oocytes as possible has the best outcome. The oocytes are fragile while in methanol. Therefore, be gentle with them during fixation. Have the slides and methanol equilibrated in a freezer prior to fixation, and return the slides with the fixative and oocytes to the freezer during fixation. We keep the slides in a 60-mm Petri dish during this procedure. If more oocytes are being fixed, 35-mm dishes containing methanol may be used. However, the oocytes tend to move around in the methanol during fixation, making it more challenging to find and remove them after the procedure. It can be done with practice, however.
9. Because methanol precipitates proteins, if methanol and protein-containing medium are mixed in the transfer pipet, it can become clogged with protein precipitate, possibly leading to the loss of the oocytes. We find it essential to evacuate the liquid out of the pipet before going into the methanol to retrieve the fixed cells. Simply evacuate the pipet before moving into the methanol, then allow the pipet to draw up methanol prior to picking up the cells. Keep the oocytes near the end of the pipet during transfer from one medium to another. If the oocytes are too far up into the pipet when fixed, they can rupture during transfer into the blocking buffer.
10. Blocking buffer is used to lower the nonspecific binding of antibodies in the cells. Holding the cells in this buffer will prevent this and reduce the background staining of the cells. We do not keep the dishes in a humidified chamber. Instead, we check on the volume of the liquid and add additional medium if necessary to prevent all of the medium from evaporating. The holding time can range from days to weeks, if desired. Just be sure to check on the volume of the liquid to avoid complete evaporation. We have found that staining mouse oocytes with this procedure produces high levels of background staining when compared to oocytes from other species. However, this can be reduced by holding the cells in the blocking buffer for a few days. For oocytes from other species, a few hours of blocking prior to incubation with the antibodies usually will suffice.
11. The times in which the cells are held in the antibody solutions have been derived empirically. We have found that overnight incubations maximize the signal and ensure full saturation of the cellular antigen. Shorter incubation times can be performed if time is of the essence, but we recommend at least 1 h in all of the solutions (antibody and wash).

12. The precise time in the wash buffer is not critical, up to a point. At least 1 h is recommended at room temperature. We have washed them for as long as overnight and found that the signal is still strong. Longer washes have shown to reduce the signal, thus we recommend washes no longer than overnight. We recommend the washes to be performed at room temperature, even if going overnight. This is especially true for mouse oocytes.
13. The time for the final wash can be as short as 1 h. If control cells are used in the experiment (which we strongly recommend), they can be viewed under the microscope briefly to assess the level of background staining. If it is low after 1 h, then the cells can be mounted on slides. We usually place the cells into the final wash buffer in the morning and then mount the cells in the afternoon. The signal should remain strong at least until the next morning. When the cells have been labeled with the secondary antibody, keep them protected from light until mounting. We cover the dishes with aluminum foil during the wash step.
14. This step requires some practice. It may be frustrating at first, but with experience, it will become routine. Performing it properly, however, will result in high-quality imaging and less frustration in the long run. Because a small number of cells are usually being observed, it helps if they are all together on the slide. Thus, once the coverslip has been applied, allow the antifade solution to slowly spread out across the area under the coverslip. If the movement is too fast, the cells will detach from the glass and move with the flow. It is not a problem for imaging the cells, just more frustrating to find the cells during imaging. The cells do not always stick to the glass, so do not be surprised if they move. We have found that within 2 d, the fluorescence signal begins to diminish, especially if the coverslips are not sealed with nail polish. Therefore, we recommend imaging the cells within 2 d (1 d preferably). Although we do not store the slides in a freezer, reports in the literature suggest this to prevent loss of the signal. We recommend imaging with fluorescence microscopy as soon as possible. Sealing the edges of the coverslip with clear nail polish is recommended, but not necessary if the slides are handled carefully at this point. Give the samples at least 1 h to dry before imaging. Overnight drying time is recommended. Be careful not to apply pressure to the coverslip before complete drying, which might cause the coverslip to move relative to the slide and destroy the sample.

References

1. Ethics Committee of the American Fertility Society (1994) Ethical considerations of assisted reproductive technologies. *Fertil. Steril.* **62,** 1S–125S.
2. Magistrini, M. and Szollosi, D. (1980) Effects of cold and of isopropyl-*N*-phenylcarbamate on the second meiotic spindle of mouse oocytes. *Eur. J. Cell Biol.* **22,** 699–707.
3. Johnson, M. H. and Pickering, S. J. (1987) The effect of dimethylsulphoxide on the microtubular system of the mouse oocyte. *Development* **100,** 313–324.
4. Critser, J. K., Agca, Y., and Gunasena, K. T. (1997) in *Reproductive Tissue Banking, Scientific Principles* (Karow, A. M. and Critser, J. K., eds.), Academic, San Diego, CA, pp. 329–357.

5. Hyams, J. S. and Lloyd, C. W. (1994) *Microtubules*, Wiley–Liss, New York.
6. Mitchison, T. and Kirschner, M. (1984) Dynamic instability of microtubule growth. *Nature* **312,** 237–242.
7. Hayden, J. H., Bowser, S. S., and Rieder, C. L. (1990) Kinetochores capture astral microtubules during chromosome attachment to the mitotic spindle: direct visualization in live newt lung cells. *J. Cell Biol.* **111,** 1039–1045.
8. Gelfand, V. I. and Bershadsky, A. D. (1991) *Annu. Rev. Cell Biol.* **7,** 93–116.
9. Summers, K. (1974) ATP-induced sliding of microtubules in bull sperm flagella. *J. Cell Biol.* **60,** 321–324.
10. Kelly, R. B. (1990) Microtubules, membrane traffic, and cell organization. *Cell* **61,** 5–7.
11. Wittmann, T., Hyman, A., and Desai, A. (2001) The spindle: a dynamic assembly of microtubules and motors. *Nature Cell Biol.* **3,** E28–E34.
12. Flemming, W. (1879) Beiträge zur Kenntnisse der Zelle und ihrer Lebenserscheinungen. *Archiv. Mikrosk. Anat.* **18,** 151–259.
13. Nurse, P. (2000) A long twentieth century of the cell cycle and beyond. *Cell* **100,** 71–78.
14. Karsenti, E. and Vernos, I. (2001) The mitotic spindle: a self-made machine. *Science* **294,** 543–547.
15. Saxton, W. M., Stemple, D. L., Leslie, R. J., et al. (1984) Tubulin dynamics in cultured mammalian cells. *J. Cell. Biol.* **99,** 2175–2186.
16. Wassarman, P. M. and Albertini, D. F. (1994) The mammalian ovum, in *The Physiology of Reproduction*, Raven, New York, pp. 79–122.
17. Kim, N. H., Moon, S. J., Prather, R. S., et al. (1996) Cytoskeletal alteration in aged porcine oocytes and parthenogenesis. *Mol. Reprod. Dev.* **43,** 513–518.
18. Messinger, S. M. and Albertini, D. F. (1991) Centrosome and microtubule dynamics during meiotic progression in the mouse oocyte. *J. Cell Sci.* **100,** 289–298.
19. Kim, N. H., Chung, H. M., Cha, K. Y., et al. (1998) Microtubule and microfilament organization in maturing human oocytes. *Hum. Reprod.* **13,** 2217–2222.
20. Kim, N. H., Cho, S. K., Choi, S. H., et al. (2000) The distribution and requirements of microtubules and microfilaments in bovine oocytes during in vitro maturation. *Zygote* **8,** 25–32.
21. Szollosi, D. (1972) in *Oogenesis* (Biggers, J. D. and Schuetz, A. W., eds.), University Park Press, Baltimore, MD, pp. 47–64.
22. Schatten, G. and Schatten, H. (1987) Cytoskeletal alterations and nuclear architectural changes during mammalian fertilization. *Curr. Topics Dev. Biol.* **23,** 23–54.
23. Nasmyth, K. (2001) Disseminating the genome: joining, resolving, and separating sister chromatids during mitosis and meiosis. *Annu. Rev. Genet.* **35,** 673–745.
24. Hewitson, L., Simerly, C., and Schatten, G. (1997) Inheritance defects of the sperm centrosome in humans and its possible role in male infertility. *Int. J. Androl.* **20,** 35–43.
25. Navara, C. S., First, N. L., and Schatten, G. (1996) Phenotypic variations among paternal centrosomes expressed within the zygote as disparate microtubule lengths

and sperm aster organization: correlations between centrosome activity and developmental success. *Proc. Natl. Acad. Sci. USA* **93,** 5384–5388.

26. Asch, R., Simerly, C., Ord, T., et al. (1995) The stages at which human fertilization arrests: microtubule and chromosome configurations in inseminated oocytes which failed to complete fertilization and development in humans. *Hum. Reprod.* **10,** 1897–1906.
27. Van Blerkom, J. (1996) Sperm centrosome dysfunction: a possible new class of male factor infertility in the human. *Mol. Hum. Reprod.* **2,** 349–354.
28. Hassold, T., Abruzzo, M., Adkins, K., et al. (1996) Human aneuploidy: incidence, origin, and etiology. *Environ. Mol. Mutagen* **28,** 167–175.
29. Hassold, T. and Hunt, P. (2001) To err (meiotically) is human: the genesis of human aneuploidy. *Natl. Rev. Genet.* **2,** 280–291.
30. Penrose, L. (1933) The relative effects of paternal and maternal age on mongolism. *J. Genet.* **27,** 219–224.
31. Morton, N. E., Jacobs, P. A., Hassold, T., et al. (1988) Maternal age in trisomy. *Ann. Hum. Genet.* **52,** 227–235.
32. Anon. (1989) Postponed childbearing—United States, 1970–1987. *MMWR* **38,** 810–816.
33. Battaglia, D. E., Goodwin, P., Klein, N. A., et al. (1996) Influence of maternal age on meiotic spindle assembly in oocytes from naturally cycling women. *Hum. Reprod.* **11,** 2217–2222.
34. Jamieson, M. E., Coutts, J. R., and Connor, J. M. (1994) The chromosome constitution of human preimplantation embryos fertilized in vitro. *Hum. Reprod.* **9(4),** 709–715.
35. Volarcik, K., Sheean, L., Goldfarb, J., et al. (1998) The meiotic competence of in-vitro matured human oocytes is influenced by donor age: evidence that folliculogenesis is compromised in the reproductively aged ovary. *Hum. Reprod.* **13(1),** 154–160.

15

Spindle Imaging in Living Mammalian Oocytes Using the Liquid-Crystal Polarized Light Microscope and Its Practical Use

Wei-Hua Wang, Jimmy Gill, and Cathy Boutin

1. Introduction

Ovulated oocytes in most mammals are arrested in metaphase II, at which point a dense array of microtubule filaments and bundles form the meiotic spindle. The spindle is vital to the alignment and separation of the chromosomes along the metaphase plate so that a proper division of the chromosome pair can occur during meiosis. Abnormalities in its integrity may be related to failed or abnormal fertilization. The architecture of the spindle's filamentous network is dynamic and can be used to understand the changing biology of the oocyte in vitro. It is also a potentially fragile structure, and, as a practical matter, techniques for visualization of the spindle can prove effective in studying dynamics of the spindle. Most conventional high-resolution methods for observing the spindle, such as immunocytochemistry and transmission electron microscopy, rely on fixation and staining procedures, so they are of limited value in clinical application and in studies of spindle dynamics.

Polarization light microscopy is a technique that relies on a material's optical properties, the birefringence, to examine microstructure *(1)*. It can be best applied to materials that are anisotropic and have two indices of refraction. Many biological materials that have a well-aligned molecular structure exhibit birefringence *(2)*. Because spindles are composed of microtubules that exhibit birefringence and most other cellular organelles and their cytoplasm do not exhibit apparent levels of birefringence *(3,4)*, the spindle in the oocytes can be detected using polarization microscopy. Polarized light microscopy employs two polarizers, the first is positioned in the light path somewhere

From: *Methods in Molecular Biology, vol. 253: Germ Cell Protocols: Vol. 1 Sperm and Oocyte Analysis*
Edited by: H. Schatten © Humana Press Inc., Totowa, NJ

before the specimen, and the second is placed in the optical pathway between the objective rear aperture and the observation tubes or camera port *(1)*. The conventional polarization microscope uses plain polarized light, and the operator must rotate the microscope stage and/or the linear analyzer in order to visualize birefringent structures of interest. As the analyzer is rotated, the structures with the property of birefringence can be detected when the molecules of that structure align with the angle of polarization of the microscope. This type of linear polarization microscopy is limited by orientation dependence *(4)*. Visualization is also limited by the range of sensitivity of the setup. Using this technology, Inoue *(5)* was able to examine mitotic spindles in living cells and, subsequently, was able to study their dynamic architecture. However, this technology has not been widely applied to mammalian embryology because of its orientation dependence and limited sensitivity in visualizing spindle structure during specimen analysis.

The liquid-crystal (LC) polarization microscope (LC-PolScope™) augments the conventional polarization microscope by integrating liquid-crystal variable retarders, electronic imaging, and digital imaging processing tools to build a highly sensitive, orientation-independent imaging system *(1,4)*. When circularly polarized light passes through the specimen, it is analyzed by a liquid-crystal variable compensator at four precalibrated polarization angles and subsequently transferred to the camera. The data from the four acquired images minus a background stack are subsequently combined by the software using specific algorithms to generate a final processed image displaying specimen retardance at each point in the images. The result of the computation is an image representing the absolute retardance magnitude and orientation measured at every pixel. Retardance is represented by gray-scale intensities that correlate directly to the level of birefringence of the sample, such as spindles. The LC-PolScope imaging system is truly orientation independent and has a sensitivity of 0.02 nm retardance. In contrast to fluorescence microscopy, LC-PolScope imaging does not require invasive preparative techniques such as fixation and staining; thus, the fine structure of the spindle can be imaged in living mammalian oocytes *(6–8)*. A user-friendly version of the LC-PolScope, called the SpindleView™ Imaging System, has been developed by Cambridge Research and Instrument Inc. (CRI. Inc.), Woburn, MA, USA; www.cri-inc.com). The system contains a set of integrated microscope peripherals including LC-based polarizing contrast optics, electronics, camera, and software. In this chapter, we will explore the setup of this imaging system, step-by-step system operation, and its current and future applications. As this is a new and promising technology, many aspects of the chapter will hinge on the possible roles that this system can play to expand the utility of the standard microscope.

2. Materials

2.1. Hardware (see *Note 1*)

1. Inverted microscope, such as Leica DM-IRB, Nikon Diaphot and TE series, Zeiss Axiovert 200, and Olympus IX series. Note: The microscope is not included with the purchase of the SpindleView System.
2. Green interference filter (*see* **Fig. 1**, A). This 546-nm green bandpass filter is a requisite of the LC polarizing compensator optics. Green light also helps to preserve the viability of the sample by reducing potentially harmful blue-UV (ultraviolet) and thermal-IR (infrared) wavelengths. This filter is inserted between the light source and the specimen, with exact positioning depending on the microscope brand and model.
3. Circular polarizer (*see* **Fig. 1**, B). This optical module is inserted between the light source and the specimen, with exact positioning depending on the microscope brand and model.
4. LC-based polarizing compensator (*see* **Fig. 1**, C). The heart of the SpindleView optical system, this LC-controlled component changes the polarization states with no moving parts, no noise and no image shift. It is connected to the SpindleView electronics box by a cable (*see* **Fig. 1**, D).
5. Optical coupler (*see* **Fig. 1**, G). The 0.45× (or 0.5×) relay coupler is placed between the camera port of the microscope and the CCD camera. This magnifies the amount of light hitting the CCD chip, thereby reducing image capture time. Exact magnification may differ based on microscope model.
6. A CCD camera (*see* **Fig. 1**, F), computer (*see* **Fig. 1**, M), and monitor. These components, along with the preinstalled frame grabber and software, permit digital image capture and control the electrooptics. Digital images simplify observation and manipulation of the oocytes and foster identification techniques, archiving, retrieving, and exporting practices.
7. SpindleView electronics box (*see* **Fig. 1**, H). This controller box controls the LC compensator optics.
8. Cables from camera to image acquisition board and from LC compensator optics to SpindleView controller box. Serial communication cable and nine-pin D-sub COM port adapter. Cable to power transform for SpindleView electronics box (*see* **Fig. 1**, I, J, K, and L).

2.2. Software

SpindleView software. CRI has integrated patented imaging software algorithms with an easy-to-follow user interface.

2.3. Accessories

1. Glass-bottomed Petri dishes (*see* **Note 2**).
2. HEPES-buffered oocyte manipulation medium.
3. Tissue-culture-grade oil.
4. Oocytes.
5. Heating stage for temperature control during imaging.

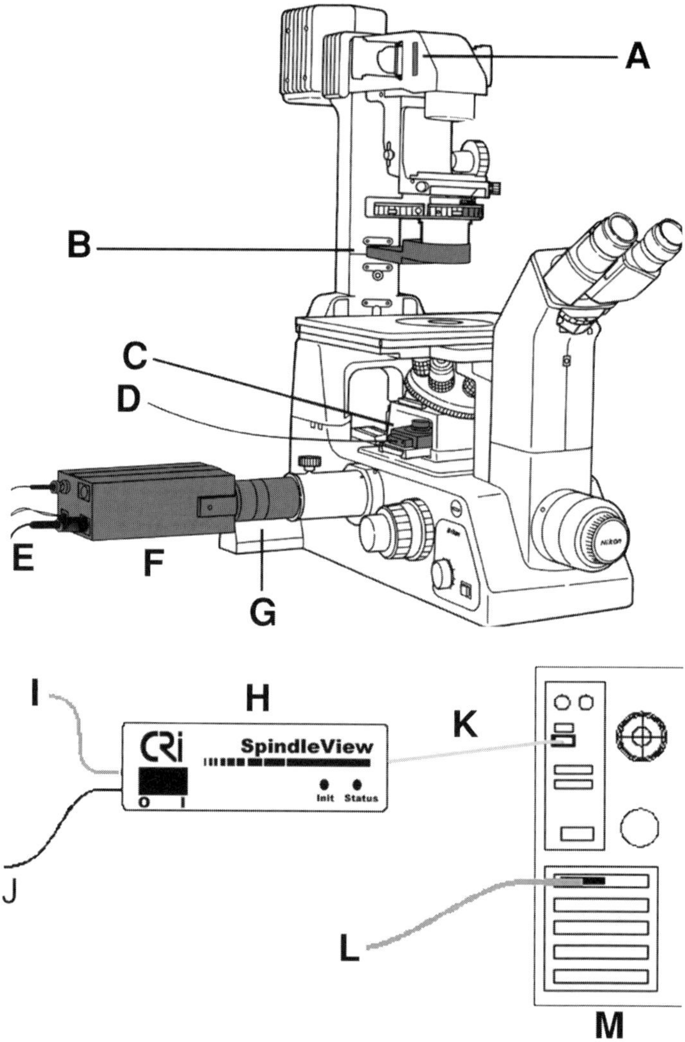

Fig. 1. The SpindleView operational schematic for imaging on a microscope. The SpindleView system is shown set up on a Nikon TE 300 inverted microscope. (A) A 546-nm interference filter in supplementary lens pocket; (B) circular polarizer in CRI Hoffman adapter; (C) LC polarizing compensator optics; (D) cable from LC polarizing compensator optics to SpindleView electronics box; (E) cable from camera to image acquisition board and serial communication cable and nine-pin D-sub COM port adapter; (F) charge-coupled device (CCD) camera; (G) 0.45× relay lens; (H) SpindleView control box; (I) cable from SpindleView electronics box to LC polarizing compensator optics; (J) cable to power transfer; (K) serial communications cable and COM port adapter; (L) cable from camera to image acquisition board (inside computer); (M) computer.

3. Methods

3.1. Set Up Hardware (Fig. 1)

1. Load PCI-bus image acquisition board to the computer according to manufacturer's instruction. It is better to have the manufacturer preinstall the PCI-bus image acquisition board prior to shipping. The board is specifically configured by CRI and any adjustments affect the system's performance.
2. Mount the CCD camera: Mate the female C-mount of the camera to the male C-mount of the relay lens. Then, insert the other side of the relay lens into the camera port of the microscope. The camera should be placed so that the sample as seen in the CameraView on the software screen, and that seen through the eyepieces of the microscope is oriented in the same direction.
3. Connect the video cables: Connect the 25-pin male D-sub connector on the image acquisition board on the back of the computer. At the opposite end of the video cable, connect the circular chrome BNC jack to the CCD camera's connector marked VIDEO. Then, connect the round eight-pin DIN connector marked AUX to the camera. Finally, connect the two stripped and tinned power leads on the cable to the 12-V ac/dc power input terminal on the back of the camera. Tighten the set screws on the camera to securely hold these power leads.
4. Install the LC polarizing compensator optics module beneath the objective lens and connect the cable from LC polarizing compensator optics to the SpindleView electronics box. Exact positioning may differ depending on microscope model.
5. Install the circular polarizer/interference filter unit:
    a. For Hoffman-based systems, install the circular polarizer/interference filter assembly onto the Hoffman condenser. Slide the adaptor ring, which contains a circular polarizer and the interference filter in a sliding frame, up from below the microscope's Hoffman HMC condenser lens. The slide mechanism should face the *rear* of the microscope. Secure the adapter to the lower part of the Hoffman HMC condenser lens by tighting the set screws on the back of the adapter ring. Be sure that when slid into place, the filter is flush with the base of the condenser.
    b. For a system without Hoffman optics, place the circular polarizer/interference filter assembly into the aperture above the condenser as noted in the installation manual. Exact positioning will depend on microscope model.
6. Connect the LC polarizing compensator cable to the 15-pin connector on the back of the SpindleView electronics box.
7. Connect the SpindleView electronics box to the host computer by attaching the RI-12 serial communication cable and nine-pin D-sub adapter to the serial COM1 port on the back of the computer.
8. Connect the SpindleView electronics box power transformer to the wall's electrical outlet.
9. Turn on the computer and then turn on the SpindleView electronics box.

3.2. Set Up the SpindleView System

1. Installation of the SpindleView Software: Systems come equipped with the drivers and software preloaded, but should you need to reinstall, follow the manufacturer's instruction.
2. Select Start/Settings/Control Panel/Display/Settings on the computer's desktop. Adjust Windows display settings to 1280 by 1024 pixels, 24-bit true color, with small fonts.
3. Launch the SpindleView software by double-clicking the SpindleView icon on the Windows desktop or selecting the SpindleView software that is called out in Start/Programs/CRI.

3.3. Spindle Imaging in Living Oocytes with SpindleView (see *Fig. 2*)

1. Turn on all power to the inverted microscope, computer, SpindleView electronics box, and heating plate.
2. Start the SpindleView software.
3. Prepare the microscope for SpindleView operation. Direct majority of the light to the camera.
4. Make droplets (5 µL each) in the glass-bottomed Petri dish; cover the droplets with tissue-culture-grade oil.
5. Transfer the control dish to the heating stage (*see* **Note 3**).
6. Select the microscope objective and focus on the edge of the media droplet.
7. Be sure that the SpindleView optics are in the light path and that any interfering optics are removed.
8. Reposition the sample so that the imaging plane is focused in the center of the media droplet.
9. Follow the instructions on the screen to quickly and effectively achieve Koehler illumination and then click the "OK" icon when done.
10. Perform automated calibration according to the step-by-step guide in the right of the screen and then click the "Done" icon.
11. Prepare sample dish by making 5-µL droplets in a new glass-bottomed Petri dish; cover the droplets with tissue-culture-grade oil.
12. Place one oocyte without cumulus cells in each droplet.
13. Transfer the dish containing the oocyte to the heating stage.
14. Position the dish so that no eggs or debris are in the central region of the CameraView window (*see* **Fig. 2**, A) in the upper right of the screen.
15. Take a background image by clicking on the Take Background (BK) icon (*see* **Fig. 2**, B).
16. Move the oocyte into the CameraView in the upper right-hand corner of the software screen.
17. Engage the eye cycle mode to locate the spindle (*see* **Fig. 2**, C). Spindle appears to flicker or flash when present (*see* **Notes 4** and **5**).
18. Take an image of the sample by clicking the SpindleView (SV) icon (*see* **Fig. 2**, D). The image will appear in the processed image window (*see* **Fig. 1**, M).
19. Take additional images with "auto-expose" off (*see* **Fig. 2**, E).

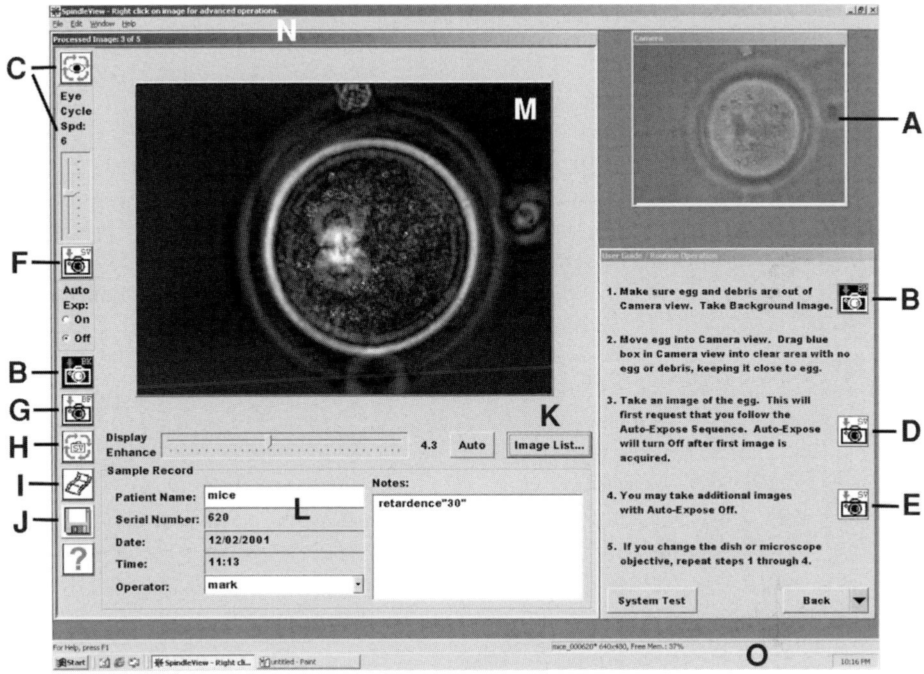

Fig. 2. A full-screen view of the SpindleView software. (A) Camera View (Live Video) window; (B and B') take Background icon; (C) Eye Cycle icon for viewing the spindle through the eyepieces, and Eye Cycle Speed Setting. (D and D') take SpindleView Image icon to take single processed image; (E) take image icon with auto-expose off; (F) display Enhance Control; (G) take Brightfield Image icon; (H) SpindleView Cycle icon; (I) take Timelapse Image icon; (J) save Image icon; (K) Image List icon; (L) Sample Record window; (M) Active Image window; (N) menu bar; (O) status bar.

20. Move dish to take image of other oocytes by clicking the SpindleView image icon (*see* **Fig. 2**, D').
21. Check image list by clicking "image list" (*see* **Fig. 2**, K) after imaging all sample.
22. Record data in Sample Record window (*see* **Fig. 2**, L).
23. Save image and data by clicking Save Image icon (*see* **Fig. 2**, J).
24. Perform any post-data processing by right-clicking on the processed images.

3.4. Time-Lapse Spindle Imaging in Living Oocytes with SpindleView (see Note 6)

1. Do **steps 1–17** in **Subheading 3.2.**
2. Click a time-lapse icon in the left side of the screen (*see* **Fig. 2**, I) and set up the time intervals for taking images (e.g., 10-min intervals) and number of images to be taken (e.g., 50).

3. Click "OK" and the images will be captured—for instance, every 10 min with termination after 50 images have been taken.
4. Save data and images by exporting as an animated GIF file (*see* **Fig. 2**, J).

4. Notes

1. Before purchasing the SpindleView system, inform the manufacturer of the specific make and model of the inverted microscope you intend to use, as the optical housings and subsequent assembly differ based on the microscope.
2. Petri dishes for imaging the spindle must be glass-bottomed, as plastic is highly birefringent.
3. It is highly recommended that the temperature of the media be tested for temperature accuracy and stability, as the meiotic spindle in most mammalian species has been shown to be highly temperature sensitive.
4. Not all oocytes exhibit spindles. This may depend on the stages of oocytes, the manipulating temperature, and/or various other factors ***(8–10)***. Approximately 20% of human oocytes that have been observed in patients consulting for infertility treatment do not exhibit a visible spindle ***(9)***. A typical and normal meiotic spindle image in a human metaphase II oocyte is shown in **Fig. 3**.
5. Rotating the oocytes with a micropipet during imaging may be necessary if you do not initially observe a spindle.
6. Practical use of imaging spindles in living oocytes with SpindleView include the following applications:
    a. Intracytoplasmic sperm injection (ICSI) by monitoring spindle position. Normal offspring have been reported in many mammals by ICSI procedure, including human beings. Traditional ICSI method dictates that the oocyte be secured to the holding pipet with the first polar body at the 6 or 12 o'clock position, so that when sperm is injected at the 3 o'clock position, the metaphase II spindle is avoided ***(11)***. However, it has been found that the first polar body is not an accurate predictor for the spindle location in either fixed ***(12)*** or live ***(6,8)*** metaphase II oocytes. Therefore, it is believed that the injection needle may damage the spindles in some oocytes during ICSI. With the aid of the SpindleView system, one can easily and assuredly avoid the spindle during ICSI. When the spindle is localized with the system, proceed to rotate oocytes to place the spindles at the 6 or 12 o'clock position and then inject sperm at the 3 o'clock position.
    b. Study spindle dynamics during oocyte maturation and activation. With the SpindleView system, the dynamics of the spindle can be analyzed noninvasively and quantitatively by measuring, in real time, the spatial patterns and temporal changes of spindle birefringence and orientation ***(10,13)***. These retardance data reflect the changes in the spindle architecture, including changes in the optical density and alignment of microtubules and changes in the alignment of microtubules. Either capture individual images or proceed to take time-lapse movies of an oocyte to monitor dynamic changes in the

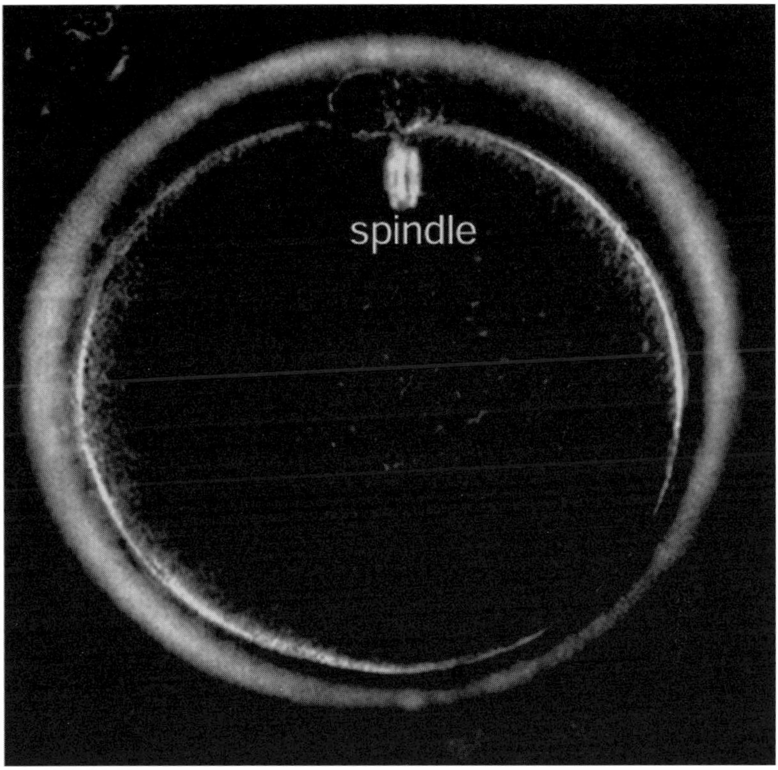

Fig. 3. A birefringent spindle imaged with the SpindleView system in a human oocyte at the metaphase II stage.

spindle. The data can be analyzed at any time in detail using data processing tools, such as pseudocolor overlay, stamp function, or line-scan plots.

c. Spindle removal (enucleation) and spindle transfer. Nuclear transfer is fast becoming an important reproductive technology. A crucial step in nuclear transfer involves enucleation of recipient oocytes. Traditional methods to remove the nucleus include the blind removal of cytoplasm close to the first polar body or oocytes if prestained with DNA-specific dyes before enucleation. As metaphase chromosomes are associated with spindle fibers, locating and then removing the spindles with the SpindleView system ensures complete removal of chromosomes with limited removal of cytoplasm *(7)*. Spindle transfer is an alternative technique to be used in nuclear transfer. For this, hold the oocytes with a holding pipet and then image the spindles. After locating the spindles, remove the spindles using an enucleation pipet.
d. Examination of spindle morphology for predicting oocyte quality. Abnormal meiotic spindles are occasionally present in human oocytes and have been found more readily in older women ***(14)***. Such spindle abnormalities include

short spindles, partially depolymerized spindles, dividing spindles, or two spindles in a single oocyte ***(15)***. To analyze such data, image the oocyte and then inseminate the oocytes and culture resulting embryos individually. Record the spindle image and subsequent embryo development, and/or implantation and fetus development. One can also feasibly combine this information with preimplantation genetic diagnosis data, to potentially understand the relationship between spindle morphology in oocytes and chromosome defects in embryos, including aneuploidy formation.

References

1. Oldenbourg, R. (1996) A new view on polarization microscopy. *Nature (Lond.)* **381,** 811–812.
2. Bragg, W. L. and Pippard, A. B. (1953) The form birefringence of micromolecules. *Acta Crystallogr.* **6,** 865–867.
3. Oldenbourg, R., Salmon, E. D., and Tran, P. T. (1989) Birefringence of single and bundled microtubules. *Biophys. J.* **74,** 645–654.
4. Oldenbourg, R. (1999) Polarized light microscopy of spindles. *Methods Cell Biol.* **61,** 175–208.
5. Inoue, S. (1953) Polarization optics studies of the mitotic spindle. I. The demonstration of spindle fibers in living cells. *Chromosoma* **5,** 487–500.
6. Silva, C. P., Kommineni, K., Oldenbourg, R., et al. (1999) The first polar body does not predict accurately the location of the metaphase II meiotic spindle in mammalian oocytes. *Fertil. Steril.* **71,** 719–721.
7. Liu, L., Oldenbourg, R., Trimarchi, J. R., et al. (2000) A reliable, noninvasive technique for spindle imaging and enucleation in mammalian oocytes. *Nature Biotechnol.* **18,** 223–225.
8. Wang, W. H., Meng, L., Hackett, R. J., et al. (2001) The spindle observation and its relationship with fertilization after intracytoplasmic sperm injection in living human oocytes. *Fertil. Steril.* **75,** 348–353.
9. Wang, W. H., Meng, L., Hackett, R. J., et al. (2001) Developmental ability of human oocytes with or without birefringent spindle imaged by Polscope before insemination. *Hum. Reprod.* **146,** 4–8.
10. Wang, W. H., Meng, L., Hackett, R. J., et al. (2001) Limited recovery of meiotic spindles in living human oocytes after cooling–rewarming observed using polarized light microscopy. *Hum. Reprod.* **16,** 2374–2378.
11. Palermo, G., Cohen, J., and Alikani, M. (1995) Intracytoplasmic sperm injection: a novel treatment for all forms of male factor infertility. *Fertil. Steril.* **63,** 1231–1240.
12. Hardarson, T., Lundin, K., and Hamberger, L. (2000) The position of the metaphase II spindle cannot be predicted by the location of the first polar body in the human oocyte. *Hum. Reprod.* **15,** 1372–1376.
13. Liu, L., Oldenbourg, R., Trimarchi, J. R., et al. (2000) Increased birefringence in the meiotic spindle provides a new method for the onset of activation in living oocytes. *Biol. Reprod.* **63,** 251–258.

14. Battaglia, D. E., Goodwin, P., Klein, N. A., et al. (1996) Influence of maternal age on meiotic spindle in oocytes from naturally cycling women. *Hum. Reprod.* **11,** 2217–2222.
15. Wang, W. H., Meng, L., Hackett, R. J., et al. (2001) Imaging metaphase II spindle in living human oocytes with the Polscope reveals a high proportion of human oocytes have abnormal spindles. *Fertil. Steril.* **75(Suppl.),** O5.

16

In Vitro Maturation and Fertilization of Pig Oocytes

Heng-Yu Fan and Qing-Yuan Sun

1. Introduction

In pigs, as in many other species, immature oocytes liberated from ovarian follicles can resume meiosis and complete maturation in culture. Although the matured oocytes can be penetrated in vitro by spermatozoa under appropriate conditions, low rates of pronuclear formation and a high incidence of polyspermy have been reported by many workers. Recently, pigs have become increasingly important in the field of biomedical research and there is increasing interest in using transgenic pigs as potential xenograf donors in the future. Because most studies that attempt to produce transgenic pigs by nuclear transfer/cloning techniques or pronuclear microinjection use matured oocytes and early embryos, respectively, it is becoming more important to produce a large number of developmentally competent oocytes and embryos for research. Furthermore, accumulating information suggests that rodents may be atypical with regard to regulating mechanisms of oocyte maturation and fertilization. The mechanisms for fertilization in domestic animals such as pig and cattle may be more similar to low vertebrates and human than the mouse *(1)*. Therefore, porcine oocytes have now become an important model for studying the molecular control of the meiotic cell cycle and signal transduction during fertilization.

At the same time, in vitro maturation and fertilization of pig oocytes is becoming a potential method for improving pig reproduction. The system for production of swine for human food continues to undergo major changes, including, in some countries, the consolidation of swine into large, integrated units. These swine operations are very receptive to the use of technologies to reduce labor costs as well as a basis of increased production efficiency *(2)*.

From: *Methods in Molecular Biology, vol. 253: Germ Cell Protocols: Vol. 1 Sperm and Oocyte Analysis*
Edited by: H. Schatten © Humana Press Inc., Totowa, NJ

In recent reviews, the relationships between oocyte maturation and male pronuclear formation ***(3)*** or polyspermy ***(4)***, and available methodology for in vitro fertilization ***(5,6)*** have been discussed. Fundamental studies on cellular and molecular factors operating during porcine oocyte maturation and fertilization provide the basis for defining the conditions for efficient embryo production. Successful methods for in vitro maturation and fertilization (IVM–IVF) of porcine oocytes ***(7–9)*** and culture of porcine embryos ***(10)*** have been developed. Here, we present a series of popular methods for IVM and IVF of pig oocytes. They are suitable for both the fundamental research of pig fertilization mechanism and the practice of pig embryo production.

2. Materials

2.1. Boar Sperm Freezing by the Pellet Method

1. Sperm washing medium ***(11)***: 14.3800 g glucose, 0.6250 g lactose, 1.1250 g Na-citrate·$2H_2O$, 0.8750 g EDTA (disodium salt), 0.3000 g $NaHCO_3$, 0.1000 g KCl, 0.0325 g penicillin G, 0.0125 g streptomycin, make up 250 mL with diluted water.
2. Freezing solution (BF5) ***(11)***: 6.000 g *N*-Tris[hydroxymethyl]-methyl-2-aminoethane sulfonic acid (TES), Tris (crystallized free base), 1.000 g, glucose, 16.000 g, Equex STM, 2.5 mL, penicillin G, 0.065 g, streptomycin, 0.050 g, egg yolk (fresh), 100 mL (add last). Add diluted water to 500 mL. Mix well (0.5–1 h) and centrifuge for 15 min at 1900*g*. Take the supernatant.
3. BF5 with 2% (v/v) glycerol: Take 250 mL of BF5 supernatant, remove 5 mL of BF5, and add 5 mL of glycerol. Mix well. Freeze both BF5 and BF5 + 2% glycerol at –20°C until used. Thaw them out gradually when you need them. First place them inside a refrigerator for about 24 h. Then, take only BF5 solution and keep at room temperature. Once it reaches room temperature, use it within a couple of hours. Leave the BF5 + 2% glycerol solution at 4°C. Once used, the remaining solutions can be refrozen and reused.

2.2. In Vitro Maturation and In Vitro Fertilization of Oocytes

1. A CO_2 incubator, a stereoscope, and an inverted phase-contrast microscope.
2. Modified TL–HEPES–PVA medium: 6.6633 g NaCl (114.00 m*M*), 0.2386 g KCl (3.20 m*M*), 0.0408 g NaH_2PO_4 (0.34 m*M*), 60% syrup Na-lactate (w/w), 1.4 mL (10.00 m*M*), 0.1017 g $MgCl_2$·$6H_2O$ (0.50 m*M*), 2.3830 g HEPES (10.00 m*M*), 0.0220 g Na-pyruvate (0.20 m*M*), 2.1860 g sorbitol (12.00 m*M*), 0.1680 g $NaHCO_3$ (2.00 m*M*), 0.2940 g $CaCl_2$·$2H_2O$ (2.00 m*M*), 0.0250 g gentamycin, 0.0650 g penicillin G, 0.1000 g poly(vinyl alcohol) (PVA). $CaCl_2$·$2H_2O$ is added last. Make up a 1000-mL solution, adjust pH to 7.3–7.4, filter, and store at 4°C.
3. In vitro maturation medium:
   a. Medium 199 with Earle's salts (Gibco; cat. no. 31100-027 or 31100-035).

b. Basic maturation medium (stock solution): TCM-199 supplemented with 0.1% PVA, 3.05 m*M* D-glucose, 0.91 m*M* sodium pyruvate, 75 μg/mL penicillin, and 50 μg/mL streptomycin.
c. Final maturation medium (working solution): Basic maturation medium supplemented with 0.57 m*M* cysteine (Sigma; cat. no. C8152), 10 ng/mL epidermal growth factor (Sigma; cat. no. E4127), 0.5 μg/mL follicle-stimulating hormone (FSH) (Sigma; cat. no. F2293), and 0.5 μg/mL luteinizing hormone (LH) (Sigma; cat. no. L5269).

4. Sperm washing medium: Dulbecco's phosphate-buffered saline (DPBS) (Gibco; cat. no. 11500-022): Without added $CaCl_2$ and supplemented with 75 μg/mL penicillin and 50 μg/mL streptomycin on the day before IVF. Adjust the pH to 7.2 and keep in a refrigerator.
5. In vitro fertilization medium (modified Tris-buffered medium [mTBM]) ***(12)***: 0.6611 g NaCl (113.1 m*M*), 0.0224 g KCl (3.0 m*M*), 0.1102 g $CaCl_2 \cdot 2H_2O$ (7.5 m*M*), 0.2423 g Tris crystallized free base (20.0 m*M*), 0.1982 g glucose (11.0 m*M*), 0.0550 g Na-pyruvate (5.0 m*M*); make up 100 mL medium. After preparation, the pH is about 10.0 (do not adjust the pH). Store at 4°C and use within 3 wk.
6. In vitro embryo development medium (BSA-free North Carolina State University 23 medium [NCSU-23 medium]) ***(13)***: 0.6355 g NaCl (108.73 m*M*), 0.0356 g KCl (4.78 m*M*), 0.0162 g KH_2PO_4 (1.19 m*M*), 0.0293 g $MgSO_4 \cdot 7H_2O$ (1.19 m*M*), 0.1000 g D-glucose (5.55 m*M*), 0.0146 g glutamine (1.00 m*M*), 0.0876 g taurine (7.00 m*M*), 0.0546 g hypotaurine (5.00 m*M*), 0.2106 g $NaHCO_3$ (25.07 m*M*), 0.0250 g $CaCl_2 \cdot 2H_2O$ (1.70 m*M*), 0.0075 g penicillin G, 0.0050 g streptomycin; make up 100 mL medium. Adjust the pH to 7.2–7.3 and filter. Store at 4°C and use within 3 wk.

2.3. Electrical Activation of Pig Oocytes

Electroporation medium: 0.28 *M* mannitol, 0.05 m*M* $CaCl_2$, 0.1 m*M* $MgSO_4$, and 0.01% (w/v) BSA. Store at 4°C and use within 1 mo.

2.4. Evaluation of Nuclear Status: Orcein Staining

1. Fixation solution: Ethanol: acetic acid = 3:1. Store at room temperature and make up fresh every month.
2. Staining solution: 1% Orcein in 45% acetic acid. Dissolve 1 g orcein (Sigma) in 45 mL acetic acid and 55 mL H_2O at 80°C and filter. Store at room temperature.
3. Washing solution: Ethanol: acetic acid: glycerol = 3:1:1. Store at room temperature.

3. Methods

3.1. Boar Sperm Freezing by the Pellet Method

1. Collect sperm-rich semen (milklike fraction) while filtering through a piece of gauze.
2. Examine sperm motility (should be >80%) and concentration (subjectively).
3. Let stand at room temperature (approx 25°C) for about 1.5–2 h.

4. Use 50-mL Falcon tubes and place 15–20 mL sperm-rich fraction and 15–20 mL (1:1) of sperm washing medium.
5. Mix gently; centrifuge at 1000*g* for 5–7 min.
6. Remove the supernatant, add BF5 medium (3–4 mL), and mix well with a pipet.
7. Calculate the sperm concentration using a hemocytometer after appropriate dilution (using 10,000 times dilution in DPBS).
8. Adjust the sperm concentration to (5–10) $\times$ 10^8/mL by adding BF5 medium.
9. Place the sperm suspension inside a glass beaker containing water at room temperature. Keep the beaker inside a refrigerator and cool down to 4°C over a period of 1.5 h (after about 45 min, check the water temperature and periodically add ice to gradually bring the temperature down to 4°C).
10. Hold the sperm suspension at 4°C for a few minutes (in order to get the internal temperature) and add an equal volume of BF5 (also cooled down to 4°C in advance) containing 2% glycerol (at 4°C). Remember to always keep the sperm suspension in the water bath kept at 4°C by adding ice.
11. Mix well and check sperm motility.
12. Freeze sperm (100 μL) on dry ice (need to make some depressions on dry ice before freezing sperm as pellets).
13. Leave the pellets on dry ice for 1–2 min and place in liquid nitrogen for long-term storage. (Fresh semen may also be used; *see* **Note 1**)

3.2. In Vitro Maturation and Fertilization of Pig Oocytes

Day 1:

1. Prepare IVM medium (8:00 AM):
   a. Place 5 mL TCM-199 into a 15-mL tube and add 0.007 g cysteine.
   b. Place 19 mL TCM-199 into a 50-mL tube; add 1 mL medium from **step a**. Then, add 20 μL EGF (final concentration 10 ng/mL). Filter the medium into a 15-mL tube.
   c. Prepare oocyte washing dishes (three 35-mm dishes in a 90-mm dish) and oocyte maturation dish (four-well multiwell dish) with 500 μL of the medium prepared in **step b** in each well. Cover the washing medium and maturation medium with paraffin oil.
   d. Place the prepared dishes in the incubator for at least 3 h before adding oocytes.
2. Aspirate gilt ovaries. Ovaries are collected from gilts at a local slaughterhouse and transported to the laboratory as soon as possible in 0.9% NaCl solution containing 75 μg penicillin G/mL and 50 μg streptomycin/mL at 37°C. Aspirate the follicular fluid from antral follicles (2–6 mm in diameter) with an 18-gauge needle affixed to a 20-mL disposable syringe and collect follicular fluid in a 50-mL tube. Discard the supernatants, and wash the sediments twice with modified TL–HEPES–PVA medium, which has been previously warmed to 37°C. Wash the obtained oocytes (*see* **Note 2**) three times with maturation medium and put them in the four-well dish (50 oocytes/well). Next, add 5 μL of both FSH and LH to each well (final concentration of 0.5 μg/mL). (*See* **Note 3**.)

3. Prepare insemination medium (2:30 PM)
   a. Take 15 mL mTBM into a 15-mL tube. Next, add 0.01 g caffeine (2 m*M*) and dissolve. Then, add 0.03 g BSA (Sigma; cat. no. A-7888).
   b. Filter the medium and cover the tube with foil that has holes poked in it to allow gas exchange. Place in the CO_2 incubator.

Day 2:

1. Make insemination drops: Make four 50-µL drops of insemination medium prepared on d 1 and cover the drops with paraffin oil.
2. Make IVF washing dishes: Put three 35-mm dishes into a 90-mm dish, place 500 µL of insemination medium into each, and cover with paraffin oil.
3. Prepare DPBS for sperm washing: Take 50 mL DPBS into a 50-mL tube. Then, add 0.05 g BSA (Sigma; cat no. A-7888). Filter and place in a 39°C non-CO_2 incubator.
4. Prepare developmental medium (NCSU-23):
   a. Put 10 mL NCSU-23 into a 15-mL tube and add 0.04 g BSA (Sigma; cat no. A-8022).
   b. After filtration, put 500 µL into each well of the four-well dishes. Cover the medium with paraffin oil. Also, prepare the washing dishes at the same time using this medium.

Day 3:

1. Oocyte preparation: Place 900 µL TL–HEPES–PVA medium into a 1.5-mL Eppendoff tube and add 100 µL hyaluronidase stock solution (30 mg/mL in TCM-199) and warm it in the non-CO_2 incubator. The expanded cumulus cells of oocytes, which have been cultured for 42–44 h, are removed by vortex for 30–60 s. Then, wash the oocytes and put them in IVF drops (30–35 oocytes/drop). Place the dishes back into the CO_2 incubator.
2. Sperm washing:
   a. Take 10 mL of DPBS (sperm washing medium) into a tube placed in a water bath at 39°C.
   b. Add a frozen sperm pellet, allow it to thaw, and then mix gently.
   c. Centrifuge three times at 1800*g* to 2000*g* for 4 min each or at 1000*g* for 5 min each.
   d. Resuspend the final pellet in a little fertilization medium (100 µL).
   e. Count the sperm concentration using a hemocytometer after 1000 times dilution.
   f. Resuspend the concentrated sperm suspension in fertilization medium in order to get (2–5) $\times$ 10^5 cells/mL. (For a particular pellet, sperm concentration should be determined after preliminary studies.) (*See* **Note 4**.)
3. In vitro fertilization: Add 50 µL of diluted sperm to each of the IVF drops that contain oocytes.
4. Place the fertilization dishes back into the CO_2 incubator for 5–6 h. Then, wash the oocytes three times in the development medium (NCSU 23 + 0.4% BSA) to

remove loosely attached sperm. Transfer the eggs into the same medium (500 μL) for further development. (*See* **Note 5**.)

3.3. Electrical Activation of Pig Oocytes

1. Remove the cumulus cells of the oocytes by a short exposure to 3 mg/mL hyaluronidase after maturation culture, and wash three times in the electroporation medium.
2. Cumulus-free oocytes that have been cultured for 42–44 h are put in a row in a stainless-steel fusion chamber and balanced for 5 min.
3. A single 80-μs pulse at 120 V/mm dc is applied to the eggs. Two minutes after activation, the eggs are then washed three times and cultured in NCSU-23 medium containing 0.4% BSA.
4. Examine the extrusion of second polar body and the formation of pronucleus 14 h after electrical stimulation.

3.4. Evaluation of Nuclear Status: Orcein Staining

1. Mount the denuded oocytes between a coverslip and a glass slide supported by four columns of a mixture of vaseline and paraffin (9:1).
2. Fix the oocytes in acetic acid: ethanol (1:3, v/v) for at least 48 h at room temperature.
3. Stain the cells with 1% orcein for 5 min.
4. Wash off the surplus orcein with washing solution and seal the slide with nail polish.
5. Examine the eggs with a phase-contrast microscope.

4. Notes

1. Fresh semen may be kept at 16–18°C and used for insemination within 1 wk.
2. Only oocytes possessing a compact cumulus and evenly granulated ooplasm should be selected for maturation culture.
3. For maturation culture, the cumulus cells of the oocytes should be kept intact because the cumulus cells are necessary for IVM of porcine oocytes. Therefore, relatively thick pipets should be used in collection and transfer of cumulus–oocyte complexes.
4. For frozen–thawed sperm: mTBM + 1 m*M* caffeine + 0.1% BSA (Sigma; cat. no. A-7888). Sperm concentration: (2–5) × 10^5/mL. For fresh semen: mTBM + 0.4% BSA and 2.5 m*M* caffeine. Sperm concentration: (1–2) × 10^4/mL.
5. Cleavage is usually examined 2 d after insemination, and blastocysts are collected 6 d after insemination.

References

1. Sun, Q. Y., Lai, L., Park, K. W., et al. (2001) Dynamic events are differently mediated by microfilaments, microtubules, and mitogen-activated protein kinase during porcine oocyte maturation and fertilization *in vitro*. *Biol. Reprod.* **64,** 879–889.
2. Day, B. N. (2000) Reproductive biotechnologies: current status in porcine reproduction. *Anim. Reprod. Sci.* **60–61,** 161–172.

3. Moor, R. M., Mattionli, M., Ding, J., et al. (1990) Maturation of pig oocytes in vivo and in vitro. *J. Reprod. Fertil.* **40,** 197–210.
4. Hunter, R. H. F. (1991) Oviduct function in pig, with particular reference to the pathological condition of polyspermy. *Mol. Reprod. Dev.* **29,** 385–391.
5. Prather, R. S. and Day, B. N. (1998) Practical considerations for the in vitro production of pig embryos. *Theriogenology* **49,** 23–32.
6. Niwa, K. (1993) Effectiveness of in vitro maturation and in vitro fertilization techniques in pigs. *J Reprod Fertil.* **48(Suppl.),** 49–59.
7. Wang, W. H., Abeydeera, L. R., Okuda, K., et al. (1994) Penetration of porcine oocytes during maturation in vitro by cryopreserved, ejaculated spermatozoa. *Biol. Reprod.* **50,** 510–515.
8. Abeydeera, L. R., Wang, W. H., Cantley, T. C., et al. (1998) Presence of beta-mercaptoethanol can increase the glutathione content of pig oocytes matured in vitro and the rate of blastocyst development after in vitro fertilization. *Theriogenology* **50,** 747–756.
9. Abeydeera, L. R., Wang, W. H., Cantley, T. C., et al. (1998) Coculture with follicular shell pieces can enhance the developmental competence of pig oocytes after in vitro fertilization: relevance to intracellular glutathione. *Biol. Reprod.* **58,** 213–218.
10. Abeydeera, L. R., Wang, W. H., Cantley, T. C., et al. (1998) Presence of epidermal growth factor during in vitro maturation of pig oocytes and embryo culture can modulate blastocyst development after in vitro fertilization. *Mol. Reprod. Dev.* **51,** 395–401.
11. Pursel, V. G. and Johnson, L. A. (1976) Frozen boar spermatozoa: methods of thawing pellets. *J. Anim. Sci.* **42,** 927–931.
12. Abeydeera, L. R. and Day, B. N. (1997) Fertilization and subsequent development in vitro of pig oocytes inseminated in a modified tris-buffered medium with frozen–thawed ejaculated spermatozoa. *Biol. Reprod.* **57,** 729–734.
13. Petters, R. M. and Wells, K. D. (1993) Culture of pig embryos. *J. Reprod. Fertil.* **48(Suppl.),** 61–73.

17

In Vitro Maturation and In Vitro Fertilization of Human Oocytes

Wei-Hua Wang and Jimmy Gill

1. Introduction

1.1. In Vitro Maturation of Human Oocytes

In vitro maturation (IVM) of human oocytes is an attractive technique for reducing the costs and averting the side effects of gonadotropin stimulation for in vitro fertilization (IVF). Edwards reported that human follicle oocytes could be matured in vitro when they were isolated from follicles and cultured in an appropriate system *(1)*. The first successful fertilization of human oocytes matured in vitro was reported in 1969 *(2)*. However, such a fertilized oocyte was not transferred back to patients until live birth of IVF of in vivo matured oocytes was achieved *(3)*. A few IVF clinics have attempted to transfer embryos resulting from in vitro matured oocytes and have succeeded in getting the patient pregnant *(4–13)*. However, only recently did IVM become clinically viable for treating infertile women in some IVF clinics *(8,10–12)*. When compared with in vivo matured oocytes, the success rate is still low. The quality of oocytes after IVM appears to be suboptimal because embryos resulting from IVM show more frequent developmental block and transfer of such embryos results in low implantation and clinical pregnancy rates. Recent research advances, especially in the patient preparation, small follicle aspiration, and the oocyte culture system make it possible to adopt this new assisted reproductive technologies (ART) in IVF clinics. Women with infertility who have polycystic ovaries or polycystic ovarian syndrome (PCOS) should be considered for IVM treatment. Other patients, such as patients who repeatedly produce poor quality embryos or patients who are poor responders to ovarian stimulation for conventional IVF, may also be considered for IVM treatment.

From: *Methods in Molecular Biology, vol. 253: Germ Cell Protocols: Vol. 1 Sperm and Oocyte Analysis*
Edited by: H. Schatten © Humana Press Inc., Totowa, NJ

The main advantage of IVM treatment is that no medications are used to stimulate multiple follicular development in the ovaries. Therefore, side effects, especially ovarian hyperstimulation syndrome and the associated costs for medications, are eliminated. Currently, IVM of human oocytes can be obtained from the following three sources.

1.1.1. Immature Oocytes from Stimulated Ovaries

When oocytes are aspirated from ovarian follicles of women who have been treated with hormones to induce multiple follicular growth, immature oocytes are occasionally retrieved. Oocytes are retrieved after human chorionic gonadotropin (hCG) administration; more than 80% of oocytes are mature, reaching the metaphase II (MII) stage. However, others are maturing or immature oocytes at the metaphase I (MI) or germinal vesicle (GV) stage. These oocytes are capable of undergoing spontaneous nuclear maturation in culture and of undergoing normal fertilization and development. Pregnancy has also been reported after the transfer of embryos resulting from immature oocytes from stimulated ovaries ***(4,7,13)***. Maturing and mature oocytes from stimulated patients usually can produce enough embryos for transfer; hence, the immature oocytes are generally not used for embryo production in most cases. Such oocytes, in fact, are good materials for research purposes ***(14–17)***. Recently, Chian used this kind of oocyte to develop IVM medium and obtained exciting results ***(17)***.

1.1.2. Immature Oocytes from PCOS Patients

It has been found that clinical IVM plays important roles for infertile treatment of PCOS patients ***(5,18,19)***. Patients with PCOS are frequently infertile. Exogenous gonadotropin stimulation in PCOS patients may increase the risk of ovarian hyperstimulation because these patients are extremely sensitive to external gonadotropins ***(20)***. Supplementation of gonadotropins, serum, and other additives, such as growth factor and estradiol, to IVM medium is necessary for both nuclear and cytoplasmic maturation ***(8,10,19)***. Priming gonadotropin (hCG) 36 h before egg retrieval significantly increases egg quality and improves subsequent embryo development and clinical pregnancy rates ***(8,21)***. A few IVF clinics have established clinical protocols for the treatment of PCOS and obtained acceptable (approx 30%) pregnancy rates ***(10–12)***.

1.1.3. Immature Oocytes from Unstimulated Ovaries

Oocytes are collected from patients undergoing infertility treatment or patients undergoing tuboplasty, cesarean section, and oophorectomy ***(19,22–24)***. For IVM of this kind of oocyte, supplementation of gonadotropins and serum, either from patients or from fetal calf, is necessary to support both nuclear and cytoplasmic maturation ***(10)***. Clinical pregnancies have also been reported and it has been found that some oocytes from tuboplasty, cesarean section, and oophorectomy can contribute to establish an egg donation program or an oocyte bank when combined with oocyte cryopreservation ***(19,22)***.

1.2. In Vitro Fertilization of Human Oocytes

In vitro fertilization means placing matured human eggs with spermatozoa in a Petri dish outside the women and then transferring the resulting embryos back to the women's uterus. It has become a standard clinical procedure for treatment of human infertility. IVF techniques were initially adopted for the treatment of women who suffered from tubal blockage. The first successful report of the birth of a baby after transferring embryos resulting from IVF in humans was in 1978 ***(3)***. After that, it has been found that a wider range of infertility problems could be treated by IVF. In recent years, remarkable advances have been made in human IVF and associated techniques, especially in the development of the embryo culture system ***(25–29)***. Currently, IVF is performed in women with most infertility problems including tubal diseases, immunologic infertility, male factor infertility, endometriosis and unexplained infertility, cervical factor infertility, polycystic ovarian disease, and preimplantation genetic diagnosis. High pregnancy rates have been reported in recent years and more than 100,000 cycles per year are performed around the world.

2. Materials

2.1. Egg Retrieval

1. 60-mm Organ culture dishes.
2. 100-mm Tissue culture dishes.
3. 17-gauge aspiration needle.
4. 10-mL Sterile tubes.
5. Sterile 5.75-in. Pasteur pipets.
6. 2-mL pipet pumps.
7. Sterile 30-gauge needles.
8. Sterile 1-cm^3 syringes.
9. Warm modified human tubal fluid (mHTF).
10. Stereomicroscope with stage warmer.

2.2. IVM of Oocytes

1. IVM medium (*see* **Table 1**).
2. Follicle-stimulating hormone (FSH) and luteinizing hormone (LH).
3. Estradiol.
4. 60-mm Organ culture dishes.
5. 70-µm-Pore size filters.
6. Tissue-culture-grade oil.
7. 40 IU/mL Hyaluronidase.

Table 1
Composition of Oocyte IVM Medium[a]

Inorganic salt	
$CaCl_2$	200.0
KCl	400.0
$MgSO_4$	98.0
NaCl	6800.0
$NaHCO_3$	1250.0
$NaH_2PO_4 \cdot H_2O$	125.0
Amino acids	
L-Alanine	8.9
L-Arginine	126.4
L-Asparagine	13.2
L-Aspartic Acid	13.3
L-Cystine	24.0
L-Glutamic Acid	14.7
L-Glutamine	292.0
Glycine	7.5
L-Histidine·HCl·H_2O	42.0
L-Isoleucine	52.4
L-Leucine	52.4
L-Lysine·HCI	72.5
L-Methionine	15.1
L-Phenylalanine	33.0
L-Proline	11.5
L-Serine	10.5
L-Threonine	47.6
L-Tryptophan	10.2
L-Tyrosine	36.0
L-Valine	46.8
Vitamins	
Biotin	0.001
D-Ca Pantothenate	1.0
Choline chloride	1.0
Folic acid	1.0
i-Inositol	2.0
Nicotinamide	1.0
Pyridoxal·HCl	1.0
Riboflavin	0.1
Thiamine·HCl	1.0
Other components	
D-Glucose	1000.0
Sodium pyruvate	110.0
Insulin	0.5
Human transferrin	5.0
Hydrocortisone	0.0004
FGF	0.0005
EGF	0.0010
Phenol Red	5.0
Penicillin G	50.0 units
Streptomycin	50.0 μg

Note: Units are milligrams per liter unless indicated otherwise.

[a]Add 10% SSS, 0.075 IU/mL FSH + LH and 1.0 μg/mL estradiol to the medium before use. *Source:* **ref.** ***17***.

Table 2
Commercial Media for IVF of Human Oocytes

Purpose of media	Commercial name and provider[a]
Egg-retrieval media	Modified HTF (Irvine Scientific) Quinn's Advantage Modified HTF (Sage BioPharma)
Sperm-washing media	HTF + 6% SSS (Irvine Scientific) Quinn's Advantage Sperm Washing Medium + 6% SSS (Sage Biopharma)
Insemination media	P-1 medium + 6% SSS (Irvine Scientific) Quinn's Advanced Fertilization Medium + 6% SSS (Sage Biopharma)
Cleavage (growth) media	P-1 medium + 10% SSS (Irvine Scientific) Quinn's Advanced Cleavage Medium + 10% SSS (Sage Biopharma)
Blastocyst media	Blastocyst Medium +10% SSS (Irvine Scientific) Quinn's Advanced Blastocyst Medium + 10% SSS (Sage Biopharma)

Abbreviation: SSS = synthetic serum substitute.

[a]Irvine Scientific, 2511 Daimler Street, Santa Ana, CA 92705-5588; Tel: 1-800-577-6097; Fax: 1-949-261-6522; website: www.irvinesci.com.

Sage BioPharma, Bedminster One, 135 US Route 202/206, Bedminster, NJ 07921-9801; Tel: 1800-368-8324 or 1-908-306-5790; Fax: 1-908-306-5777; website; www.sagebiopharma.com.

2.3. Sperm Preparation

1. Sterile specimen cup.
2. 9-in. Conical tubes.
3. Sterile snap-cap tubes.
4. 9-in. Pasteur pipets.
5. Glass slides.
6. Glass coverslips.
7. 5-mL serological pipet.
8. Rubber bulb.
9. Pipet tips.
10. Hemocytometer.
11. Slide warmer.
12. Phase-contrast microscope.
13. Centrifuge.
14. Sperm-washing medium (*see* **Tables 2** and **3**).
15. Mini-vortex.
16. Chymotrypsin, 5 mg/vial (Sigma).
17. Water bath.

Table 3
Compositions of Commercial Media for IVF of Human Oocytes

Component (m*M*)	Irvine Scientific			Sage Biopharma		
	HTF	mHTF	P-1	HTF[a]	mHTF[b]	Cleavage medium[c]
NaCl	101.6	101.6	101.6	113.8	110.0	117.1
KCl	4.69	4.69	4.69	4.7	4.7	4.7
$MgSO_4 \cdot 7H_2O$	0.20	0.20	0.20	0.20	0.20	0.20
KH_2PO_4	0.37	0.37	—	0.01	0.01	—
$CaCl_2 \cdot 2H_2O$	2.04	2.04	2.04	—	—	—
Ca-lactate (L)	—	—	—	2.04	2.04	2.04
$NaHCO_3$	25.0	4.0	25.0	17.9	4.0	17.9
HEPES	—	21.0	—	—	21.0	—
Glucose	2.78	2.78	—	2.78	2.78	0.1
Na-pyruvate	0.33	0.33	0.33	0.33	0.33	0.33
Na-lactate(D/L)	21.4	21.4	21.4	—	—	—
Na-citrate	—	—	0.15**	0.0005	0.0005	0.0005
Alanyl-glutamine	—	—	—	1.0	1.0	1.0
Aspartic acid	—	—	—	0.1	0.1	0.1
Asparagine	—	—	—	0.1	0.1	0.1
Glycine	—	—	—	0.1	0.1	0.1
Proline	—	—	—	0.1	0.1	0.1
Serine	—	—	—	0.1	0.1	0.1
Taurine	—	—	0.05	0.1	0.1	0.1
EDTA	—	—	—	0.01	0.01	0.01
Penicillin G*	100	100	—	—	—	—
Streptomycin sulfate**	50	50	—	—	—	—
Gentamicine**	—	—	10	0.01	0.01	0.01
Phenol Red***	10	10	5	3	3	3

Note: Concentrations are millimoles except *IU/mL, **mg/mL, and ***mg/L.
[a]Commercial name is Quinn's Advantage Fertilization Medium.
[b]Commercial name is Quinn's Advantage Modified HTF Medium.
[c]Commercial name is Quinn's Advantage Cleavage Medium.

2.4. Insemination of Oocytes

1. Sterile pipet tips.
2. Alcohol prep pad.
3. Fertilization medium (*see* **Tables 2** and **3**).
4. 60-mm Organ culture dishes.
5. Tissue-culture-grade oil.

2.5. Fertilization Evaluation

1. Stripping pipets.
2. Growth medium (**Tables 2** and **3**).
3. Stereomicroscope with stage warmer.
4. Phase-contrast microscope.

2.6. Embryo Transfer

1. 35-mm Tissue culture dishes.
2. 1-cm^3 Syringe.
3. Sterile gloves.
4. Transfer catheter.

3. Methods

3.1. Immature Oocyte Retrieval and IVM of Oocytes

3.1.1. IVM of Oocytes from Stimulated Ovaries

Immature oocytes from women undergoing controlled ovarian stimulation and intracytoplasmic sperm injection (ICSI) cycles with standard stimulation protocols are the primary materials for IVM.

1. Perform oocyte retrieval with ultrasound guidance 36 h after hCG administration (for details, *see* **Subheading 3.2.1.**).
2. Culture the collected oocyte–cumulus complexes in fertilization medium supplemented with 6% synthetic serum substitute (SSS) for 5–6 h in a CO_2 incubator (37°C with high humidity).
3. Remove cumulus cells in a mHTF containing 40 IU/mL hyaluronidase by mechanical pipetting until all oocytes are completely denuded of their cumulus cells.
4. Check oocytes stages (*see* **Fig. 1**) under the inverted microscope to assess oocyte maturity. Mature (MII) oocytes are determined by the presence of a first polar body extrusion and are used for ICSI immediately.
5. Wash the immature oocytes twice with IVM medium and then culture individually.
6. Record oocyte stage and start IVM.
7. Determine the maturity of the oocytes (*see* **Fig. 1**) 6 h after IVM for MI stage oocytes and 24 h and 48 h after IVM for all oocytes.
8. Inseminate mature (MII) oocytes by routine IVF or ICSI.

3.1.2. IVM of Oocytes from Unstimulated Ovaries

1. Aspirate oocytes from follicles with ultrasound guide or directly aspirate from follicles on the ovary surface.
2. Search the oocyte–cumulus complex under a stereomicroscope.
3. Wash the oocyte–cumulus complex with IVM medium (*see* **Table 1**) and culture in the medium for 44–48 h.
4. Determine oocyte maturity and inseminate all mature oocytes through IVF or ICSI.

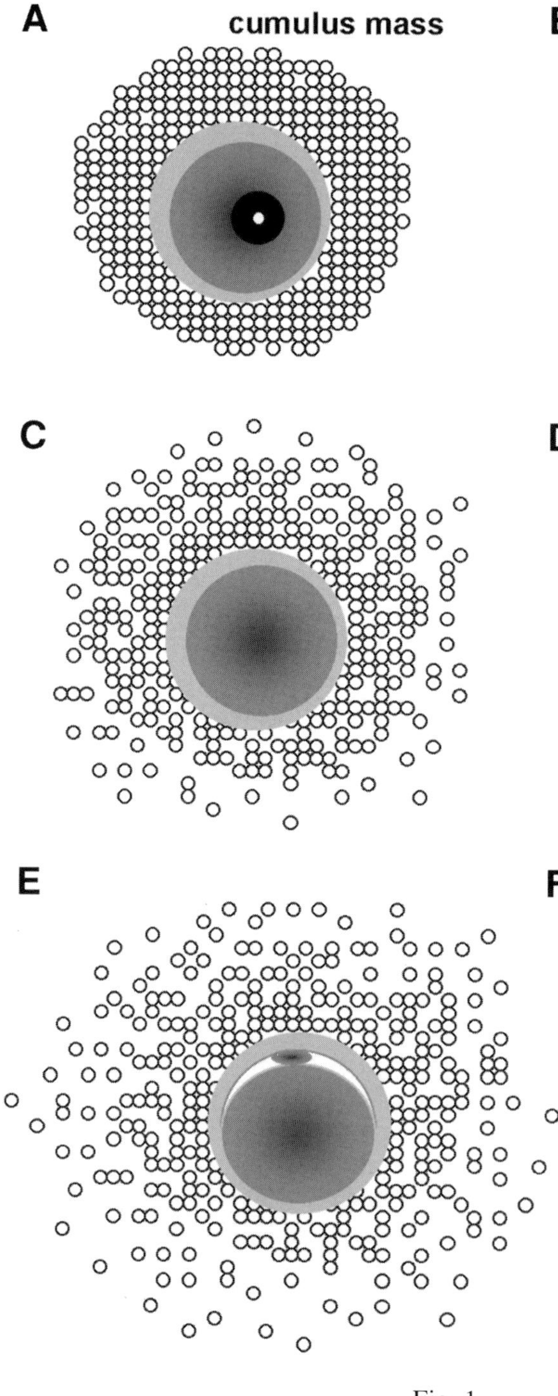

Fig. 1.

3.1.3. IVM of Oocytes from PCOS Patients

3.1.3.1. Patient Preparation

1. The treatment cycle is initiated by the administration of progesterone for 10 d.
2. On d 2 or 3 following the onset of menstrual bleeding, the patient undergoes a baseline ultrasound scan.
3. Perform the second ultrasound scan on d 6–8 and withdraw blood for serum, which is used for oocyte maturation.
4. Oocyte retrieval is performed in the follicular phase between d 10 and d 14 of the cycle. Prime 10,000 IU hCG 36 h before oocyte retrieval.

3.1.3.2. Day Before Egg Retrieval

1. Do a quality control check of CO_2 content and the temperatures of the incubator (*see* **Note 1**).
2. In the incubator, place two tightly capped flasks (30–40 mL) of mHTF per patient.
3. Prepare IVM medium. Add 1 mL of patient's (maternal) serum (56°C, 30-min heating inactivation) or SSS to 9 mL oocyte IVM medium (*see* **Table 1**). Add 0.075 IU/ml FSH + LH and 10 µg/mL estradiol to the medium.
4. Prepare IVM dishes: Add 1 mL of IVM medium into the center well and 2 mL into the moat in the organ culture dish; then, cover the dishes with tissue-culture-grade oil.
5. Equilibrate the dishes in a 5% CO_2 incubator overnight.

3.1.3.3. Day of Egg Retrieval

1. Aspirate follicles using transvaginal ultrasound-guided technique with a specially designed 17G single-lumen aspiration needle with an aspiration pressure of 7.5 kPa. Collect follicular fluid in tubes containing prewarmed mHTF.
2. Place follicular fluid into a 70-µm-pore size filter that has been rinsed with flushing medium. Wash the filtrates with medium by vigorous pipetting using a 10-mL serological pipet.
3. Resuspend the retained cells in the medium and isolate oocytes under a stereomicroscope. An alternative method is to isolate oocytes directly under the

Fig. 1. *(previous page)* Schematic morphology of cumulus–oocyte complexes and oocyte stages during maturation: **(A)** immature oocyte at the germinal vesicle stage; the cumulus and corona cells tightly enclose the oocyte; **(B)** a large clear nucleus (GV) is present after removing the cumulus cells; **(C)** immature oocyte at metaphase I stage; cumulus cells are moderate expanded, the oocyte is surrounded with a tight layer of corona cells, and the corona is enclosed tightly with a few cumulus cell layers; **(D)** the oocyte does not have a GV or polar body after removing the cumulus and corona cells; **(E)** mature oocyte with cumulus cells being well expanded and corona cells being visible around the oocyte; **(F)** a polar body is present in the perivitelline space after removing cumulus and corona cells. GV: germinal vesicle. Pb: polar body. ZP: zona pellucida.

stereomicroscope in the same way as mature oocyte collection if there is not much blood in follicular fluid.
4. Repeat this procedure until all dishes have been searched.
5. Once the retrieval is finished, rinse all oocytes in the moat of a maturation dish and then place oocytes into the center well.
6. Record the time in patient's chart; this is the start of IVM (*see* **Note 2**).

3.1.3.4. Days After IVM

1. Prepare a 40 IU/mL hyaluronidase solution in mHTF.
2. Make a hyaluronidase drop in a 60-mm culture dishes in the center of dish and mark it; make several washing droplets with mHTF for washing and cover the droplets with warm oil.
3. After 24 h of culture, denude all eggs from the cumulus cells. For this, put oocyte–cumulus complexes in a hyaluronidase drop for 45 s and then transfer oocytes to a washing drop. Remove cumulus mechanically using different sizes of denuding pipets.
4. Identify the oocyte stages (*see* **Fig. 1**). If oocytes are MII, inseminate these oocytes by IVF or ICSI. If the oocytes are immature, put these oocytes back into the IVM medium for further maturation (*see* **Note 3**).
5. Check oocytes every 12 h until 48 h after IVM. Inseminate all matured oocytes.

3.2. Maturing and Mature Egg Retrieval from Stimulated Ovaries (see Note 4)

3.2.1. Patient Preparation

Controlled ovarian stimulation is used to obtain a number of oocytes for insemination. Ovulation induction is a safe and effective means of restoring fertility in many women. It is important to carefully select treatment regimes based on the diagnosis. It usually involves stimulating the ovary to produce more eggs and may be accomplished with a number of different medications. The stimulatory protocol may be different between different disorders.

3.2.1.1. Injectable Gonadotropins Therapy Protocol

1. On d 3 of the menstrual cycle, perform an ultrasound scan and blood work for hormones.
2. Start FSH injection on d 3 if everything is normal.
3. Perform an ultrasound scan and blood work again after 5 d of FSH injection. Continue FSH injection and perform ultrasound scan and blood work every other day until hCG administration.
4. When at least two follicles reach 17 mm in diameter, administer 10,000 IU hCG subcutaneously to trigger ovulation.
5. Retrieve eggs 35–37 h after hCG injection.

3.2.1.2. Gonadotropin-Releasing Hormone Agonist Therapy Protocol

3.2.1.2.1. Long Protocol

1. Start treatment with a gonadotropin-releasing hormone agonist (GnRHa) (0.5 mg/d, subcutaneous injection) on d 21 of the menstrual cycle and continue until the day of hCG administration.
2. Check estradiol levels 14 d after the first GnRHa injection.
3. Reduce GnRHa dose to 0.2 mg/d until hCG injection if estradiol levels are below 100 pg/mL.
4. Perform a vaginal ultrasound scan and start follicular stimulation the following day by giving 225 IU of Gonal-F daily until egg retrieval.
5. Perform vaginal ultrasound scan on d 6 of Gonal-F administration and continue every other day to determine the number and size of follicles.
6. When at least two follicles reach 17 mm in diameter, administer 10,000 IU hCG subcutaneously to trigger ovulation.
7. Retrieve eggs 35–37 h after hCG injection.

3.2.1.2.2. Short Protocol

1. Start treatment with GnRHa and Gonal-F at the same time on d 2 of the patient's menses. These two hormones are administrated daily until two dominant follicles reach 17 mm in diameter.
2. Administer 10,000 IU hCG to trigger ovulation.
3. Retrieve eggs 35–37 h after hCG injection.

3.2.2. Day Before Egg Retrieval

1. Do a quality control check of the CO_2 content and the temperatures of the incubator.
2. Place two tightly capped flasks (30–40 mL) of mHTF in the incubator.
3. Make insemination medium and dishes. Insemination medium: Add 1.2 mL SSS to 18.8 mL fertilization medium (*see* **Tables 2** and **3**) in a 50-mL flask. Insemination dishes: Pipet 1 mL of insemination medium into the center well and 2.0 mL of insemination medium into the moat of 60-mm organ culture dishes with a center well. Cover both the center well and the moat with oil.
4. Prepare sperm-washing medium: Add 1.8 mL SSS to 28.2 mL sperm-washing medium (*see* **Tables 2** and **3**) in a 50-mL flask.
5. Place these media and dishes in the incubator.

3.2.3. Day of Egg Retrieval

1. Place several sterile 100-mm tissue culture dishes on the slide warmer. It is always best to pour the trap into a warm dish.
2. Pour follicular fluid into a sterile 100-mm tissue culture dish. Rinse the trap with about 2 mL of mHTF and pour this final rinse into the tissue culture dish.
3. Search the dishes for oocytes under the microscope until all of the dishes have been searched.

4. Strip away the cumulus cells surrounding the oocytes using two sterile 1-cm^3 syringes attached to two 30G needles if excessive blood or debris is present once the retrieval is finished. For this, use one needle to hold the granulosa cells and the other needle to cut the cells away from the oocyte.
5. Rinse all oocytes in the moat of an insemination dish and then place oocytes into the center well (maximum of 10 eggs/well).
6. Place the dishes containing the eggs into the incubator until insemination (5 h post-retrieval) (*see* **Note 3**).

3.3. Sperm Preparation (Swim-up)

On the day of egg retrieval, have the male partner produce a semen sample in a sterile sample cup. The male partner should be told to abstain from ejaculation for 48–72 h before the swim-up. The purpose of an insemination swim-up is to isolate the majority of motile sperm in the male partner's ejaculate and rid the sample of most of the dead sperm and debris. This motile sample is then used to inseminate the female partner's eggs. Sometimes, cryopreserved semen is used. Take the following steps to process the sample.

3.3.1. Fresh Sperm Preparation

1. Place the cup containing semen at room temperature to liquefy for at least 45 min.
2. After 45 min, check to see if the sample has liquefied; do this by drawing semen up into a sterile 9-in. Pasteur pipet. If the sample is difficult to draw up or stringy when emitted, let the sample sit for another 15 min.
3. If the sample has not liquefied after 60 min, use a 18-gauge needle attached to a 10-mL syringe to aspirate the sample up the syringe several times (do this slowly so as not to create any bubbles) or use chymotrypsin to help liquefy the sample. To make the chymotrypsin solution, add 1 mL of sperm-washing medium to a vial of chymotrypsin until dissolved. Add the solution to the sperm sample and let it sit at room temperature for 5 min.
4. Measure the volume of semen with a 5-mL serological pipet after the sample has liquefied.
5. Place two drops of semen in warmed slides for motility assessment.
6. Add two 25-µL semen sample to test tubes for concentration assessment.
7. Divide the sperm sample into four conical tubes (centrifuge tubes) and add sperm-washing medium equal to two times the volume of semen to each tube.
8. Vortex the tubes gently and centrifuge twice at 200*g*, each for 5 min.
9. Once the tubes stop spinning, remove supernatant with a sterile 9-in. Pasteur pipet; make sure the pellet remains.
10. Slowly add 250 µL of sperm-washing medium to each tube and loosely cap them. Place the rack of tubes in the incubator for 75 min to allow the sperm to "swim-up."
11. After 75 min, remove tubes from the incubator. Remove the supernatant from all tubes with a sterile 9-in. Pasteur pipet and place in a small sterile snap-cap tube. Determine sperm motility and concentration.

12. Calculate the insemination volume. The concentration for insemination is 135,000 motile sperm per milliliter.
13. Place sperm sample back in the incubator until insemination (5 h postretrieval).

3.3.2. Frozen Semen Thawing and Preparation

1. Prepare water bath and set the temperature at 37°C.
2. Locate frozen sperm; the log book tells in which tank, row, and cane the sperm is stored.
3. Once the vial is removed from the liquid-nitrogen storage tank, place vial in a 37°C water bath for 10 min.
4. Determine the volume of the sample and calculate sperm concentration and motility. If the motile count is not high enough to inseminate all eggs, thaw a second vial.
5. Add sample to a conical tube. If the sample volume is greater than 4 mL, divide the sample between two tubes. Slowly add three times the volume of sperm-washing medium to the sample.
6. Resuspend the sperm and the medium with a sterile 9-in. Pasteur pipet.
7. Centrifuge at 200*g* for 5 min.
8. Finish preparation of sample as in **Subheading 3.3.1.**, **steps 9–13**.

3.4. Insemination of Oocytes (see Note 5)

Oocytes are inseminated 5 h postretrieval with prepared human sperm.

1. Set the pipetman for the correct amount of sperm (in microliters) necessary for 135,000 motile sperm per milliliter.
2. Wipe the end of the pipetman with an alcohol prep pad and let it air-dry.
3. Remove sperm from the incubator. Always double check to verify that you have the correct sperm.
4. Agitate the sperm by gently tapping the capped tube.
5. Ensure that all of the oocytes are present in the insemination dish before insemination.
6. Place the pipet tip into the insemination dish away from the oocytes and slowly inject the sperm.
7. Return the dish to the incubator (*see* **Note 3**).
8. Record the date and time of insemination on the patient's chart (*see* **Note 2**).

3.5. Fertilization Assessment (see Note 3)

The process of checking for fertilization of the eggs involves the removal (or stripping) of the cumulus cells surrounding the egg, followed by a microscopic evaluation to determine the fertilization status 12–20 h after insemination. Once the cumulus cells are removed, it is easy to determine whether or not two pronuclei have occurred. The two pronuclei appear between 12 and 20 h after insemination and then disappear with syngamy.

3.5.1. Day Before Fertilization Assessment

1. Make 10 mL of 10% growth (cleavage) medium in a 50-mL flask by adding 1 mL SSS to 9.0 mL growth medium (*see* **Tables 2** and **3**).
2. Make growth dishes with growth medium. In 35-mm dishes, place up to six 25-µL drops of growth medium and fully cover with oil. Make one washing dish using a 60-mm dish. Place up to eight 50-µL drops in the dish and fully cover with oil.
3. Place the medium and dishes to the incubator.

3.5.2. Day of Fertilization Assessment

1. Perform daily quality control measures prior to fertilization assessment.
2. Make a sterile set of stripping pipets. These pipets are made using presterilized 9-in. Pasteur pipets and pulled using the flame of an alcohol lamp.
3. Take out the first insemination dish.
4. Strip the granulosa and cumulus cells off the oocytes until the pronuclei are visible. Do the stripping under low magnification on the dissecting microscope so that you can watch the oocyte as it comes out of the pipet.
5. Check fertilization status as shown in **Fig. 2**, wash the oocytes three or four times, place one oocyte in each 25-µL drop of growth medium covered by oil, and label it on the bottom of dishes as 1, 2, 3, 4, etc., and write down the fertilization results in the patient's chart as 0 PN (pronucleus), 1 PN, 2 PN, 3 PN, or whatever you find.
6. Check all oocytes with no pronucleus to see how many of them have a polar body. Record the number in the patient's chart.
7. Discard all aneuploid (≥3 PN) and degenerated oocytes, which are considered nonviable.
8. Culture embryos in growth medium until embryo transfer.

Fig. 2. *(opposite page)* Schematic morphology of human oocytes after fertilization: **(A)** unfertilized oocyte, and the oocyte is still at metaphase II stage and one polar body is in the perivitelline space; **(B)** oocyte has one pronucleus with one polar body present in the perivitelline space; this is an abnormal oocyte; **(C)** oocyte has one pronucleus with two polar bodies being in the perivitelline space, possibly delayed fertilization; it is necessary to examine the oocyte again within 12–24 h; if two pronuclei form, this oocyte is normal; **(D)** a normal fertilized oocyte, two pronuclei are in the cytoplasm and two polar bodies in the perivitelline space; more than 95% of human fertilized oocytes are normal; **(E,F)** three pronuclear oocytes with one or two polar bodies being in the perivitelline space; these oocytes are polyspermy and aneuploid embryos usually result from this kind of oocytes; 3–5% of fertilized oocytes in human form three or more pronuclei; all three pronuclear oocytes should be discarded. Pb: polar body. ZP: zona pellucida. PN: pronucleus.

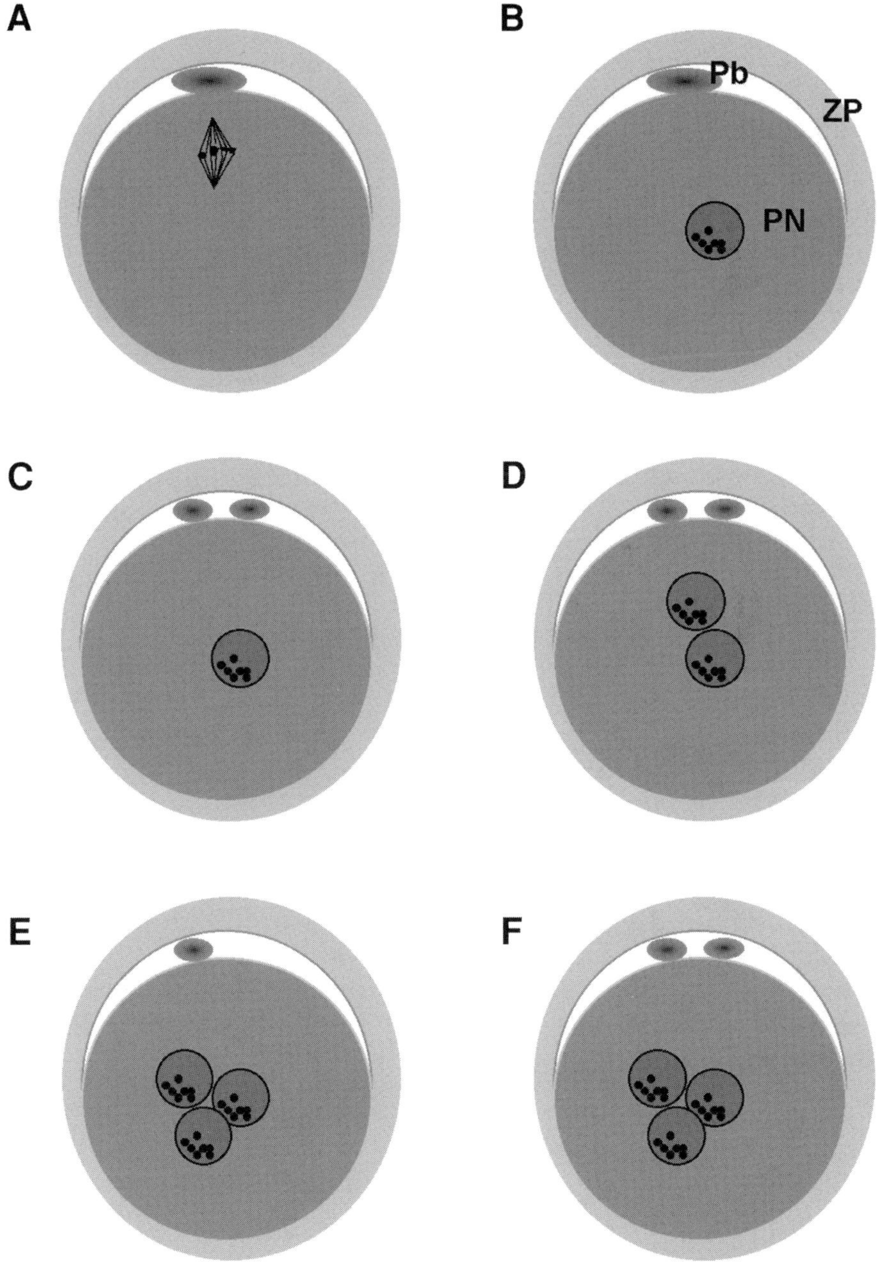

Fig. 2.

3.6. Embryo Evaluation and Embryo Transfer (see Note 3)

Once human oocytes have been successfully fertilized and divided in vitro, it is necessary to transfer them back to the patient. Embryos are at 2–4 cell stages at d 2, 6–10 cell stages at d 3, or blastocyst stage at d 5 or 6. Determine the transfer day based on clinical protocols and the number of embryos the patient has.

1. Check the embryo transfer consent to determine the number of embryos a patient wants to transfer.
2. Select the best quality of embryos before physician is ready for transfer.
3. When patient is ready for embryo transfer, carefully remove the dish containing embryos from the incubator and place on the stage of the dissecting microscope.
4. Using a sterile pulled Pasteur pipet, round up all of the embryos in the transfer dish.
5. Take a 35-mm tissue culture dish and place it on the stage of the microscope alongside the dish containing embryos. Make a drop of 100 μL of growth medium in the 35-mm dish.
6. Draw up all of the embryos from the transfer dish into the pipet. Gently expel the embryos into the small drop of medium so that all of the embryos remain close together in the drop.
7. Put on sterile gloves.
8. Pick up the catheter, attach 1-cm^3 syringe to the catheter, and wash the catheter for five times with the growth medium.
9. Detach the syringe from the catheter, and fill with medium from the transfer dish. Remove all air from the syringe; then, reattach to the catheter and push the plunger all the way down to fill the catheter with medium.
10. With one hand, grasp the catheter with thumb and index finger, about 1-in. away from the end. Use the other hand to grasp the 1-cm^3 syringe.
11. First draw approx 0.5 cm of air, and then place the open end of the catheter next to the embryos in the drop of medium. Draw up the embryos in 0.5 cm of medium or less. This should be done while looking through the microscope to be certain that all of the embryos are within the catheter.
12. After drawing up the embryos, draw up 0.5 cm of air followed by 0.5 cm of medium.
13. Carefully walk the catheter into the transfer room.
14. Hand off the catheter to the IVF physician. The physician will insert the catheter through the patient's cervix and expel the embryos into the uterus.
15. Return the catheter to the laboratory after the physician transfers the embryos.
16. Remove the syringe from the catheter, making sure the end of the catheter is over an empty 35-mm dish.
17. Attach a syringe filled with medium to the end of the catheter and expel the contents into the dish.
18. Attach a second syringe to the catheter and expel the content. Remove the syringe, draw up air, and reattach the syringe to the catheter. Depress the plunger again, making sure that all of the contents are expelled.

19. Examine the end of the catheter under the microscope. Pay close attention to any cervical mucous, which may contain embryos. Attach a third syringe to the catheter sheath and expel the contents into a dish.
20. Thoroughly search both of the dishes to make certain that no embryos have been left behind in the catheter.
21. If any embryos have been left behind, they should be reloaded into a new catheter, as explained above.
22. At this point, if no embryos can be found, transfer is complete and one IVF cycle is done. If there are extra embryos, they should be cryopreserved for future use.

4. Notes

1. Equipment, culture media, plasticware, catheters, and gloves should be tested to see whether they have any toxic effects on embryos. All new brands of items that contact gametes must be subjected to quality control testing before use. For quality control testing, mouse embryo culture and human sperm survival test are the most popular tests in many IVF laboratories. At least 75% of mouse embryos must reach blastocysts in order to pass the mouse embryo assay or 85% of human sperm must have motility as compared with control.
2. As with any procedure that is done in IVF laboratory, great care must be taken to document every step on the correct forms.
3. It is extremely important to minimize pH and temperature fluctuations during gamete in vitro manipulation. Therefore, perform the procedure as quickly as possible and then return the dishes to the incubator.
4. The protocols for ovarian stimulation are dependent on the physician's experience, patient's age, diagnosis, and hormone levels. Sometimes, changes are necessary during stimulation.
5. Fertilization failure in IVF happens in some patients. Most fertilization failure is related with semen quality, and sometimes it is related with oocyte quality, such as immature oocyte. Therefore, in order ensure a successful fertilization, a complete semen analysis, including sperm morphology, is necessary before IVF treatment.

References

1. Edwards, R. G. (1965) Maturation in vitro of human ovarian oocytes. *Lancet* **ii,** 926–929.
2. Edwards, R. G., Bavister, B. D., and Steptoe, P. C. (1969) Early stages of fertilization in vitro of human oocytes matured in vitro. *Nature* **221,** 632–735.
3. Steptoe, P. C. and Edwards, R. G. (1978) Successful birth after IVF. *Lancet* **ii,** 366.
4. Veeke, L. L., Wortham, J. W., Jr., Witmyer, J., et al. (1983) Maturation and fertilization of morphologically immature human oocytes in a program of in vitro fertilization. *Fertil. Steril.* **39,** 594–602.
5. Trounson, A., Wood, C., and Kausche, A. (1994) *In vitro* maturation and fertilization and developmental competence of oocytes recovered from untreated polycystic ovarian patients. *Fertil. Steril.* **62,** 353–362.

6. Cha, K. Y., Koo, J. J., Ko, J. J., et al. (1991) Pregnancy after in vitro fertilization of human follicular oocytes collected from nonstimulated cycles, their culture in vitro and their transfer in a donor oocyte program. *Fertil. Steril.* **55,** 109–113.
7. Liu, J., Katz, E., Garcia, J. E., et al. (1997) Successful *in vitro* maturation of human oocytes not exposed to human chorionic gonadotropin during ovulation induction, resulting in pregnancy. *Fertil. Steril.* **67,** 566–568.
8. Chian, R. C., Buckett, W. M., Too, L. L., et al. (1999) Pregnancies resulting from in vitro matured oocytes retrieved from patients with polycystic ovary syndrome after priming with human chorionic gonadotropin. *Fertil. Steril.* **72,** 639–642.
9. Jaroudi, K. A., Hollanders, J. M., Elnour, A. M., et al. (1999) Embryo development and pregnancies from in-vitro matured and fertilized human oocytes. *Hum. Reprod.* **14,** 1749–1751.
10. Chian, R. C., Buckett, W. M., Tulandi, T., et al. (2000) Prospective randomized study of human chorionic gonadotrophin priming before immature oocyte retrieval from unstimulated women with polycystic ovarian syndrome. *Hum. Reprod.* **15,** 165–170.
11. Cha, K. Y., Han, S. Y., Chung, H. M., et al. (2000) Pregnancies and deliveries after in vitro maturation culture followed by in vitro fertilization and embryo transfer without stimulation in women with polycystic ovary syndrome. *Fertil. Steril.* **73,** 978–983.
12. Yoon, H. G., Yoon, S. H., Son, W. Y., et al. (2001) Pregnancies resulting from in vitro matured oocytes collected from women with regular menstrual cycle. *J. Assist. Reprod. Genet.* **18,** 325–329.
13. Kim, B. K., Lee, S. C., Kim, K. J., et al. (2000) In vitro maturation, fertilization, and development of human germinal vesicle oocytes collected from stimulated cycles. *Fertil. Steril.* **74,** 1153–1158.
14. Combelles, C. M., Cekleniak, N. A., Racowsky, C., et al. (2002) Assessment of nuclear and cytoplasmic maturation in in-vitro matured human oocytes. *Hum. Reprod.* **17,** 1006–1016.
15. Wang, W. H., Meng, L., Hackett, R. J., et al. (2001) Limited recovery of meiotic spindles in living human oocytes after cooling–rewarming observed using polarized light microscopy. *Hum. Reprod.* **16,** 2374–2378.
16. Wang, W. H. and Keefe, D. L. (2001) Prediction of chromosome misalignment among in vitro matured human oocytes by spindle imaging with the Polscope. *Fertil. Steril.* **78,** 1077–1081.
17. Chian, R. C. and Tan, S. L. (2002) Maturational and developmental competence of cumulus-free immature human oocytes derived from stimulated and intracytoplasmic sperm injection cycles. *Reprod. Biomed.* Online **5,** 125–132.
18. Nagele, F., Sator, M. O., Juza, J., et al. (2002) Successful pregnancy resulting from in-vitro matured oocytes retrieved at laparoscopic surgery in a patient with polycystic ovary syndrome: case report. *Hum. Reprod.* **17,** 373–374.
19. Cha, K. Y. and Chian, R. C. (1998) Maturation in vitro of immature human oocytes for clinical use. *Hum. Reprod. Update* **4,** 103–120.

20. Goldzieher, J. W. and Green, J. A. (1962) The polycystic ovary. I. Clinical and histologic features. *J. Clin. Endocrinol. Metab.* **22,** 325–338.
21. Mikkelsen, A. L. and Lindenberg, S. (2001) Benefit of FSH priming of women with PCOS to the in vitro maturation procedure and the outcome: a randomized prospective study. *Reproduction* **122,** 587–592.
22. Cha, Y. K. (1992) In vitro fertilization using immature follicular oocytes harvested from ovary tissue, in *Progress in Infertility* (Behrman, S. J., Paatton, G. W., Jr. and Holtz, G., eds.), Little, Brown and Co., New York, Vol. 7, pp. 99–112.
23. Cha, K. Y., Do, B. R., and Chi, H. J. (1992) Viability of human follicular oocytes collected from unstimulated ovaries and matured and fertilized in vitro. *Reprod. Fertil. Dev.* **4,** 685–701.
24. Barnes, F. L., Kausche, A., Tiglias, J., et al. (1996) Production of embryos from in vitro-matured primary human oocytes. *Fertil. Steril.* **65,** 1151–1156.
25. Quinn, P., Kerin, J. F., and Warnes, G. M. (1985) Improved pregnancy rate in human in vitro fertilization with the use of a medium based on the composition of human tubal fluid. *Fertil. Steril.* **44,** 493–498.
26. Quinn, P., Moinipanah, R., Steinberg, J. M., et al. (1995) Successful human in vitro fertilization using a modified human tubal fluid medium lacking glucose and phosphate ions. *Fertil. Steril.* **63,** 922–924.
27. Carrillo, A. O., Lane, B., Pridman, D. D., et al. (1998) Improved clinical outcomes for in vitro fertilization with delay of embryo transfer from 48 to 72 hours after oocyte retrieval: use of glucose- and phosphate-free media. *Fertil. Steril.* **69,** 329–334.
28. Coates, A., Rutherford, A. J., Hunter, H., et al. (1999) Glucose-free medium in human in vitro fertilization and embryo transfer: a large-scale, prospective, randomized clinical trial. *Fertil. Steril.* **72,** 229–232.
29. Gardner, D. K., Vella, P., Lane, M., et al. (1998) Culture and transfer of human blastocysts increases implantation rates and reduces the need for multiple embryo transfer. *Fertil. Steril.* **69,** 84–88.

18

In Vitro Maturation and Fertilization of Canine Oocytes

Karine Reynaud, Marie Saint-Dizier, and Sylvie Chastant-Maillard

1. Introduction

1.1. Generalities on Bitch Cycle and Ovarian Function

In vitro maturation of canine oocytes was first described in 1976 *(1)*, but reproductive biotechnologies applied to canine species developed mostly during the last 10 yr. Few studies have been conducted on dogs because the availability of biological material (bitch ovaries in particular) is limited, whereas bovine, porcine, ovine, or equine ovaries can be found in a slaughterhouse and smaller animals (mice, rats) can be bred easily and at low cost.

The dog (*Canis familiaris*) is a monoestral polyovulatory nonseasonal species. The cycle is divided in three phases: heat period, divided into proestrus (3–20 d, the bitch is not receptive to mating) and estrus (1–10 d, the bitch is receptive to mating), diestrus (2 mo; also called metestrus, corresponding to pseudopregnancy or pregnancy with secretion of progesterone by the corpora lutea), and anestrus (3–9 mo) (for review, *see* **ref. *2***).

Reproductive endocrinology during the cycle in this species is unusual: During proestrus, estradiol concentration increases progressively, but during estrus, the ratio between estradiol and progesterone changes. Estradiol decreases at the beginning of estrus, whereas progesterone starts to increase a few days before ovulation (at the moment of luteinizing hormone [LH] peak). This preovulatory luteinization, with a progesterone level between 5 and 7 ng/mL at the moment of ovulation, is typical of the bitch. As a consequence, oocytes present in their preovulatory follicles are exposed to increasing levels of progesterone before ovulation, which is spontaneous, occurring 1–3 d after LH surge (sometimes more). Duration of the persistence of corpora lutea is then more or less similar whether the bitch is

From: *Methods in Molecular Biology, vol. 253: Germ Cell Protocols: Vol. 1 Sperm and Oocyte Analysis*
Edited by: H. Schatten © Humana Press Inc., Totowa, NJ

pregnant or not (2 mo) and luteolysis is followed by a particularly long period of ovarian inactivity of 3–9 mo.

1.2. Manipulation of Cycle and Ovaries

Active bitch ovaries on natural cycles are scarce material (one follicular phase on average every 6 mo). Unfortunately, induction of artificial cycles (follicular growth) and superovulation are difficult in the bitch and imply heavy protocols. Protocols with estrone, equine chorionic gonadotrophin (eCG) + human chorionic gonadotrophin (hCG), estradiol, and hCG *(3)* or with a dopamine agonist like cabergoline *(4)* have been tested, but often with limited results.

1.3. Oocyte Collection

The simplest way to gain access to canine oocytes is ovary collection after routine ovariectomy. However, it is classically recommended to perform ovariectomy during the anestrus phase. Ovariectomy during proestrus and estrus can appear surgically more difficult because of a higher vascularization of the tractus; furthermore, neutering during metestrus can induce overt pseudopregnancy and persistent lactation.

In bovine, oocytes are now widely collected in vivo through ultrasonography or endoscopy-guided follicular puncture (ovum pick-up). This technique classically allows one to obtain an average of five oocytes two times a week during months, without hormonal treatment. Regrettably, the endoscopic method is difficult to transpose in the bitch because of the presence of an adipous ovarian bursa enclosing ovaries; ultrasonography-guided puncture remains to be tested, but whichever method is chosen, the limit remains in the scarcity of follicular phases.

1.4. Particularities of Oocyte Maturation in the Bitch

In the bitch, oocyte maturation is very different from that observed in other species (mouse, bovine, porcine, ovine, feline, etc.). In the vast majority of mammals, nuclear maturation occurs in the preovulatory follicle and oocytes are ovulated at the metaphase II stage and are already fertilizable. On the contrary, in the bitch, oocytes are ovulated at the immature germinal vesicle (GV) stage and reach metaphase II stage after more or less 48 h spent in the oviduct. When oocytes are collected from large follicles, bitch oocytes need at least 48 h to reach the in vitro metaphase II stage (only 18–24 h in mouse or bovine oocytes). Furthermore, in vitro maturation is really difficult in the bitch and researchers failed to produce numerous metaphase II oocytes and, as a consequence, a high number of canine embryos. Indeed, in vitro oocyte maturation rates can vary from 0% to 39% (metaphase I to II stages pooled), with a mean of 10–20%. Oocytes collected from preovulatory follicles exhibit a higher maturation rate (up to 32% of metaphase II), but, as we discussed earlier, they are scarce material, explaining why the large majority of works are conducted with oocytes from anestrus bitches. In addition to the role of the follicular stage, a possible positive role of sperm penetration in the oocyte maturation has been suggested ***(5)***.

Another difficulty is that canine oocytes (like porcine or feline oocytes) present a dark cytoplasm, because of high lipid content, which interferes with visibility and the determination of the nuclear stage after maturation. It explains why researchers often present maturation results with a high rate of oocytes at a "not determined" nuclear stage. Different staining methods have been tested to evaluate the oocyte nuclear stage more exactly, such as aceto/orcein or Hoechst 33258 ***(6)***. However, as described below, we think that confocal microscopy is even a better method to evaluate the nuclear stage with a higher accuracy (Saint-Dizier et al., in press).

2. Materials

2.1. Ovaries and Oocytes Collection

1. Phosphate-buffered saline (PBS): 0.01 *M* phosphate buffer, 2.7 m*M* potassium chloride, and 0.127 *M* sodium chloride, pH 7.4 (PBS Tablets; Sigma, cat. no. P-4417).
2. Dissection medium: M199 with 25 m*M* HEPES and Earle's salts supplemented with 20% heat-inactivated fetal calf serum (FCS), penicillin G (100 μg/mL), and streptomycin sulfate (100 IU/mL) (Sigma) and filtered (0.2-μm sterile filter). Decomplement the serum by warming it at 56°C for 30 min; then, aliquot and store at –20°C.
3. Isotherm bottle; Falcon tubes, 50 mL; scalpel or razor blades (*see* **Fig. 1A**); large, plastic Petri dishes (90 mm in diameter); warming plate; sterile hood.
4. Mouth pipet (*see* **Fig. 1B**,**C**): This useful tool can be made in the lab using mouth tips (or classic yellow or blue tips for a micropipet), two different sizes of rubber tube (0.5 and 1 cm in diameter), a 0.2-μm filter (classically used to sterilize culture medium), and a Pasteur pipet. The diameter of the pipet hole should be adjusted by heating the extremity on a flame and pulling it before cutting at the desired diameter.

2.2. Oocyte Selection

1. Stereomicroscope equipped with a heating stage. The heating stage can be purchased from different companies (e.g., Olympus, stereomicroscope SZX + heating plate MATS-SZX; see also the website http://www.tokaihit.com/).
2. Organ culture dish (Falcon 3037) and four-well dishes (Nunc).

2.3. In Vitro Maturation

1. Maturation medium: M199 medium without HEPES, supplemented with 20% FCS and filtered (0.2-μm sterile filter).
2. Incubator with a humidified atmosphere of 5% CO_2 in air, at 38°C.

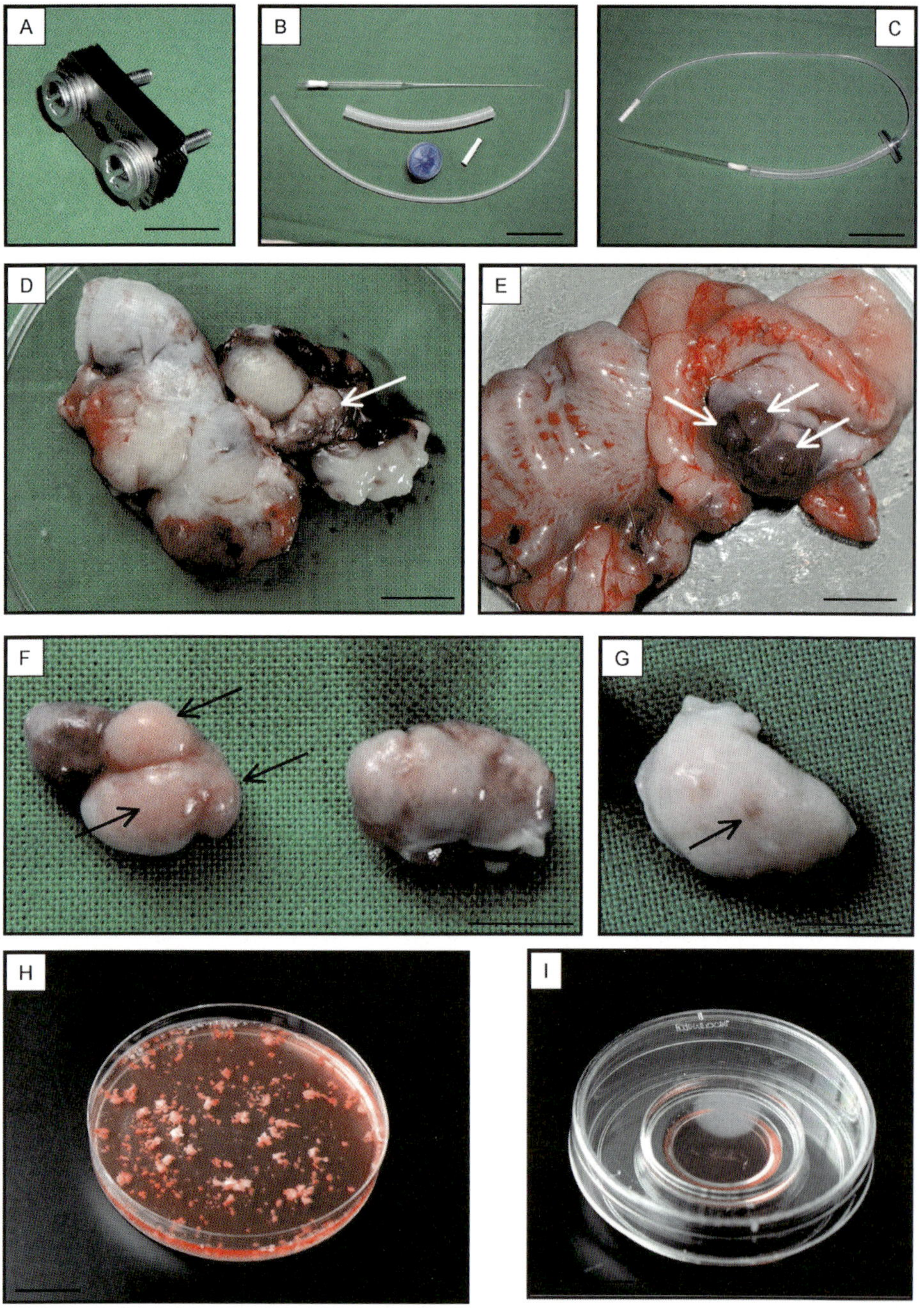

Fig. 1. Technical and biological material for canine oocyte collection. Ovaries are sliced using razor blades **(A)** and collected with a mouth pipet **(B,C)**. Before beginning the slicing, ovaries should be separated from the ovarian bursa and the fat.

2.4. Evaluation of Oocyte Maturation

2.4.1. Denudation

1. Pronase (Roche-Boehringer; cat. no. 165921) solution, 7 mg/mL in M199/HEPES
2. FCS and small incubator at 38°C (not gassed).

2.4.2. Fixation

1. Solution 4% paraformaldehyde (PAF): In a fume hood, weigh 4 g of paraformaldehyde powder into a beaker and add 100 mL PBS. Place the beaker in a larger beaker of water and heat to 60°C (always in the fume hood). Do not boil. Stop the procedure when the solution is clear; let it cool and store in the refrigerator. The solution can be kept for 1 mo at 4°C or can also be aliquoted and frozen at –20°C.
2. 2% paraformaldehyde, 0.25% Triton X-100: 5 mL of 4% PAF + 5 mL PBS + 25 µL Triton X-100 (Sigma).
3. 4% paraformaldehyde, 0.25% Triton X-100: add 25 µL of Triton X-100 in 10 mL of 4% PAF.
4. PBS, 1% bovine serum albumin (PBS/BSA): Dissolve 1 g of BSA fraction V (Sigma) in 100 mL PBS.
5. 0.5% Triton X-100: Dissolve 500 µL Triton X-100 in 100 mL PBS/BSA.

2.4.3. Staining Procedure for DNA (Hoechst 33342)

Make a stock solution with 1 mg/mL Hoechst 33342 (2'-[4-ethoxyphenyl]-5-[4-methyl-1-piperazinyl]-2,5'-bi-1*H*-benzimidazole; Sigma). Because Hoechst powder is toxic and light sensitive, do not weigh the powder, but add PBS directly in the tinted flask. Make aliquots, protect them from the light, and store at –20°C. Prepare the working solution: Dissolve 1 µL (1 µg) of Hoechst stock solution in 1 mL of PBS/BSA and protect this solution from the light using aluminum foil.

2.4.4. Staining Procedure for DNA/Actin

Dissolve 2 µL of ethidium homodimer-2 (EH-2; Molecular Probes, cat. no. E-3599) and 20 µL of Oregon Green 488 phalloidin (Molecular Probes; cat. no. O-7466) in 1 mL of 0.25% Triton X-100/PBS/BSA. EH-2 stains DNA,

---

(Fig. 1 caption continued) **(D,E)** The bitch tractus (proximal part of the uterus, ovarian bursa) and ovary during anestrus **(D)** and preovulatory **(E)** phases. **(F,G)** Ovaries separated from the ovarian bursa and the fat. **(F)** Metestrus ovaries with corpora lutea (arrows) and **(G)** anestrus ovaries, on which antral follicles, as shown by the arrow, are rarely observed. These ovaries are then sliced **(H)** and good quality oocytes are selected and transferred in an organ culture dish **(I)**. Scale bars represent 1 cm **(F,G)**, 2 cm **(A, D, E, H, I)**, and 5 cm **(B,C)**.

whereas phalloidin stains F-actin protein (polymerized). Final working concentrations are 4 units/mL (0.13 μ*M*) and 4 μ*M* for EH-2 and Oregon Green phalloidin, respectively.

2.4.5. Staining Procedure for DNA/Tubulin

1. Propidium iodide (10 μg/mL).
2. Blocking solution: 0.5% Triton X-100 in PBS/BSA, 10% goat serum.
3. The first antibody is a monoclonal anti-α-tubulin (IgG1 isotype, clone DM1A; Sigma) raised against chicken microtubules and is diluted 1/200 in 0.25% Triton X-100/PBS/BSA. The secondary antibody is a goat polyclonal anti-mouse IgG antibody, labeled with Alexa Fluor 488 (Molecular Probes) and diluted 1/200 (10 μg/mL). We also tested another antibody: monoclonal antibody anti-α-tubulin (IgG_1 isotype, clone B-5-1-2, Sigma), which crossreacts also with canine oocytes.
4. Chamber slides (Labteck 177380).
5. Dako Pen, delimiting pen for immunocytochemistry (Dako S-2002).
6. Mounting medium for fluorescence (Vectashield, Vector Laboratories).

2.5. In Vitro Fertilization

1. M199 medium without HEPES, supplemented with 20% FCS and filtered (0.2-μm sterile filter).
2. Hemocytometer.

3. Methods

3.1. Preparation of Culture Dishes

1. The day before oocyte collection, prepare the culture dishes in the laminar-flow hood. A mean of one four-well dish (2 mL of culture medium) per ovary is necessary. Put 500 μL of culture medium in each well and fill in the space between the wells with 2 mL of sterilized water to avoid dehydration.
2. Let the culture dish equilibrate for one night in the gassed incubator.
3. Prepare the dissection medium (50 mL per ovary) and let it equilibrate at 38°C.

3.2. Ovaries Dissection and Oocytes Collection

1. The morning of ovariectomies, prepare the Petri dishes with 25 mL of dissection medium per dish and let them warm on the hot plate. Prepare the selection dishes (1 mL of dissection medium in the organ culture dish). Prepare the isotherm bottle with water at 38°C and Falcon tubes with PBS.
2. Ovaries are collected by routine ovariectomies (*see* **Notes 1** and **2**) and are transported to the laboratory in PBS in the tubes at 38°C (to avoid temperature shock). Ovaries should be processed as soon as possible (oocytes should be placed in culture 30 min to 1 h after ovary collection).

3. They are placed in dissection medium on the warming plate at 38°C, the ovarian bursa is dissected (*see* **Fig. 1D**), and the fat is removed (*see* **Note 3**). The size of the ovary depends on the dog breed (from 10 mm × 5 mm to 30 mm × 15 mm). Just before ovulation, the mean size of preovulatory follicles (which are four or five per ovary) is 7 mm (*see* **Fig. 1E**). Most of the time, ovaries are in metestrus or anestrus and no large antral follicles are visible (*see* **Note 4** and **Fig. 1F,G**).
4. When the bursa is removed, put the ovary in another Petri dish with fresh dissection medium and start the slicing (*see* **Fig. 1H**). To slice the ovaries and release cumulus–oocyte complexes (COCs), we use razor blades (*see* **Fig. 1A**), but a single razor or scalpel blade can also be used.

3.3. Oocyte Selection

1. After dissection, good quality oocytes are selected under a stereomicroscope, on a heating stage to avoid heat loss and minimize temperature variation. They are collected using a mouth pipet (*see* **Note 5** and **Fig. 1C**) and are transferred in an organ culture dish in new dissection medium (*see* **Fig. 1I**). Good quality oocytes should have uniformly dark cytoplasm (in some, GV can be visible as a clearer round zone; *see* **Fig. 2O**) with many lipid droplets and should be surrounded with at least two layers of granulosa cells (*see* **Fig. 2K**). The oocyte diameter varies from 70 to 130 μm in the bitch, but the best in vitro maturation (IVM) rate will be obtained with largest oocytes (>110 μm) *(7)*. The zona pellucida is thick in this species. Bad oocytes should be discarded: diameter smaller than 100 μm, denuded or with incomplete layers of cumulus cells, with a pale/vacuolized cytoplasm and/or heterogeneous cytoplasmic color, or already lysed (zona pellucida is damaged/ruptured or zona pellucida content appears flat and lighter) (*see* **Fig. 2J**). The yield of the dissection/selection steps is highly variable from bitch to bitch. The average number of oocytes collected is around 20 per ovary (from 10 to 100), but only 30–50% will be good quality oocytes after selection.

3.4. In Vitro Maturation

1. Oocytes are matured in M199 medium supplemented with 20% FCS (*see* **Notes 6** and **7**), for 72 h (*see* **Note 8**). Twenty oocytes are taken from the selection dish, washed in the first and the second wells of the equilibrated (38°C and 5% CO_2) culture dish, and then 10 oocytes are placed in the third and fourth wells (*see* **Notes 9** and **10**).
2. After 72 h of culture, oocyte maturation should be evaluated (determination of nuclear maturation stage).

3.5. Evaluation of Oocyte Maturation

With bitch oocyte, after maturation, none or a limited cumulus expansion is noted, but granulosa cells are fixed to the bottom of the dish (*see* **Fig. 2L**). The degeneration rate is very high (20–25%), even if good quality oocytes have been selected 72 h earlier.

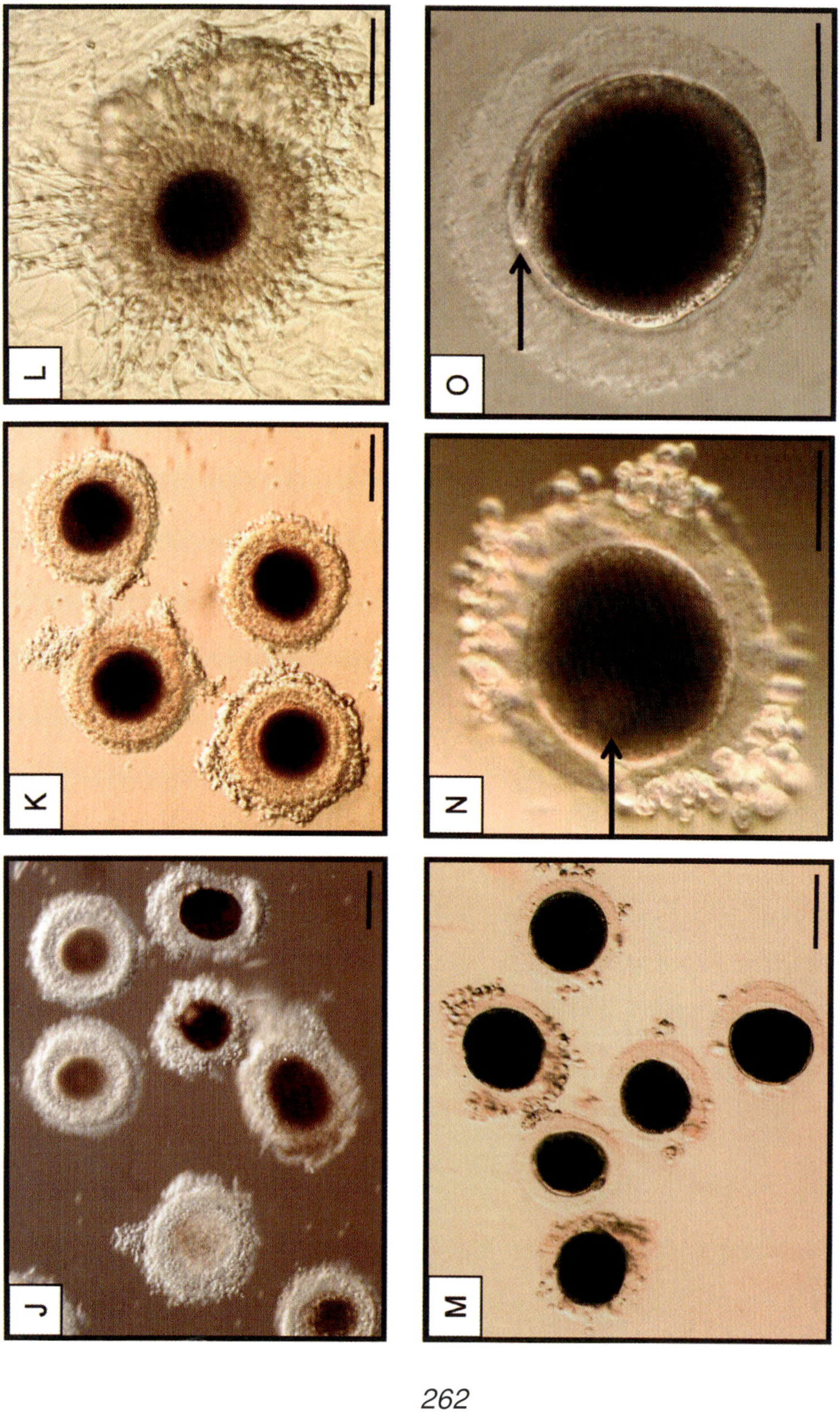

Fig. 2.

3.5.1. Decoronization

1. Before the evaluation of their nuclear stage, oocytes have to be denuded. This step of denudation is particularly difficult with canine oocytes, and even more difficult with in vivo than in vitro matured oocytes. Actually, cumulus cells are deeply linked to the oocyte, via their transzonal projections (through the zona pellucida). For instance, even when the zona pellucida is totally digested using enzymes, some granulosa cells can sometimes be still observed, attached on the oocyte!

   Hyaluronidase treatment, classically used in species like mouse or rabbit, is inefficient in canine oocytes. The best method is gentle pipeting, using a mouth pipet. However, sometimes, cumulus cells remain around the oocyte. In these cases, an enzymatic method, using pronase, can be applied (*see* **Note 11**).
2. After thawing, centrifugate pronase solution (5 min, 3000*g*) and prepare a four-well dish with one well of 0.7% pronase (500 µL) and three wells of M199/10% FCS. Equilibrate this dish at 38°C in the ungassed incubator.
3. Incubate oocytes for 30 s in the first well (on the hot plate at 38°C) and then rinse them in the second and third wells (*see* **Note 12**). Finally, leave them in the fourth well for 15 min, at 38°C in the ungassed incubator.

3.5.2. Fixation

1. After 15 min, denude oocytes by gentle pipeting and then fix them in two steps: 30 min in 2% paraformaldehyde/0.25% Triton X-100/PBS for 30 min at 38°C (some proteins, such as tubulin, are highly sensitive to temperature changes) (*see* **Note 13**) followed by 30 min in 4% paraformaldehyde/0.25% Triton X-100/PBS at 4°C.
2. Keep the oocytes in PBS/BSA (*see* **Note 14**) in four-well dish at 4°C, with 2 mL of distilled water between the wells to prevent dehydration, until the staining procedure.
3. Before staining, oocytes can be checked using optical microscopy (*see* **Fig. 2M**), and in some cases, the nuclear stage can be visible when a clear circular zone is visible in the cytoplasm (GV stage; *see* **Fig. 2N**) or when polar body has been extruded (*see* **Fig. 2O**).

Fig. 2. Poor quality oocytes **(J)** with light cytoplasm; those lysed or too small (<100 µm) are discarded and good quality oocytes **(K)** are placed in the culture. After 48 h, almost no cumulus expansion is observed, but numerous granulosa cells are plated on the bottom of the culture dish **(L)**. At the end of the culture (48 or 72 h), oocytes are denuded **(M)**. The nuclear stage can sometimes be observable under the stereomicroscope: **(N,O)** Two oocytes at the germinal vesicle and metaphase II (with the visible polar body) stages, respectively. Scale bars represent 50 µm **(N, O)** and 100 µm **(J, K, L, M)**.

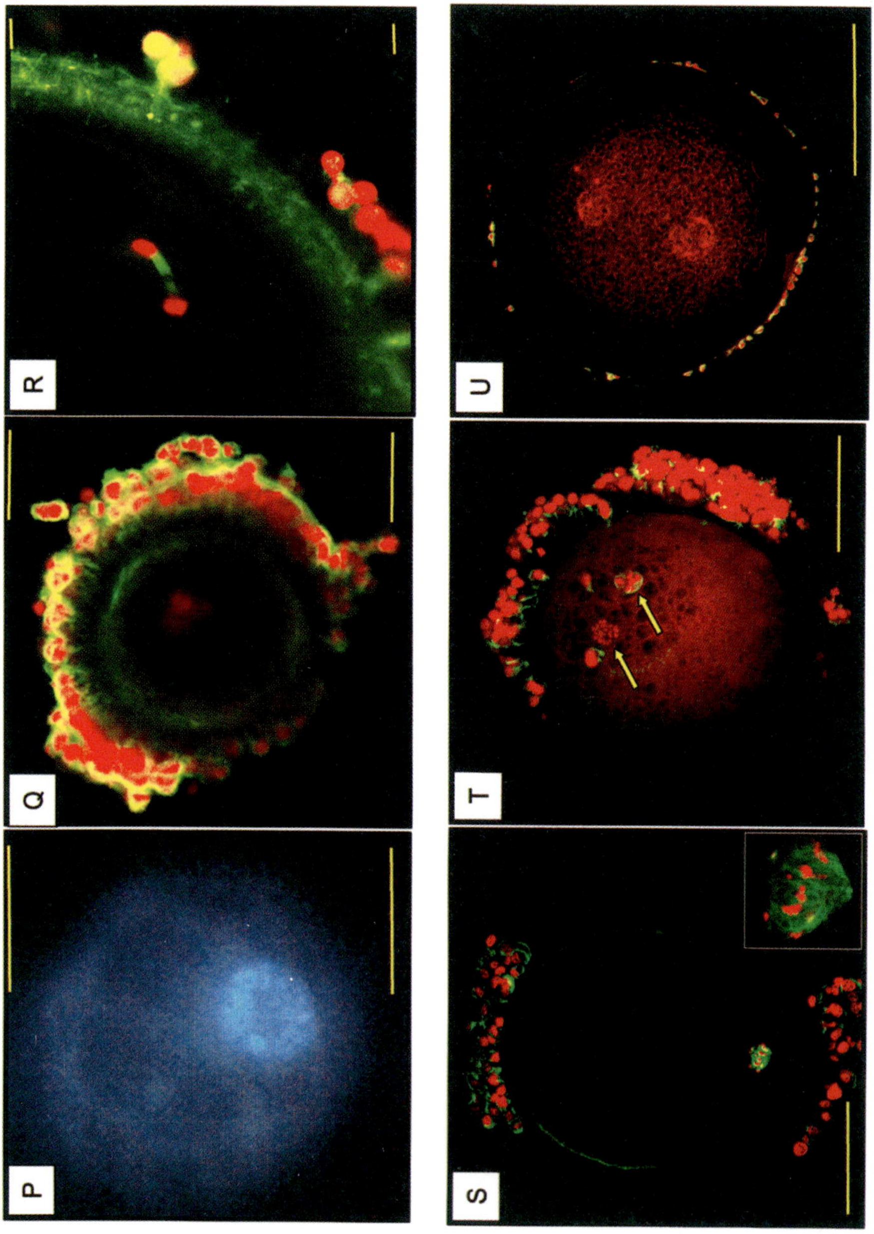

Fig. 2.

3.5.3. Staining Procedure for DNA (Hoechst)

1. Rinse the oocytes in PBS and incubate them for 30 min in Hoechst 33342 in a four-well dish, protected from light by using aluminum foil. During the incubation time, the dish may be gently shaken on an agitator.
2. Rinse in PBS for 10 min and observe under ultraviolet (UV) light (*see* **Note 15**).

3.5.4. Staining Procedure for DNA and Actin (see **Note 16**)

Microfilaments are cytoskeletal elements from actin molecules. Actin filaments are most highly concentrated in the cellular cortex, just beneath the plasma membrane. Thus, detection of actin filaments allows us to visualize the shape and the limits of the cells (oocyte and cumulus cells). Actin can be detected using phallotoxins, isolated from the *Amanita phalloides* mushroom, which bind specifically to F-actin in small and large filaments. These phallotoxins can be conjugated to fluorescent dyes such as Texas Red or Oregon Green (used in our method).

1. Prepare a four-well dish with 500 μL of 0.5% Triton X-100 in PBS/1% BSA in the first well, 500 μL of phalloidin Oregon green/EH-2/PBS/BSA solution (*see* **Notes 17** and **18**) in the second well, and 500 μL PBS/BSA in the third and the fourth wells. Add 2 mL water between the wells to avoid dehydration and protect from the light using aluminum foil (*see* **Note 19**).
2. First, permeabilize the oocytes in 0.5% Triton/PBS/BSA for 15 min.
3. Then, transfer in EH-2/phalloidin green/0.25% Triton X-100 for 30 min at room temperature (*see* **Note 20**).
4. Rinse two times 10 min in PBS/BSA. Oocytes can be stored in PBS/BSA, protected from the light, at 4°C for 2–3 d without significant loss of staining.

3.5.5. Staining Procedure for DNA and Tubulin

1. First, permeabilize the oocytes in 0.5% Triton X-100/PBS/BSA/10% goat serum (*see* **Note 21**) for 1 h.
2. Rinse in PBS/BSA (5 min) and then incubate with the first antibody (anti-α-tubulin; dilution 1/200) for 2 h (*see* **Note 22**) at room temperature.

Fig. 2. *(continued)* **(T)** An oocyte at germinal vesicle stage, observed after Hoechst staining. However, confocal microscopy is the best way to observe nuclear stage. Different stages such as germinal vesicle **(Q)**, telophase **(R)**, metaphase I **(S)**, and metaphase II **(T)** can be observed. **(S)** A magnification of the spindle can be seen on the bottom right-hand side of the photograph. Chromosomes, aligned on the metaphase plate and stained using ethidium homodimer, are visible. Arrows on **(T)** indicate metaphase II plate on the left and polar body on the right, and small nuclei of granulosa cells are also visible. **(U)** A canine zygote at the two-pronuclei stage, after fertilization. Scale bars represent 5 μm **(R)** and 50 μm **(P, Q, S, T, U)**.

3. Rinse the oocytes (three times 10 min in PBS/BSA, to avoid background) before incubation with the secondary antibody (anti-mouse IgG conjugated to Alexa Fluor, dilution 1/200) (*see* **Note 23**).
4. Incubate 2 h and then rinse extensively the oocytes (two times 20 min) to avoid background induced by unbound fluorescent secondary antibody.

3.5.6. Observation by Confocal Laser Scanning Microscopy

1. Mount the oocytes on microscope slides: make a circle (1 cm in diameter) with a Dako pen, to delimitate the position of the oocytes on the slide. Then, create a space between the slide and the coverslip using nail polish, put one drop of mounting medium (*see* **Note 24**) on the slide, and place the oocytes in this drop.
2. Observe the oocytes using a confocal microscope equipped with argon/krypton (or helium) lasers. For actin/DNA staining, absorption/emission wavelengths are 498/624 and 496/520 nm for EH-2 and Oregon Green, respectively. As a consequence, both fluorochromes can be excited with the argon-ion laser. For tubulin staining, absorption/emission wavelengths of Alexa Fluor are 495/519 nm.

3.6. In Vitro Fertilization

1. Dog semen should be collected from two different dogs with known good in vivo fertility, using manual stimulation (*see* **Note 25**). Three different fractions are obtained (presperm, sperm-rich, and prostatic phases). The prostatic phase contains few spermatozoa and for in vitro fertilization (IVF), only the second one (sperm-rich fraction) is used.
2. Assess this ejaculated semen for sperm motility using optical microscopy and use only samples with >80% motile spermatozoa.
3. Evaluate sperm concentration using a hemocytometer (*see* **Note 26**).
4. Transfer matured oocytes in wells with fresh maturation medium and incubate with spermatozoa at a final concentration of 10^6 motile spermatozoa/mL (*see* **Note 27**).
5. Leave sperm in contact with oocytes for 24 h (*see* **Note 28**).
6. The day after, rinse oocytes/eggs, transfer them to fresh culture medium, and culture for 24–48 h more (*see* **Note 29**).

   The mean fertilization rate (evaluated by DNA and tubulin staining) is relatively low. This can be the result of the very low oocyte maturation efficiency (few metaphase II), of abnormal sperm capacitation, or of the high rate of polyspermy (*see* **Note 28**). As in humans, Intracytoplasmic sperm injection has been tested to improve the fertilization rate, but even with this method, very few pronuclei and no embryonic development have been obtained (*8*; *see* **Note 30**).

3.7. In Vitro Embryonic Development

The presence of male and female pronuclei after insemination has been reported by several authors ***(3,5,9,10)*** and the rate of two pronuclei embryos varies from 8% to 37%. The first embryonic cleavages can also be observed

(3,11) and Yamada et al. reported that 33% of the matured and fertilized oocytes developed to the 1- to 8-cell stage. However, further in vitro embryonic development has rarely been reported in the canine species and, in 2000, Otoi et al. obtained one blastocyst. Because of this low number of successful embryonic development observed in our lab and in others, the kinetics of in vitro canine embryonic development cannot be precise to date. Transfer of in vitro embryos to recipient bitches can be performed, even if bitches are highly difficult to synchronize and if embryonic mortality is high. Pregnancy after transfer of in vitro canine embryos has been reported for the first time in 2001 ***(12)***; 169 oocytes were collected and, after in vitro maturation and fertilization, 90 "oocytes/embryos" were transferred in one bitch. A pregnancy with three conceptus was obtained, but unfortunately, fetal development, followed by ultrasonography, stopped after 20 d of pregnancy.

3.8. Conclusions

Research on in vitro maturation in canine species started mainly during the last 10 yr. As a consequence, much information on ovarian physiology, follicular function, and oocyte features (quality, meiotic maturation, fertilization) in the bitch are lacking. Further investigations are required to optimize the maturation and fertilization rate and to control the estrus cycle of bitches. The development of these techniques will help obtain more embryos and develop research on embryo physiology/quality/survival. When these reproductive biotechnologies are available, they will be widely applied to preserve and save endangered breed and canid species.

4. Notes

1. When ovariohysterectomy is performed because of pyometra, ovaries are not suitable for IVM. Macroscopic examination of these ovaries often reveals an abnormal aspect (numerous cysts). Moreover, the uterine bacterial infection may contaminate the ovaries at the time of dissection and then favor bacterial development in culture dishes. Finally, our own experience indicates that very few oocytes can be obtained from this type of ovary.
2. Most of the ovaries are in the anestrus stage and oocytes collected from these ovaries originate from small follicles, and nuclear competence (aptitude to resume meiosis) of these oocytes may be limited. However, it is sometimes possible to set up protocols of ovarian stimulation and obtain oocytes originating from larger/preovulatory follicles ***(3)***. In that study, 32% of metaphase II oocytes were obtained after 72 h of culture, and some embryos (unfortunately blocked between two-cell and eight-cell stages) were observed.
3. The presence of fat during ovary cutting will interfere with visibility during oocyte selection.

4. If ovaries at the preovulatory stage can be obtained, it is possible to puncture each preovulatory follicle by using a 1-mL syringe with a 26G needle and to collect oocyte and follicular fluid.
5. The mouth pipet represents a real efficient tool for manipulating oocytes and embryos. The Pasteur pipet can be sterilized and different sizes of pipet can be made, depending on the experiment. This system allows denudation of oocytes and the manipulation of a small volume of liquid during staining procedure.
6. Numerous supplementations have already been tested by different authors:
   - 10 or 20% FCS ***(1,3,6,13)***.
   - BSA supplementation (0.3% or 4%) was demonstrated to increase GVBD rate ***(6)***.
   - Gonadotrophin supplementation (FSH and/or LH) ***(13,14)***.
   - 10% estral bitch serum and 20 μg estradiol ***(9)***.

   The effects of steroid addition (estradiol, progesterone, or estradiol + progesterone) were tested ***(15)***, but none of these hormonal treatments improved oocyte maturation. The progesterone was added to mimic in vivo conditions, because in the bitch, as described in **Subheading 1.**, progesterone levels start to increase before ovulation (preovulatory luteinization). However, to date, no beneficial effect has been evidenced on oocyte maturation rates, and progesterone may act on spermatozoa by inducing acrosome reaction ***(16)***.

   Other culture media have been tested too, such as synthetic oviductal fluid (SOF) medium, which did not improve either GVBD or MI–MII rate ***(17)***, or modified-TYH (Krebs–Ringer bicarbonate solution) medium ***(3)***. However, because numerous parameters (natural or induced cycles, BSA or FCS supplementation) were different in these works, it remains difficult to evaluate separately the effects of oocyte origin or medium quality.

   To mimic in vivo conditions, research has been conducted on oviductal environment also. Different authors ***(9,17)*** tested oocyte culture in the presence of canine oviductal cells but did not observe positive effects. Another team ***(18)*** tested a system in which they cultured canine oocytes in the isolated and ligated oviduct. They obtained 32% of metaphase I to metaphase II stages after only 30 h.
7. Optimal concentrations of FCS have been reported to be 10% or 20%, but only M199 + 20% FCS allows for an increase of maturation rate after 96 h of maturation ***(6)***. A lower FCS concentration (5%) increases oocyte degeneration rate.
8. The time of maturation should be at least 24 h, but the maximum rate of metaphase II is obtained after 48–72 h. A longer time of maturation does not improve significantly the rate and oocytes tend to degenerate.
9. Some authors culture oocytes in wells under oil, because the presence of oil prevents dehydration and contamination. In our lab, we do not to use oil because (1) the 2 mL water we add between the wells is sufficient to prevent dehydration and (2) some compounds of culture medium can be lipophilic and be absorbed by the oil. For example, steroids (added to the culture medium or produced by the oocytes/cells in culture) are highly lipophilic. This should be kept in mind if steroid assays are performed after culture under oil.

10. The majority of researchers working on canine oocyte maturation use culture in droplets or wells, with groups of oocytes in 400–500 µL of medium ***(15)***. However, other culture systems have been tested, such as a system with ligated oviduct ***(18)***. This system is ingenious and gives good results (32% of MI–MII after 30 h) but is difficult to set up: It requires bitch oviducts (meaning ovariohysterectomy) and is difficult to standardize (oviductal features vary substantially according to the cycle stage), whereas the culture medium is easy to standardize to compare the effects of a treatment.
11. The classical method of oocyte denudation in human, bovine, or mouse oocytes use hyaluronidase enzyme. However, with canine oocytes, we did not obtain any satisfactory results. We tested hyaluronidase type I-S and VIII at two different concentrations (30 IU/mL and 300 IU/mL for 5 min or 15 min, as used with rabbit oocytes) and hyaluronidase had no effect or transformed the cumulus in a gelatinous mass, which is really difficult to handle. We also tested the effect of trypsine/EDTA and did not observe better results.
12. Pronase is a nonspecific protease, with various types of endopeptidase and exopeptidase. The incubation time in pronase should be very precise (not more than 1 min); the enzyme will completely digest the oocyte if the incubation is too long (3 or 5 min, for instance). The optimum temperature for pronase activity is in the range 40–60°C, so oocytes should be kept at 38°C during the pronase reaction. Pronase denudation can also be used after fixation. After this fixation protocol, the incubation time in pronase will be longer to obtain denudation (2 min with our protocol).
13. Triton X-100 in the fixation solution will permeabilize the membranes and allow the penetration of the fixative.
14. The presence of BSA prevents the oocytes from sticking on the plastic of the well and reduces nonspecific background staining.
15. Hoechst staining can be useful to determine the nuclear stage without oocyte destruction. In this case, we use the same UV lamp, but with a special filter: only 1/100 of the light is transmitted to the oocyte and some assay can be done on the oocyte thereafter, because it remains alive.
16. DNA and actin staining can also be applied on embryos, to check their quality (embryo/cell shapes, total cell number).
17. As described on the Molecular Probes website (www.probes.com), ethidium homodimers 1 and 2 (EH-1 and EH-2) strongly bind to double-strand DNA, single-strand DNA, RNA, and oligonucleotides with a large fluorescence enhancement (>30-fold). However, the DNA affinity of EH-2 is about twice that of EH-1, and EH-2 is also about twice as fluorescent bound to double-strand DNA than to RNA. We tested this point because we compared staining obtained with EH-2 and propidium iodide: EH-2 staining localization on oocyte DNA is more specific, whereas propidium iodide induces more diffuse staining in the cytoplasm. Moreover, because ethidium homodimers are cell impermeant, they can be used, as propidium iodide, to detect loss of membrane integrity (dead-cell indicator).

18. As described in **Subheading 1.**, canine oocytes are full of lipid droplets. This induces more background in the oocyte cytoplasm after DNA staining (even with EH). As a consequence, concentration of DNA dye should not be too high.
19. Fluorescent stainings are highly photosensitive: Light exposure will induce bleaching and fluorescence will be irreversibly lost. As a consequence, each step with fluorochrome should be performed in dishes protected from the light using aluminum foil.
20. The incubation time can vary from 20 min to 2 h and, generally, any incubation temperature between 4°C and 37°C is suitable.
21. Classically in immunocytochemistry, serum of the animal that produces the secondary antibody is added to the blocking solution.
22. The first antibody can also be incubated overnight at 4°C.
23. Alexa Fluor is a conjugate developed by Molecular Probes, which is more stable (lower bleaching) than fluorescein isothiocyanate.
24. Mounting medium for fluorescence can be bought as a commercial preparation (e.g., VectaShield from Vector Laboratories) or prepared in the lab with PBS/glycerol (4/1) supplemented with DABCO. DABCO (1,4-diazabicyclo [2.2.2]octane) is an antifade reagent added to the medium to retard photobleaching of fluorescent dyes.
25. We always use fresh semen to perform IVF. We tested IVF with frozen semen, but motility is difficult to preserve after thawing and is lost after a few hours.
26. Some authors performed sperm capacitation in specific media:
    - Capacitation during 4 h in the canine capacitation medium (CCM) ***(12,19,20)***.
    - Krebs–Ringer bicarbonate solution with 4% BSA for 5 h ***(3)***. Capacitation was performed for 7 h.
    - Brackett–Oliphant medium with BSA, heparin, and caffeine ***(11)***.

    The presence of calcium and a high bicarbonate concentration seem important. Furthermore, follicular fluid induces capacitation. In our lab, we tested CCM and TALP media (widely used for bovine sperm), but the fertilization rate was not improved and we often observed spermatozoa agglutination.
27. Concentration: Spermatozoa concentration used by other authors varies from 5×10^5/mL ***(9)*** to 1×10^6/mL ***(3,12)***.
28. Depending on the authors, sperm can be coincubated with oocytes from 6 h ***(11)*** to 12 h ***(12)*** or 20 to 24 h ***(3,5)***. In the dog, a high rate of polyspermy is observed (from 33% to 59%). This does not appear to be the result of a prolonged time of coincubation, and a high polyspermy rate (50%) is observed as early as 4 h after the beginning of fertilization ***(3)***.
29. B2 culture medium has also been tested for embryo culture, but it did not improve the development significantly.
30. Fulton et al. ***(8)*** reported for the first time ICSI in dogs: After 48 h of maturation, they injected 38 metaphase II oocytes, but they observed both male and female pronuclei only in three of them (8%). They did not describe any further embryo development.

Acknowledgments

We wish to thank Dr. Alain Fontbonne and his team for help concerning estrus cycle follow-up, and the team of reproductive surgery in the Veterinary School of Alfort. This work was supported by the National Veterinary School of Alfort (ENVA) and National Institute for Agronomic Research (INRA).

References

1. Mahi, C. A. and Yanagimachi, R. (1976) Maturation and sperm penetration of canine ovarian oocytes in vitro. *J. Exp. Zool.* **196,** 189–196.
2. Johnston, S. D., Root-Kustritz, M. V., and Olson, P. N. S. (2001) *Canine and Feline Theriogenology.* W. B. Saunders, Philadelphia.
3. Yamada, S., Shimazu, Y., Kawaji, H., et al. (1992) Maturation, fertilization, and development of dog oocytes in vitro. *Biol. Reprod.* **46,** 853–858.
4. Verstegen, J. P., Onclin, K., Silva, L. D., et al. (1999) Effect of stage of anestrus on the induction of estrus by the dopamine agonist cabergoline in dogs. *Theriogenology* **51,** 597–611.
5. Saint-Dizier, M., Renard, J. P., and Chastant-Maillard, S. (2001) Induction of final maturation by sperm penetration in canine oocytes. *Reproduction* **121,** 97–105.
6. Hewitt, D. A., Watson, P. F., and England, G. C. (1998) Nuclear staining and culture requirements for in vitro maturation of domestic bitch oocytes. *Theriogenology* **49,** 1083–1101.
7. Otoi, T., Fujii, M., Tanaka, M., et al. (2000) Oocyte diameter in relation to meiotic competence and sperm penetration. *Theriogenology* **54,** 535–542.
8. Fulton, R. M., Keskintepe, L., Durrant, B. S., et al. (1998) Intracytoplasmic sperm injection (ICSI) for the treatment of canine infertility. *Theriogenology* **48,** 366.
9. Nickson, D. A., Boyd, J. S., Eckersall, P. D., et al. (1993) Molecular biological methods for monitoring oocyte maturation and in vitro fertilization in bitches. *J. Reprod. Fertil. Suppl.* **47,** 231–240.
10. Shimazu, Y. and Naito, K. (1996) Both male and female pronuclei formation in canine oocytes inseminated at germinal vesicle stage. *J. Mammal. Ova. Res.* **13,** 122–124.
11. Otoi, T., Murakami, M., Fujii, M., et al. (2000) Development of canine oocytes matured and fertilised in vitro. *Vet. Rec.* **146,** 52–53.
12. England, G. C., Verstegen, J. P., and Hewitt, D. A. (2001) Pregnancy following in vitro fertilisation of canine oocytes. *Vet. Rec.* **148,** 20–22.
13. Cinone, M., Ghneim, A., Caira, M., et al. (1992) Collection and maturation of oocytes in the bitch. Proceedings of the 12th International Congress on Animal Reproduction, Vol. 4, pp. 1767–1769.
14. Olson, M. A., Anderson, A. C., Amodeo, D., et al. (2001) Resumption of meiosis in canine oocytes cultured with or without bovine serum albumin and gonadotropins. *Theriogenology* **55,** 489.
15. Hewitt, D. A. and England, G. C. (1997) Effect of preovulatory endocrine events upon maturation of oocytes of domestic bitches. *J. Reprod. Fertil. Suppl.* **51,** 83–91.

16. Sirivaidyapong, S., Bevers, M. M., Gadella, B. M., et al. (2001) Induction of the acrosome reaction in dog sperm cells is dependent on epididymal maturation: the generation of a functional progesterone receptor is involved. *Mol. Reprod. Dev.* **58,** 451–459.
17. Hewitt, D. A. and England, G. C. (1999) Synthetic oviductal fluid and oviductal cell coculture for canine oocyte maturation in vitro. *Anim. Reprod. Sci.* **55,** 63–75.
18. Luvoni, G. C., Chigioni, S., Allievi, E., et al. (2002) In vitro maturation of canine oocytes in isolated oviduct. Proceedings of the Third EVSSAR (European Veterinary Society for Small Animal Reproduction) Congress, p. 139.
19. Mahi, C. A. and Yanagimachi, R. (1978) Capacitation, acrosome reaction, and egg penetration by canine spermatozoa in a simple defined medium. *Gamete Res.* **1,** 101–109.
20. Hewitt, D. A. and England, G. C. (1999) Culture conditions required to induce capacitation and the acrosome reaction of canine sperm in vitro. *Vet. Rec.* **144,** 22–23.

19

Detection and Measurement of Membrane-Bound Protein Tyrosine Kinases in the Zebrafish Egg

Wenjun Wu and William H. Kinsey

1. Introduction

The study of the biochemical events involved in fertilization and egg activation has historically been directed toward marine invertebrate and amphibian eggs *(1)*, although recent progress has been made in the analysis of a few enzymes that are abundant in mammalian eggs *(2,3)*. The zebrafish system is also a promising model that shares several advantages with marine invertebrate eggs, yet has the advantage of being a vertebrate. For example, the zebrafish eggs are reasonably clear optically and can be obtained in quantities suitable for biochemical analysis. They can be fertilized synchronously and will develop rapidly *(4)*. The zebrafish system also benefits from the fact that the DNA sequence homology with mammals is very high, so that tools developed in mammalian systems can be applied to zebrafish eggs with a reasonable expectation of success. Finally, the potential for genetic analysis of the fertilization process could ultimately provide novel insights into the signaling mechanisms used at fertilization.

Despite the above-listed advantages, the zebrafish egg does have some characteristics that complicate biochemical analysis. The relatively large amount of yolk and internal membranous organelles in zebrafish eggs has made detection of relatively low-abundance signal transduction enzymes difficult. The immunoprecipitation steps usually required for detection of Src family kinases and other signaling enzymes do not work well on egg homogenates or detergent extracts of whole zebrafish eggs. Similar problems in the *Xenopus* oocyte were overcome by subcellular fractionation *(5)*. The objective of this chapter is to describe recently developed subcellular fractionation techniques

From: *Methods in Molecular Biology, vol. 253: Germ Cell Protocols: Vol. 1 Sperm and Oocyte Analysis*
Edited by: H. Schatten © Humana Press Inc., Totowa, NJ

that have enabled us to overcome the interference by egg yolk components and successfully quantitate the Fyn tyrosine kinase as well as other PTKs and PTPases. These methods represent a refinement of our previous studies ***(6)*** and could be easily adapted to study other low-abundance enzymes present in the membrane fraction of eggs.

2. Materials

2.1. Buffers

1. Hank's BSA: 137 m*M* NaCl, 5.4 m*M* KCl, 0.25 m*M* Na_2HPO_4, 1.37 m*M* $CaCl_2$, 1.0 m*M* $MgSO_4$, 4.2 m*M* $NaHCO_3$, pH 7.2, 5 mg/mL BSA.
2. High-performance liquid chromatography (HPLC) buffer A: 4% (w/v) NaH_2PO_4, 0.1% (v/v) morpholine adjusted to pH 2.1 with CF_3COOH.
3. Immunoprecipitation buffer: 150 m*M* NaCl, 10 m*M* Tris-HCl, pH 7.5, 1 m*M* EDTA, 1 m*M* EGTA, 0.1 mg/mL NaN_3, 1 m*M* of 2-mercaptoethanol, 10 μg/mL Aprotinin (Sigma-Aldrich, St. Louis, MO), and 1% (v/v) NP-40.
4. Kinase assay buffer: 12.5 m*M* HEPES, pH 7.5, 10 m*M* $MgCl_2$, 2.5 m*M* of 2-mercaptoethanol, 0.1 m*M* sodium orthovanadate, to which was added 1 m*M* peptide-5 (YGEVYEGVFKK), and 10 μ*M* [γ-^{32}P]ATP (300 mCi/μmol).
5. Phosphatase buffer: 50 m*M* HEPES, pH 7.5, 10 m*M* $MgCl_2$, 2.5 m*M* of 2-mercaptoethanol.
6. Sperm extender solution: 10 m*M* HEPES, 80 m*M* KCl, 45 m*M* NaCl, 45 m*M* NaOAc, 0.4 m*M* $CaCl_2$, 0.2 m*M* $MgCl_2$, pH 7.2.
7. TKM buffer: 50 m*M* Tris-HCl, pH 7.5, 25 m*M* KCl, 5 m*M* $MgCl_2$, 1 m*M* EGTA, 0.1 mg/mL NaN_3 (added as a peroxidase inhibitor), 1 m*M* of 2-mercaptoethanol, 10 μg/mL Aprotinin.

2.2. Zebrafish Culture Conditions and Egg Collection

Wild-type zebrafish between 3 and 6 mo of age were maintained in an array of 10- and 15-gallon aquariums containing reverse-osmosis purified water supplemented with 0.059 g/L synthetic sea salts (Marine Enterprises, Baltimore, MD). The aquariums are connected to a central canister filtration system (Aquaclear) containing a filtering matrix and 2 lb of charcoal. An ultraviolet (UV) sterilizing unit is plumbed in line, and the temperature maintained at 28°C and pH 6.5–7.0. Wide-spectrum fluorescent lighting (Coralife Trichromatic; Energy Savers Unlimited, Carson, CA) is provided on a 13-h on/11-h off light/dark cycle. The diet consists of Tetramarine flake food (Tetra Sales, Blacksburg, VA) supplemented daily with freshly hatched, live brine shrimp (Aquatic Lifeline Inc., Salt Lake City, UT). (*See* **Note 1**.)

Once sexually mature and in the presence of male fish, females grow mature eggs continuously and spawn regularly. The fish are maintained as mixed male and female populations until 2 wk before eggs are required, at which time the females are segregated to allow the accumulation of mature eggs.

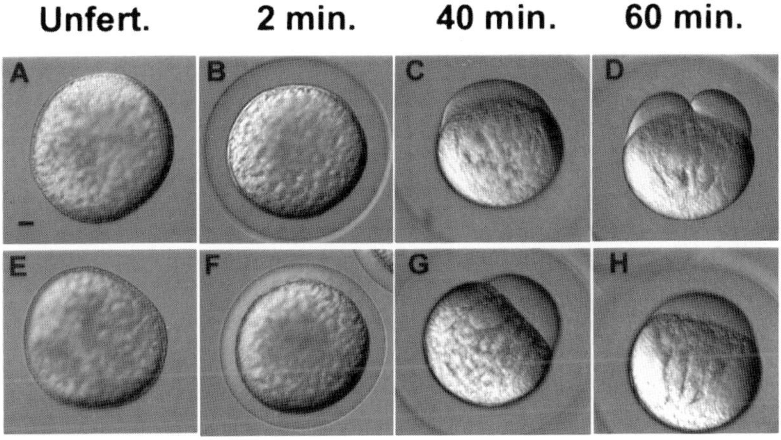

Fig. 1. Fertilization and parthenogenic activation of the zebrafish egg. The typical morphology of zebrafish eggs shed into Hank's BSA is shown in **(A)** and **(E)**. The chorion remains tightly associated with the egg plasma membrane. Once fertilized by the addition of sperm and aquarium water, the chorion elevates, as seen in **(B)**. Chorion elevation also occurs in response to addition of aquarium water alone **(F)** although it seems to occur faster if fertilization has occurred. Cytoplasmic streaming results in accumulation of yolk-free cytoplasm as the blastodisc **(C,G)**. Cleavage is complete when the two cells are seen to be separated by their own cell membrane, although both are continuous with the yolk mass in this meroblastic cleavage **(D)**. Cleavage does not occur in eggs activated by aquarium water alone **(H)**. Bar in **(A)** = 80 μm.

Once fully gravid, the females can be squeezed for egg collection every 3 wk for up to 6 mo (*see* **Note 2**). However, if eggs are not collected regularly, the males must be reintroduced to prevent accumulation of degenerating eggs. For sperm and egg collection, fish are anesthetized with 0.02% tricane (Sigma-Aldrich, St. Louis, MO) and gently squeezed between gloved fingers to express eggs or sperm. Sperm were suspended in sperm-extension buffer *(7)*, which maintains them in an immotile state, and stored on ice for up to 2 h. Eggs were stored in Hank's BSA at 25°C *(8)* and used within 30 min of collection.

3. Methods

3.1. In Vitro Fertilization

1. Fertilization is accomplished by mixing 2 μL of sperm suspension (2 mg/mL sperm protein containing approx 1.23×10^6 sperm) with approx 1000 eggs suspended in 500 μL of Hank's BSA, then adding 2 mL of aquarium water to activate sperm motility and allow fertilization to proceed.
2. Chorion elevation (9) occurs within about 45 s in fertilized eggs as well as eggs activated in the absence of sperm by hypotonic conditions (*see* **Fig. 1B,F**).

Cytoplasmic streaming leading to blastodisc formation *(10)* occurs by 30–40 min (*see* **Fig. 1C**,**G**), and cleavage is complete by 50–60 min (*see* **Fig. 1D**,**H**).

3. Eggs activated in the absence of sperm do not undergo cleavage, so the first reliable evidence of the frequency of successful fertilization is at the cleavage stage, and it is important that a group of 10–20 eggs be allowed to develop for a few hours so that the success rate of fertilization can be determined.

3.2. Preparation of Membrane Fractions

1. The immunoprecipitation steps normally required to assay specific signal transduction PTKs and PTPases are usually not successful when performed on detergent extracts of whole eggs or embryos. We have found, however, that these enzymes can easily be detected in the purified membrane fractions of eggs or embryos *(6)*. The following procedure has been developed to maximize recovery of Fyn kinase activity and can be applied to eggs, zygotes, or early gastrula embryos. Samples as small as 150 eggs can be used if 0.5-mL ultracentrifuge tubes (5 × 41 mm) are available, whereas larger samples of up to 2000 eggs require a 5-mL ultracentrifuge tube (13 × 50 mm).
2. Unfertilized eggs are used as a suspension in Hank's BSA. Fertilized zygotes are in a mixture of aquarium water and Hank's BSA. The eggs or zygotes are transferred to a glass Potter–Elvhjem homogenizer (0.5–5 mL) and all buffer is aspirated with a fine plastic pipet tip.
3. Fertilized samples are washed with 10 vol of aquarium water to remove any unbound sperm. The eggs or zygotes are then mixed with 1.5 vol of ice-cold 70% sucrose (*see* **Note 3**) (all sucrose solutions are in TKM buffer) and homogenized by hand until all cells are disrupted.
4. The homogenate, which now contains 42% sucrose, is then layered on the bottom of an ultracentrifuge tube (0.5–5.0 mL). The homogenate is then overlain with steps of 35% sucrose and 23% sucrose. A 12% sucrose layer described in our earlier publication *(6)* has been omitted because it separates only a small amount of membrane material.
5. The gradients are then centrifuged in an SW50.1 rotor at 100,000*g* for 3 h, and the material at each sucrose interface is collected. The bottom 200 μL are collected as the cytosol fraction, and the underlying pellet is also recovered.
6. Fractions are diluted with 5–10 vol of TKM buffer and centrifuged at 100,000*g* for 45 min to pellet the membranes. The fractions are resuspended in a small volume of TKM, and protein content is determined by BCA assay (Pierce, CO).

3.3. Characterization of Membrane Fractions

1. The results of a typical preparation are presented schematically in **Fig. 2**. Plasma membrane vesicles are distributed as a white band at the 23/35% interface and in an additional yellowish band at the 35/42% interface. Quantitation of ouabain-inhibitable Na^+/H^+ ATPase activity indicates that plasma membrane vesicles are present in both membrane fractions, although most highly enriched in the material at the 23/35% interface.

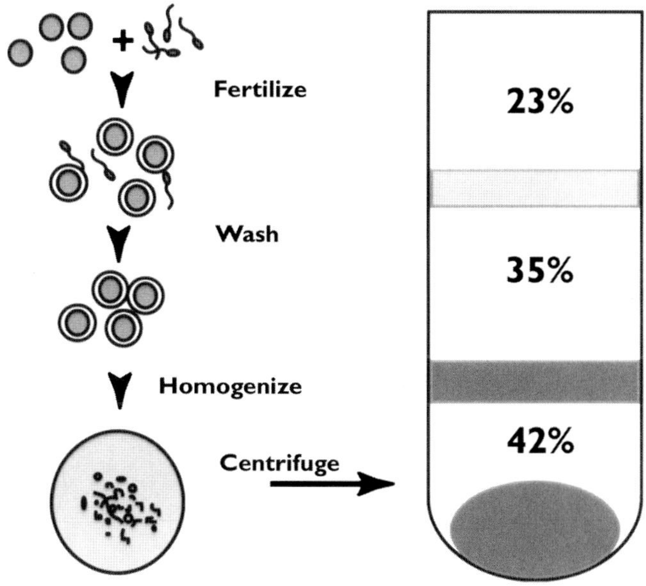

Fig. 2. Subcellular fractionation of the zebrafish egg. Eggs are mixed with sperm and activated with aquarium water. The excess sperm are washed out and the zygotes are suspended in 1.5 vol of 70% sucrose and then homogenized. The homogenate is overlain with 35% and 23% sucrose and then centrifuged at 100,000*g* as described in **Subheading 3.** The membrane fractions are found at the 42/35% interface and the 23/35% interface. A large pellet of yolk, chorions, and so forth is present at the bottom of the tube.

2. The membranes at the 35/42% interface include more cytoplasmic organelles than in the 23/35% interface, as indicated by the higher levels of marker enzymes for lysosomes and mitochondria. However, the 35/42% interface also contains significant Fyn kinase activity (*see* **Subheading 3.4.**). Therefore, experiments designed to detect very low-abundance plasma membrane kinases should take advantage of the higher purity of the membranes at the 23/35% interface, whereas experiments designed to estimate the total kinase activity associated with plasma membranes should combine the membranes at the 23/35% interface with those at the 35/42% interface.

3.4. Detection of Membrane-Associated PTKs Such as Fyn Kinase

1. Detection of Src-family PTKs is done by immune complex kinase assays in which the kinase is first purified by immunoprecipitation and then detected in a phosphorylation assay in which the kinase either undergoes autophosphorylation or phosphorylates a peptide substrate. The light and heavy membrane fractions are first solubilized by the addition of an equal volume of 2X immunoprecipitation

buffer, and after 10 min agitation on a rotator at 4°C, the extract is centrifuged in a microfuge at 10000*g* for 10 min to pellet the detergent-insoluble material.

2. The detergent-soluble extract is divided into experimental and control samples for immunoprecipitation. Fyn kinase can be immunoprecipitated using the rabbit anti-Fyn 3 antibody (Santa Cruz Biotechnology, Santa Cruz, CA) at a concentration of 0.2 µg/mL, and the peptide antigen (Fyn3)P is also available from Santa Cruz and can be used as a blocking peptide to demonstrate the specificity of the antibody. Typically, 100-µL samples of detergent extract containing either 0.2 µg/mL anti-Fyn3 antibody or 0.2 µg/mL anti-Fyn3 antibody + 5 µ*M* (Fyn3)P are incubated on a rotator at 4°C for 2 h.
3. The immune complexes are then collected by incubation with 25 µL of a 10% suspension of protein A–agarose for 30 min and washed by centrifugation. We use two washes with immunoprecipitation buffer and a third wash with kinase buffer.

3.4.1. Autophosphorylation Assays

1. Autophosphorylation assays are the most sensitive way to detect a Src-family kinase because the kinase phosphorylates itself with very high specific activity [γ-^{32}P] ATP (*see* **Note 4**). The assay is an indicator of the amount of active kinase molecules, but one has to be careful about possible contamination by serine–threonine kinases, which may phosphorylate proteins in the 50- to 60-kDa range and would give a false result. The assays are performed by incubating the immunoprecipitate with carrier-free [γ-^{32}P]ATP (4500 mCi/µmol) (ICN Biomedicals, Irvine, CA) for a very short incubation period. Typically, we add 20 µL of kinase buffer to the 5 µL immunoprecipitate and start the reaction by adding 5 µL of [γ-^{32}P]ATP.
2. The reaction is carried out for 1 min at 25°C and stopped by the addition of an equal volume of 2X gel sample buffer followed by immediate heating to 90°C for 5 min.
3. Sodium dodecyl sulfate–polyacrylamide gel electrophoresis (SDS-PAGE) is performed on a 12.5% gel and the Coomassie-stained gel is treated with 1 *M* KOH at 50°C for 30 min to hydrolyze most of the P-Ser and P-Thr residues that could result from contamination by Ser/Thr kinases. The gel is then refixed in 10% acetic acid, dried, and used for autoradiography.
4. The results of a typical autophosphorylation assay are seen in **Fig. 3**. When fertilized zygotes (2.5 min postinsemination) are simply homogenized in TKM buffer and fractionated into cytosol and particulate fractions by centrifugation at 100,000*g* for 1 h, almost all of the Fyn activity was immunoprecipitated from the particulate fraction (lane A) as evidenced by a 59-kDa band detected in the autoradiograph (arrow). This band was not detected in immunoprecipitates prepared in the presence of the (Fyn3)P peptide antigen (*see* **Fig. 3**, lane E). Because the cytosol seems to contain very little Fyn kinase (lane B), we focused our efforts on the membrane fractions, which should be highly enriched in PTK activity. Analysis of individual membrane fractions (lanes C and D) indicates that Fyn kinase

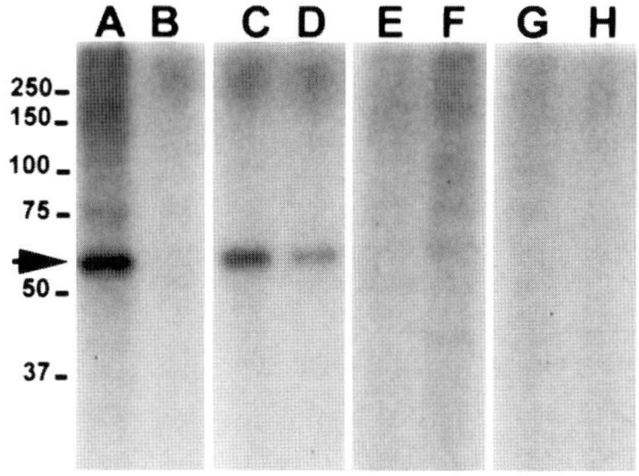

Fig. 3. Detection of Fyn kinase activity in the zebrafish zygote. Zebrafish zygotes (1000) collected 2.5 min postinsemination were homogenized in TKM buffer and centrifuged at 100,000*g* to obtain crude particulate **(A,E)** and cytosol **(B,F)** fractions. A second group of 1000 was used to prepare membranes as described in **Subheading 3.**, which were collected from the 23/35% interface **(C,G)** or the 35/42% interface **(D,H)**. Detergent extracts were prepared and one-half of each sample was incubated with anti-Fyn 3 IgG **(A–D)** or with anti-Fyn 3 IgG + 5 μ*M* (Fyn3)P as a control **(E–H)**. Immune complexes were recovered by adsorption to protein A–agarose, washed, and incubated in an autophosphorylation assay as described in **Subheading 3.** The SDS gel was stained with Coomassie blue, treated with 1 *M* KOH, refixed, dried, and used to expose X-ray film with an intensifying screen (BioMax MS, Eastman Kodak, Co., Rochester, NY) for 48 h.

activity is immunoprecipitated from both the 23/35% and the 35/42% interface, although it is most highly enriched in the 23/35% fraction.

3.4.2. Peptide Phosphorylation Assays

1. Assays in which the immunoprecipitated Fyn phosphorylates a synthetic peptide are more complicated, but provide superior quantitation because they are done under conditions where the substrates are not limiting. If the peptide contains tyrosine as the only phosphorylation site, contamination by serine–threonine kinases is not a concern. Immunoprecipitates prepared as in **Subheading 3.4.1.** are incubated with kinase buffer containing [γ-^{32}P]ATP and carrier ATP to bring the ATP concentration up to 10 μ*M* (300 mCi/μmol).
2. A peptide substrate such as peptide 5 (YGEVYEGVFKK) is added at 1–5 m*M*, which is a compromise between an attempt to get the concentration above the K_m and the poor solubility of some of these peptides. We usually carry the reaction out with the microfuge tube on its side to enhance substrate availability and incubate for 10 min at 25°C.
3. The reaction is stopped by the addition of 250 μL of 7% TCA. The reaction products are filtered through a 0.5-mL centrifugal filter (Ultrafree-MC, Millipore

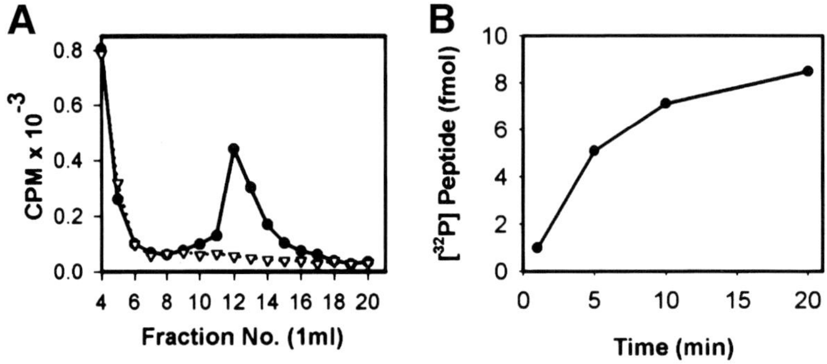

Fig. 4. Detection of Fyn kinase activity by peptide phosphorylation assay. Detergent extracts were prepared as in **Fig. 3** and aliquots representing 600 eggs were incubated with the anti-Fyn3 antibody or anti-Fyn3 antibody plus 5 μ*M* (Fyn3)P. The immunoprecipitates were incubated with kinase assay buffer containing 1 m*M* peptide-5 and [γ^{32}P]ATP (300 mCi/μmol) at 10 μ*M*. Samples representing the immunoprecipitate from 150 zygotes were removed after different incubation times and analyzed by RP-HPLC. **(A)** The elution of the phosphorylated peptide from control and anti-Fyn immunoprecipitates. The anti-Fyn3 (●) immunoprecipitate was able to phosphorylate the peptide which elutes in fractions 12 and 13, whereas control immunoprecipitates (▽) had no detectable kinase activity. **(B)** The effect of reaction time on peptide phosphorylation.

Corp., Bedford, MA) containing 300 μL Bio-Rad AG1 X8 (Bio-Rad Laboratories, Hercules, CA) to remove unreacted ATP and precipitated protein.

4. The phosphopeptide can then be analyzed by reverse phase (RP)-HPLC. The results of a typical assay analyzed on a Zorbax C-8 column are seen in **Fig. 4A**, where it is clear the Fyn immunoprecipitate phosphorylated the peptide and the control immunoprecipitate did not. Summation of the cpm under the peak less the cpm from the same fractions in the control sample yields the net product formed, which is typically in the 2- to 3-fmol range for samples immunoprecipitated from 150 eggs. Analysis of the effect of reaction time on product formation indicates that under these conditions, the reaction does not remain linear for very long (*see* **Fig. 4B**) and reaction times longer than 10 min should be avoided.

3.5. Detection of Membrane-Associated PTPases Such as rPTPa

1. Receptor protein tyrosine phosphatases are often involved in the same signaling pathways as Src-family PTKs and can be detected in immune-complex assays in which a [^{32}P] peptide is used as a substrate (*see* **Note 5**) ***(11)***. We have prepared such a substrate by using a constitutively active GST-Fyn (GST-Fyn-^{Y}532^{F}) to phosphorylate the tyrosine-containing synthetic peptide 5 to a specific activity of 6000 cpm/pmol ***(12)***.

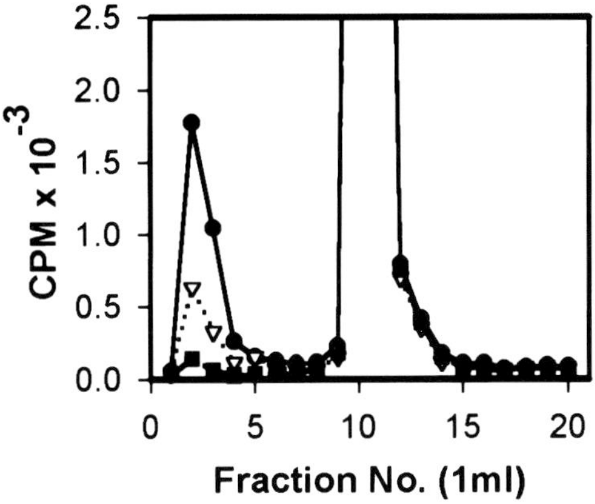

Fig. 5. Detection of rPTPα by peptide dephosphorylation assay. Membranes were prepared from 150 (2.5-min) zygotes and the detergent-soluble extract was incubated with anti-human–rPTPa (1:100) (●) or with normal rabbit serum (▽) followed by adsorption to protein A–agarose. The immunoprecipitates suspended in phosphatase buffer containing the [^{32}P]peptide were incubated on a rotator for 1 h in a hybridization oven set at 28°C. A separate assay was conducted with no immunoprecipitate (■) to control for spontaneous degradation resulting from contaminating PTPases or proteases. The free phosphate released by PTPase action elutes in fractions 2–3 and the intact [^{32}P]peptide elutes in fractions 10–11.

2. The rPTPα protein can be immunoprecipitated from detergent extracts of zygote membranes using the same methodology as described for Fyn kinase except that sodium vanadate is omitted from all steps. We have used a polyclonal anti-human–rPTPα antibody ***(13)*** (a gift from Jan Sap at New York University Medical Center, New York) to immunoprecipitate this enzyme from zebrafish zygote membranes. The immune complexes are washed twice with immunoprecipitation buffer and once with phosphatase buffer.
3. The pellet is suspended in 100 µL of phosphatase buffer containing 100 pmol of the [^{32}P] peptide after incubation for 1 h at 28°C with constant agitation.
4. The reaction was stopped by the addition of 500 µL of HPLC buffer A and analyzed by chromatography on a Zorbax (DuPont) C8 column equilibrated in buffer A and eluted with a 0–70% linear gradient of acetonitrile in buffer A. Fractions were collected, neutralized, and counted in a scintillation counter. PTPase activity results in the release of $^{32}PO_4$ from the [^{32}P] peptide. The highly charged $^{32}PO_4$ elutes in the void volume of the column, and the remaining [^{32}P] peptide is retarded on the column. The results of a typical PTPase assay from 150 zygotes are presented in **Fig. 5**, where it can be seen that PTPase activity in the rPTPa immunoprecipitate releases a significant amount of $^{32}PO_4$ eluting in fractions 2–3, whereas the intact [^{32}P] peptide elutes in fractions 10–11. The control immunoprecipitate performed with normal rabbit IgG releases little $^{32}PO_4$ and a zero

Table 1
Distribution of Marker Enzymes Among Membrane Fractions from Zebrafish Zygotes

Fraction	Protein (μg) per 1000 eggs	Na^+/H^+ ATPase	NADPH–Cyt C reductase	Succinate dehydrogenase	Acid phosphatase
23/35 int.	144 ± 39	821 ± 204	318 ± 84	204 ± 83	848 ± 389
35/42 int.	2160 ± 756	252 ± 106	433 ± 121	268 ± 97	1108 ± 396

Note: Membranes were prepared from groups of 1000 fertilized eggs (2.5 min postfertilization) and aliquots were analyzed for protein and marker enzyme activity. Enzyme assays were performed as previously described ***(6)*** and represent the average of three preparations ± SD. Values are expressed as nmols/h/mg protein except for NADPH–Cyt C reductase, which is expressed as μmols/h/mg protein.

enzyme control, demonstrating that spontaneous hydrolysis of the [^{32}P] peptide does not occur under these conditions.

In conclusion, analysis of signal transduction enzymes by immune complex assay is facilitated by subcellular fractionation to remove the yolk and other interfering egg components. Samples of eggs as small as 100 and as large as 1000 can be used at developmental stages between the one cell zygote and early gastrula. To date, we have been able to detect Fyn kinase and Yes kinase as well as rPTPα in zebrafish zygotes. This methodology together with the potential for use of genetic mutations in signal transduction genes make the zebrafish egg a useful system for the study of egg activation and early development.

4. Notes

1. Fish maintenance: We keep approx 250 fish at low density (20–25 fish per 14-gal tank). However, we feed them approx 2 g of flake food plus brine shrimp hatched from 0.5 g of dried eggs each day. This high level of feeding maximizes egg production but results in rapid acidification of the aquarium system. It is important to adjust the pH daily with Na_2CO_3 and wash the filter media each week.
2. Handling gametes: Unfertilized eggs are susceptible to rough handling and can undergo parthenogenic activation or partial lysis if they are exposed to too much agitation or are picked up with a pipet that is too narrow. Sperm can be activated by exposure to aquarium water; therefore, we rinse the anesthetized male fish with a few drops of sperm-extender buffer before collecting sperm.
3. Membrane preparation: When homogenizing the egg suspension, it is helpful to invert the homogenizer several times to speed up mixing of the viscous sucrose with the eggs. Homogenization should be done slowly to minimize bubbles that cause protein denaturation.
4. Kinase assays: ^{32}P is a strong β-emitter and heavy plexiglass shielding should be used as much as possible.

5. Phosphatase assay: Although the [^{32}P] peptide substrate works quite well with immunoprecipitates, care should be taken when using it for assays of membrane fractions or cell extracts that have substantial peptidase activity. In these cases, one must use a PTPase inhibitor such as sodium vanadate or phenylarsine oxide to demonstrate that the [^{32}P] peptide is being hydrolyzed by a phosphatase rather than by peptidases.

References

1. Whitaker, M. and Swann, K. (1993) Lighting the fuse at fertilization. *Development* **117**, 1–12.
2. Moore, H. D. (2001) Molecular biology of fertilization. *J. Reprod. Fertil.* **57(Suppl.)**, 105–110.
3. Carroll, J. (2001) The initiation and regulation of Ca^{2+} signalling at fertilization in mammals. *Semin. Cell Dev. Biol.* **12,** 37–43.
4. Hart, N. H. (1990) Fertilization in teleost fishes: mechanisms of sperm–egg interactions. *Int. Rev. Cytol.* **121,** 1–66.
5. Kamsteeg, E. J. and Deen, P. M. (2001) Detection of aquaporin-2 in the plasma membranes of oocytes: a novel isolation method with improved yield and purity. *Biochem. Biophys. Res. Commun.* **282,** 683–690.
6. Wu, W. and Kinsey, W. H. (2001) Fertilization triggers activation of Fyn kinase in the zebrafish egg. *Int. J. Dev. Biol.* **44,** 837–841.
7. Lee, K. W., Webb, S. E., and Miller, A. L. (1999) A wave of free cytosolic calcium traverses zebrafish eggs on activation. *Dev. Biol.* **214,** 168–180.
8. Westerfield, M. (1993) *The Zebrafish Book*, University of Oregon Press, Eugene.
9. Hart, N. H. and Yu, S. F. (1980) Cortical granule exocytosis and cell surface reorganization in eggs of Brachydanio. *J. Exp. Zool.* **213,** 137–159.
10. Beams, H. W., Kessel, R. G., Shih, C., et al. (1985) Scanning electron microscopy on blastodisc formation in zebrafish. *J. Morphol.* **184,** 41–49.
11. Zheng, X. M. and Shalloway, D. (2001) Two mechanisms activate PTPa during mitosis. *EMBO J.* **20,** 6037–6049.
12. Wu, W. and Kinsey, W. H. (2002) Role of PTPase(s) in regulating Fyn kinase at fertilization of the zebrafish egg. *Devel. Biol.* **247,** 286–294.
13. Su, J., Muranjan, M., and Sap, J. (1999) Receptor protein tyrosine phosphatase α activates Src-family kinases and controls integrin-mediated responses in fibroblasts. *Curr. Biol.* **9,** 505–511.

20

Single-Cell Method for Estimation of Protein Tyrosine Kinases in the Zebrafish Egg

Elinor MacGregor and William H. Kinsey

1. Introduction

The zebrafish system has many features that are ideal for the study of signal transduction enzymes involved in fertilization. However, one problem that the zebrafish egg shares with eggs of many species is that most of the signal transduction proteins are already synthesized and stored in the egg before ovulation. This makes it difficult to modify the proteome of the egg by transfection or knockdown methods and leaves microinjection of exogenous proteins as the only alternative. In many cases, the researcher must test the effect of an exogenous molecule such as a dominant-negative fusion protein or a blocking antibody on fertilization and development. Although the zebrafish egg is amendable to microinjection, the number of eggs that can be injected is usually insufficient for the type of immunoprecipitation analysis used to detect most protein tyrosine kinases (PTKs). Methods for measurement of serine–threonine kinases have been developed for individual *Xenopus laevis* ***(1)*** or mammalian eggs ***(2–4)***; however, these kinases are present in higher abundance than most PTKs. In an effort to find a solution to this problem, we have begun to develop methods to measure PTK activity in single zebrafish eggs. The general approach has been to prepare a particulate fraction from a single egg and then solubilize this crude fraction in a nonionic detergent and do an autophosphorylation assay using high-specific-activity [γ-^{32}P]ATP. When the reaction products are resolved on sodium dodecyl sulfate (SDS) gels and treated with strong alkali, autoradiography provides a crude measure of the PTK activity in the egg. This method does not use immunoprecipitation to select for a particular kinase because the large amount of yolk makes immunoprecipitation

From: *Methods in Molecular Biology, vol. 253: Germ Cell Protocols: Vol. 1 Sperm and Oocyte Analysis*
Edited by: H. Schatten © Humana Press Inc., Totowa, NJ

very unreliable. Without an immunoprecipitation step, we are only able to estimate total PTK activity. Despite this limitation, this methodology allows us to determine the effect of an injected protein or other reagent on the PTK activity in a single fertilized egg.

2. Materials

2.1. Buffers

1. Hank's BSA: 137 m*M* NaCl, 5.4 m*M* KCl, 0.25 m*M* Na_2HPO_4, 1.37 m*M* $CaCl_2$, 1.0 m*M* $MgSO_4$, 4.2 m*M* $NaHCO_3$, pH 7.2, 5 mg/mL bovine serum albumin (BSA).
2. TKM buffer: 50 m*M* Tris-HCl, pH 7.5, 25 m*M* KCl, 5 m*M* $MgCl_2$, 1 m*M* EGTA, 0.1 mg/mL NaN_3 (added as a peroxidase inhibitor), 1 m*M* of 2-mercaptoethanol, 10 µg/mL Aprotinin.
3. 2.5X Kinase buffer with NP-40: 31.25 m*M* HEPES, 25 m*M* $MgCl_2$, 6.3 m*M* mercaptoethanol, 1% NP40.
4. γ-Labeled ATP: [γ-^{32}P]ATP (3000 mCi/µmol) (ICN Pharmaceuticals, Irvine, CA).
5. Alkaline hydrolysis buffer: 1 *M* KOH.
6. Gel destain buffer: 10% (v/v) acetic acid.
7. Blue sensitive X-ray film (Molecular Technologies, St. Louis, MO).

2.2. Equipment for Egg Handling

1. Pipet tips used for homogenization. One-milliliter pipet tips (Fisher RediTip) used as a homogenization vessel are dipped in two part epoxy glue to plug the opening. The 0.2-mL yellow pipet tips with beveled ends (Fisher RediTip 1-200 µL) are used to crush the egg . Microseque pipet tips (Bio-Rad) are used to remove excess water within a homogenization vessel and to transfer the egg homogenate to the centrifuge tube.
2. Costar polymerase chain reaction (PCR) tubes of 200-µL capacity (Thermowell tubes) are used as centrifuge tubes by floating in the buckets of a swinging-bucket ultracentrifuge rotor (Beckman Ti-55).

3. Methods

3.1. Fertilization and Preparation for Analysis

Individual eggs are moved to a dish with 50 µL of Hank's BSA; then 2–5 µL of sperm are added with gentle mixing. One milliliter of water is added to activate the sperm and trigger fertilization, which occurs rapidly (chorion elevation is evident within 60 s).

3.2. Homogenization and Centrifugation

1. At 2 min postinsemination, transfer the egg in a minimal amount of water to the epoxy plugged blue pipet tip and use a microseque pipet to remove all of the water from the vicinity of the egg and to push the egg to the bottom of the tube (*see* **Fig. 1** and **Fig. 2**).

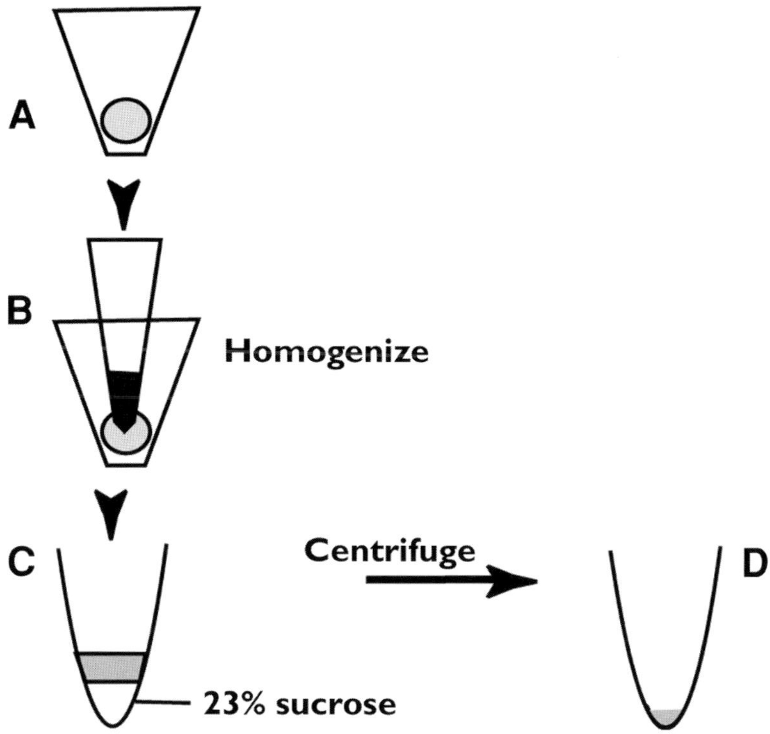

Fig. 1. Schematic representation of single-egg kinase assay. A single egg or fertilized zygote is placed in a homogenizing vessel and all extra water is removed. The egg is then crushed with a yellow pipet tip containing 12.5% sucrose in 1X kinase buffer. The homogenate is then transferred to a PCR tube containing 23% sucrose and centrifuged in the ultracentrifuge. The supernate is removed and the particulate fraction is solubilized in kinase buffer containing 1% NP-40 and [γ-^{32}P]ATP.

2. Crush the egg with a standard yellow 200-µL pipet tip with a beveled opening loaded with 5 µL of 12.5% sucrose in 2.5X kinase buffer (*see* **Note 1**). Pipet the crushed egg up and down several times to mix it with the sucrose.
3. Transfer (overlay) the homogenate to a 200-µL PCR tube containing a 2.5-µL pad of 23% sucrose in 2.5X kinase buffer (marked interface with waterproof ink). Finally, overlay the homogenate with enough 1X kinase buffer to fill the tube to within 3 mm of the top.
4. Trim the hinge and tab of the 200-µL PCR tube with a razor blade; then, the tube is placed in a bucket from a SW50-Ti ultracentrifuge rotor (Beckman Instruments), which is filled with ice-cold water. The tube should float in a more or less vertical orientation. Weigh and balance the rotor buckets; then, centrifuge at 100,000*g* for 30 min to pellet the particulate fraction.

Fig. 2. Homogenizing vessel containing a single unfertilized zebrafish egg. An egg positioned in the homogenizing vessel made from an epoxy-plugged blue 1-mL pipet tip is easily visualized under a dissecting microscope. The water surrounding the egg is removed just prior to homogenization.

3.3. Phosphorylation Reaction

1. Aspirate the 1X kinase buffer and the 12.5% sucrose from each centrifuge tube.
2. Prepare ATP mix by mixing 1 vol of 2.5X kinase buffer with NP-40 with 1.5 vol of [γ-^{32}P]ATP (3000 mCi/µmol).
3. Add 2.5 µL of the ATP mix to the 2.5-µL sample in the centrifuge tube. Pipet up and down carefully to solubilize the particulate fraction as much as possible without creating bubbles, which can denature the proteins (*see* **Note 2**).
4. Incubate the reaction for 1 min at 25°C; then, stop by mixing with 5 µL of 2X Laemmli gel sample buffer. Denature the samples by heating at 90°C for 5 min.
5. SDS-PAGE (polyacrylamide gel electrophoresis) is carried out on a 10% gel is stained with Coomassie blue, destained with acetic acid, washed with water, and incubated in 1 *M* KOH at 55°C for 1 h to hydrolyze P-Ser and P-Thr residues. The gel is then washed with gel destain solution and dried. Audoradiography is carried out using blue sensitive X-ray film and a Kodak Biomax MS intensifying screen.
6. The results of a typical autophosphorylation reaction are shown in **Fig. 3**. Several radiolabeled bands are evident at molecular weights ranging from 30 to 200 kDa,

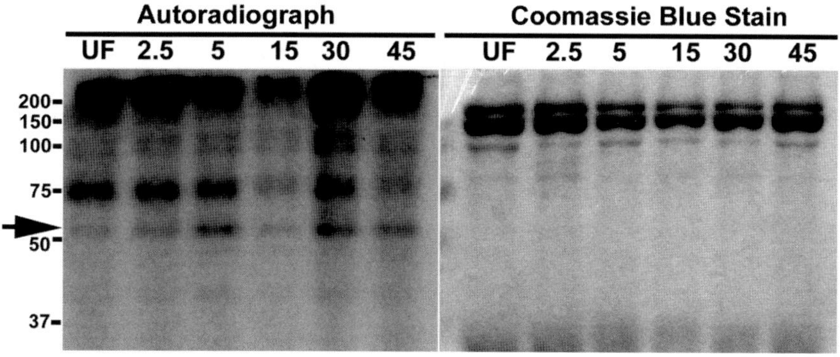

Fig. 3. SDS-PAGE analysis of ^{32}P-labeled proteins following in vitro kinase assay. Samples prepared from eggs taken before and at various times (in minutes) after fertilization were incubated in kinase reactions, as described in **Subheading 3.** The products of the kinase reaction were resolved on a 10% SDS gel and processed with KOH as described. Following autoradiography for 24 h, ^{32}P-labeled proteins are evident at 200, 90, 75, 60 (arrow), and 40 kDa.

which represent P-Tyr-containing proteins that resist alkaline hydrolysis with KOH. These ^{32}P-labeled bands may represent proteins phosphorylated by Src-family PTKs or autophosphorylation of diverse PTKs of different molecular weights. Autophosphorylated Src, Yes, and Fyn would be expected to migrate with the 60-kDa band (arrow), and it can be seen that the kinase activity at 60 kDa increased significantly after fertilization. Examination of the Coomassie stained gel (*see* **Fig. 3**) reveals the heavily stained yolk bands between 160 and 120 kDa, whereas other cellular proteins are barely visible because yolk accounts for most of the protein in these eggs. Visualization of the yolk proteins is useful because it allows one to detect differences in protein loading among lanes of the gel. For example, the amount of protein in the lane containing the 15-min sample is underloaded compared to the other lanes.

7. Quantitation of kinase activity. An estimate of the relative kinase activity within the Src-family kinase band can be made by densitometric scanning of the autoradiograph followed by quantitation of pixel density using NIH-Image software (Scion Corp., Fredrick, MD). The pixel density is expressed relative to that in the unfertilized sample. It is also possible to derive a correction factor for protein loading errors by scanning the dried, Coomassie stained gel, and quantitating pixel density of the yolk bands. The relative density of the yolk protein bands can be used to normalize the autoradiograph density.
8. Specificity of the kinase assay. In order to confirm that the phosphoproteins that migrate in the 60-kDa band were phosphorylated by tyrosine protein kinase activity, we tested the effect of a synthetic PTK inhibitor PP2 on the phosphorylation of proteins in the 60-kDa band. As seen in **Fig. 4**, PP2 inhibited phospho-

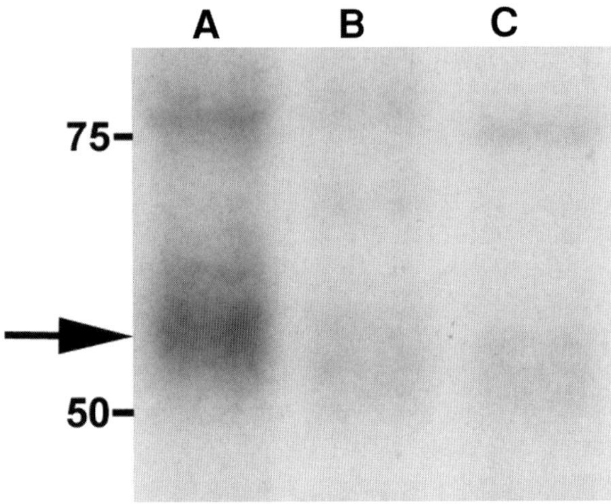

Fig. 4. Inhibition of kinase activity by PP2. Kinase reactions using single eggs collected at 5 min postinsemination were carried out in with no additions **(A)** or in the presence of 25 n*M* PP2 **(B)** or 50 n*M* PP2 **(C)**. The reaction products were analyzed by SDS-PAGE and autoradiography with 24 h exposure.

rylation of the 60-kDa proteins at concentrations between 25 and 50 n*M*, which is near the K_i reported for PP2 inhibition of Fyn kinase *(5)*.

3.4. Conclusions

The studies presented here demonstrate the feasibility of detecting Src-family kinase activity in particulate fractions prepared from single zebrafish eggs. The method is not as specific as the immune-complex kinase assays described in Chapter 19 and in **ref. *6***, but does offer the advantage of analyzing a single egg. This opens up the possibility designing experiments in which individual eggs are microinjected with test compounds and then assayed for Src-family kinase activity. Future experimentation may ultimately develop more specific methods utilizing this versatile experimental model.

4. Notes

1. Manipulating eggs: The eggs are transferred with a standard yellow 200-μL pipet tip cut off with a razor blade to a diameter slightly larger than the egg. All manipulations should be observed with the aid of an illuminated magnifying glass mounted on a movable support.
2. Because [γ-^{32}P]ATP is a strong β-emitter, we use two layers of (3/8 in. thick) plexiglass shielding to protect the user. The tubes containing radioisotope are handled only with forceps and kept in a plexiglass (1/4 in. thick) rack as much as possible.

References

1. Sohaskey, M. L. and Ferrell, J. E., Jr. (2002) Activation of p42 mitogen-activated protein kinase (MAPK), but not c-Jun NH(2)-terminal kinase, induces phosphorylation and stabilization of MAPK phosphatase XCL100 in *Xenopus* oocytes. *Mol. Biol. Cell* **13(2),** 454–468.
2. Motlik, J., Sutovsky, P., Kalous, J., et al. (1996) Co-culture with pig membrana granulosa cells modulates the activity of cdc2 and MAP kinase in maturing cattle oocytes. *Zygote* **4,** 247–256.
3. Yu, H. Q., Bou, S., Chen, D. Y., et al. (2002) Phosphorylation of MAP kinase and p90rsk and its regulation during in vitro maturation of cumulus-enclosed rabbit oocytes. *Zygote* **10,** 311–316.
4. Fan, H. Y., Li, M. Y., Tong, C., et al. (2002) Inhibitory effects of cAMP and protein kinase C on meiotic maturation and MAP kinase phosphorylation in porcine oocytes. *Mol. Reprod. Dev.* **63,** 480–487.
5. Hanke, J. H., Gardner, J. P., Dow, R. L., et al. (1996) Discovery of a novel potent, and Src family-selective tyrosine kinase inhibitor. Study of Lck and Fyn-dependent T cell activation. *J. Biol. Chem.* **271,** 695–701.
6. Wu, W. and Kinsey, W. H. (2001) Fertilization triggers activation of Fyn kinase in the zebrafish egg. *Int. J. Dev. Biol.* **44,** 837–841.

21

Activity of MAPK/p90rsk During Fertilization in Mice, Rats, and Pigs

Qing-Yuan Sun and Heng-Yu Fan

1. Introduction

Mammalian oocytes enclosed in the ovary are arrested at the diplotene stage of the first meiotic prophase, which is also termed the germinal vesicle (GV) stage. Following proper stimulation (gonadotropin in mammals), the fully grown oocytes reinitiate meiosis. The resumption of meiotic maturation is characterized by germinal vesicle breakdown (GVBD), followed by chromatin condensation, microtubule organization, and emission of first polar body. Then, the oocyte meiosis arrests again at metaphase II (MII) *(1)*, which is released by fertilization or parthenogenetic stimulation and followed by the completion of second meiotic division.

Mitogen-activated protein kinases (MAPK), which are also termed extracellular-regulated kinases (ERK), are Ser/Thr protein kinases that require dual phosphorylation on threonine and tyrosine residues to become fully activated. Two isoforms of MAPK, ERK1 (p44) and ERK2 (p42), are expressed and seem to play a pivotal role in meiosis. The core of the MAPK signaling pathway is a cascade of protein kinase phosphorylation. MAPK kinase (MAPKK, otherwise known as MEK), the direct activator of MAPK, is a dual-specificity kinase that phosphorylates MAPK on threonine and tyrosine residues in the catalytic domain. MEK is also phosphorylated (activated) by an upstream activator termed MAPKKK. MAPKKK has several members and Mos, a product of the c-*mos* proto-oncogene, is one of them. Mos, a 39-kDa Ser/Thr protein kinase, is a germ-cell-specific upstream kinase of MAPK. c-*Mos* mRNA is stored as maternal information in the growing oocytes, and its translation initiates MAPK cascade phosphorylation during oocyte maturation *(2)*. The first detected physiological substrate of MAPK in oocytes is a 90-kDa protein kinase, $p90^{rsk}$

From: *Methods in Molecular Biology, vol. 253: Germ Cell Protocols: Vol. 1 Sperm and Oocyte Analysis*
Edited by: H. Schatten © Humana Press Inc., Totowa, NJ

(ribosome S6 kinase). $p90^{rsk}$ is activated with phosphorylation on a serial of threonine and tyrosine residues. In the oocytes of *Xenopus*, starfish, mouse, and rat, the activation of $p90^{rsk}$ is MAPK1/2 activity dependent ***(3)***.

In mammalian oocytes studied so far (mouse, rat, pig, cattle, goat, and horse), MAPK is activated shortly after or at the same time as the appearance of GVBD. In bovine ***(4)*** and porcine ***(5)*** oocytes, MAPK migrates to the GV before GVBD. In mouse ***(6)*** and pig ***(7)*** oocytes, MAPK distributes to the spindle poles at metaphase and migrates to the center of the spindle at anaphase. With the polar body emission, MAPK is associated with the cytokinetic ring. It has been found that functions of microtubules and chromosomes are regulated by MAPK. MAPK can induce chromosome condensation and spindle organization in mouse oocytes even if the activity of MPF is inhibited ***(8,9)***. All the results suggest that MAPK participates in the organization and maintenance of the metaphase spindle in oocytes. In vertebrates, maturing oocytes produce a cytostatic factor (CSF), which causes oocyte to arrest at metaphase II. Mos–MAPK–$p90^{rsk}$ is known to be a critical component of CSF. The MAPK activity remains high until pronucleus formation (i.e., the S-phase of the first mitotic cell cycle) and MAPK activity does not appear in the following mitotic cell cycles ***(2)***. It seems that MAPK is an inhibitor of nuclear membrane construction, and the inactivation of MAPK is the premise for the nuclear membrane formation.

2. Materials

2.1. Manipulation of the Eggs

1. Special equipment: A CO_2 incubator, a stereoscope, and an inverted phase-contrast microscope.
2. Dulbecco's modified Eagle medium (DMEM) (Gibco, Grand Island, NY).
3. M2 medium: 1.65 g M2 medium powder (available from Sigma), 0.4349 g sodium lactate (60% syrup), 0.0035 g sodium bicarbonate, 0.006 g penicillin G potassium, and 0.005 g streptomycin sulfate. Add water to 100 mL and adjust pH to 7.4. Filter with 0.22 μm filtermats and store at 4°C (up to 1 mo).
4. M16 medium: 1.15 g M16 medium powder (available from Sigma), 0.4349 g sodium lactate (60% syrup), 0.2101 g $NaHCO_3$, 0.0313 g taurine, 0.006 g penicillin G potassium, and 0.005 g streptomycin sulfate. Add water to 100 mL and adjust pH to 7.4. Filter with 0.22-μm filtermats and store at 4°C (up to 1 mo).
5. Pig egg manipulation media: see **Chapter 16**, pp. 227–233.

2.2. Western Blot Analysis of MAP Kinase and p90rsk

1. Special equipment: Mini-protein sodium dodecyl sulfate–polyacrylamide gel electrophoresis (SDS-PAGE) and electrophoretical transfer system (Bio-Rad), X-ray exposure box, and X-ray film.
2. 10% SDS: 10 g SDS, add double-distilled water (DDW) to 100 mL, and dissolve in a 60°C water bath (store at room temperature, use for up to 6 mo).

3. 10% Ammonium persulfate (AP): 0.1 g AP, add DDW to 1 mL (made before use).
4. 10% TEMED: 0.1 mL TEMED, add DDW to 1 mL (store at 4°C in dark, use for up to 1 mo).
5. 1.5 *M* Tris-HCl, pH 8.8 (4X running gel buffer): 36.3 g Tris-HCl (formula weight [FW] 121.1), add 150 mL DDW, adjust pH to 8.8 with HCl, add DDW to 200 mL (store at 4°C at dark, use for up to 3 mo).
6. 0.5 *M* Tris-HCl, pH 6.8 (4X stacking gel buffer): 6.0 g Tris-HCl, add 80 mL DDW, adjust pH to 6.8 with HCl, add DDW to 100 mL (store at 4°C in dark, use for up to 3 mo).
7. 30.8% Acrylamide bisacrylamide: 30 g acrylamide (FW 71.08), 0.8 g bisacrylamide (FW 154.2), add DDW to 100 mL (store at 4°C in dark, use for up to 3 mo).
8. SDS sample buffer (5 mL, 2X): 0.5 mL of 1 *M* Tris-HCl (pH 6.8), 0.5 mL β-mecaptoethanol, 1 mL glycerol, 2 mL of 10% SDS, 1 mg bromophenol blue, 2 mL DDW.
9. Gel preparation recipe:

	Separating gel		Stacking gel
	5 mL	10 mL	5 mL
DDW	2.0	4.0	2.1
30% Acrylamide	1.65	3.3	0.5
1.5 *M* Tris-HCl (pH 8.8)	1.25	2.5	—
0.5 *M* Tris-HCl (pH 6.8)	—	—	0.38
10% SDS	0.05	0.1	0.03
10% Ammonium persulfate	0.05	0.1	0.03
10% TEMED	0.02	0.04	0.03

10. Running buffer stock solution (1000 mL, 10X): 30.2 g Tris base, 188 g glycine, and 10 g SDS. The pH should be 8.3 after preparation, and no adjustment of pH is necessary. Dilute 1:10 in H_2O before use.
11. Transfer buffer stock solution (1000 mL, 10X): 29.277 g glycine, 58.128 g Tris base, and 3.7 g SDS; add water to 1000 mL.
12. Transfer buffer working solution (800 mL): 39 m*M* glycine, 48 m*M* Tris, 0.037% SDS, and 20% methanol (80 mL stock solution, 160 mL methanol, add water to 800 mL). This solution should be prepared on the day of electrophoresis and cooled to 4°C before use.
13. Tris-buffered saline (TBS): 20 m*M* Tris-HCl, 137 m*M* NaCl, pH 7.4 (can be stored as 10X buffer).
14. TBST: 20 m*M* Tris-HCl, 137 m*M* NaCl, 0.1% Tween-20, pH 7.4.
15. Blocking buffer: TBST containing 5% skimmed milk.
16. Antibody dilution buffer: TBST containing 0.5% skimmed milk (for Ab1) or TBST containing 0.25% skimmed milk (for Ab2).
17. Striping buffer (10 mL): 70 μL β-mecaptoethanol (14.3 *M*), 2 mL of 10% SDS, 1.25 mL of 0.5 *M* Tris-HCl (pH 6.7), and 6.68 mL distilled H_2O. Make this fresh for each experiment.

18. Antibodies: Polyclonal rabbit anti-mouse ERK2 antibody, mouse anti-p-ERK1/2 antibody, and rabbit anti-mouse p90rsk antibody (Santa Cruz Biotechnology, Santa Cruz, CA). The antibodies can be stored at 4°C for 1 yr. Appropriate horseradish peroxidase (HRP)- or fluorescein isothiocyanate (FITC)-conjugated secondary antibodies are available from various companies. The second antibodies can be stored at –20°C in the dark for at least 1 yr.
19. Developing solution: 0.5 mL of 20X LumiGLO, 0.5 mL of 20X peroxide, and 9.0 mL water. Prepare just prior to use.
20. Blotting membrane: This protocol has been optimized for nitrocellulose membrane.

2.3. Confocal Microscopy

1. Special equipment: Confocal laser scanning microscope. This equipment is produced by several companies, such as Nikon, Leica, Bio-Rad, Zeiss, and Olympus. The Leica TCS-4D confocal laser scanning microscope has worked very well in the authors' hands.
2. Fixation solution 1: 3% Formaldehyde and 2% sucrose in PBS (pH 7.4). Store at 4°C for up to 1 wk.
3. Fixation solution 2: 4% Paraformaldehyde (*see* **Note 1**) in phosphate-buffered saline (PBS) (pH 7.4). Prepare before use.
4. PBS: 0.01 *M* Na_2HPO_4/KH_2PO_4, 0.15 *M* NaCl/KCl, (8 g NaCl, 0.2 g KCl, 1.15 g Na_2HPO_4 [or 2.9 g $Na_2HPO_4 \cdot 10H_2O$], and 0.2 g KH_2PO_4; add water to 1000 mL), adjust pH to 7.4. Store at room temperature. Discard if solution becomes cloudy or precipitated.
5. Incubation buffer (ICB): 0.5% Triton X-100 in 20 m*M* HEPES, pH 7.4, 3 m*M* $MgCl_2$, 50 m*M* NaCl, 300 m*M* sucrose, and 0.02% NaN_3. Store at 4°C for up to 1 yr.
6. DABCO: 0.4 g Triethylenediamine, 2 mL of 0.2 *M* Tris-HCl (pH 7.4), glycerol 18 mL (keep at 4°C or –20°C in dark for up to 1 yr).
7. Washing solution: 0.1% Tween-20, 0.01% Triton X-100 in PBS (pH 7.4). Store at 4°C for up to 1 mo.
8. ICB: 100 m*M* KCl, 5 m*M* $MgCl_2$, 3 m*M* EGTA, and 20 m*M* HEPES, pH 6.8, in H_2O.
9. Propidium iodide (PI) (100X): Dissolve at a concentration of 1 mg/mL in PBS. Store at –20°C in the dark.

2.4. In Vitro MAPK Assay

1. H1 kinase buffer: 80 m*M* glycerophosphate, 20 m*M* EGTA, 15 m*M* $MgCl_2$, 1 m*M* dithiothreitol (DTT), 10 µg/mL pepstatin, and 500 n*M* cAMP-dependent protein kinase inhibitor. MBP, ATP, and γ-^{32}P-ATP are available from many companies.
2. Lysate buffer: 15 m*M* EGTA, 1% Nonidet P-40, 60 m*M* sodium β-glycerol phosphate, 30 m*M* *p*-nitrophenyl phosphate, 25 m*M* MOPS (pH 7.2), 15 m*M* $MgCl_2$, 0.2 m*M* Na_3VO_4, 1 m*M* DTT, 2 µg/mL leupeptin, 2 µg/m aprotinin, 1 µg/mL pepstain, 1 m*M* phenylmethylsulfonyl fluoride, 50 µ*M* *p*-aminobenzoic acid.

2.5. Reagents Regulating MAPK Activity in Oocytes

1. U0126 [1,4-diamino-2,3-dicyano-1,4-bis(2-aminopgenylthio) butadiene]: MEK inhibitor, available from Sigma or Calbiochem Chemical Co. U0126 can be diluted in dimethyl sulfoxide (DMSO). Store at a concentration of 10 m*M* as the stock solution and store at –20°C in the dark.
2. U0124 [1,4-diamino-2,3-dicyano-1,4-bis(methylthio) butadiene]: A useful negative control for MEK inhibitor U0126. It does not inhibit MEK activity at a concentration of 100 μ*M* (available from Calbiochem Chemical Co.).
3. PD98059 (2'-amino-3'-methoxyflavone): Selective and cell-permeable inhibitor of MEK.
4. Okadaic acid (OA): Phosphatase 1 and 2A inhibitor, available from Sigma or Calbiochem Chemical Co. OA can be diluted in DMSO at a concentration of 2 m*M* as the stock solution and stored at –20°C in the dark.

3. Methods

3.1. Preparation of Samples

Because a large number of eggs are needed for analysis of MAPK activity, superovulation of animals (mouse or rat) is necessary before collection of eggs.

1. Collection of GV oocytes: GV stage oocytes are collected from ovaries of 4- to 6-wk-old animals at 48 h after the females are intraperitoneally injected with 10 IU pregnant mare's serum gonadotrophin (PMSG) (20 IU PMSG for rats). GV-intact follicular oocytes are released from the large antral follicles by puncturing with a needle in M2 medium (Sigma) with 60 μg/mL penicillin and 50 μg/mL streptomycin. All cultures are maintained in M2 medium at 37°C in a humidified atmosphere of 5% CO_2. The cumulus cells surrounding the oocytes are removed by repeated pipetting before maturation culture.
2. Collection of MII eggs: Metaphase II-arrested eggs are obtained from mice of the same strain. Females are superovulated by intraperitoneal injection of 10 IU of PMSG (20 IU PMSG for rats), and 48 h (54 h for rats) later, they are injected with 10 IU of human chorionic gonadotrophin (hCG) (20 IU hCG for rats). Mice are sacrificed and oviducts are removed at 15 h post-hCG injection. Cumulus masses are collected by tearing the oviduct with a pair of fine forceps in M2 medium. The cumulus cells surrounding the eggs are removed by a brief exposure to 300 IU/mL hyaluronidase and repeated pipetting, followed by three washes in M2 medium.
3. Fertilization in vitro: In vitro fertilization of mouse eggs is performed using 1×10^6/mL motile cauda epididymal sperm, which have been previously capacitated in M16 medium with 2.5 m*M* taurine for 1 h. Zona pellucida (ZP)-free eggs are used to minimize the lag period of sperm–egg interaction and to achieve synchronous fertilization. The emission of the second polar body and the formation of pronuclei are observed with an inverted microscope. The eggs are collected at different stages for confocal microscopy or Western blot analysis.

For in vitro fertilization of rat, ZP is removed with acid Tyrode's solution (pH 2.5). On the day of oocyte collection, male rats are sacrificed, and the cauda epididymides are dissected from animals and minced slightly into 2 mL of DMEM containing 20 mg/mL bovine serum albumin (BSA), and then placed in a 5% CO_2 air incubator for 10 min at 37°C to allow spermatozoa to swim up. Top sperm suspension showing vigorously progressive motility and containing about 1.5×10^7 cells/mL are collected. For sperm capacitation, the sperm suspension is kept in the concentrated form in a 5% CO_2 air incubator for 3 h. Capacitated spermatozoa are diluted at 1:10 before insemination. The emission of the second polar body and the formation of the pronuclei are observed with an inverted microscope. The fertilized eggs are collected at various stages for confocal microscopy or Western blot analysis.

4. In vitro maturation and fertilization of pig oocytes: see **Chapter 16**, pp. 227–233.

3.2. Western Blot Analysis of MAP Kinase and p90rsk

Both the expression and phosphorylation of MAPK and p90rsk can be analyzed by Western blot. The electrophoretic mobility of a certain protein in SDS-PAGE is determined by its relative molecular weight, and the relative molecular weights of MAPK and p90rsk increase when they are phosphorylated. In addition, the phosphorylation of MAPK and p90rsk can be analyzed in a single membrane, because there is considerable difference in their molecular weights (42/44 ku vs 90 ku), which results in electrophoretic mobility difference between these two proteins. Furthermore, the membrane can be reprobed for other protein after stripping off banding antibodies after enhanced chemiluminescence (ECL) detection. By detecting more than one protein on a single membrane and reprobing the membrane for another antibody, the investigators can obtain more information from one set of samples. The obtained results are highly reliable because they are derived from the analysis of the same sets of samples, which have been manipulated by the same experimental conditions. In the method introduced here, phosphorylation of p90rsk is assessed by examining its electrophoretic mobility shift on SDS-PAGE and phosphorylation of ERK1/2 is evaluated by both mobility shift and a specific antibody against phospho-MAPK.

1. Collect 20 or 30 oocytes (*see* **Note 2**) in an Eppendorf tube with as little liquid as possible (not to exceed 1.5 µL), add 6.5 µL double-strength sample buffer, and 5 µL ddH_2O (*see* **Note 3**), and then freeze the samples at less than –20°C (–80°C is recommended). To prevent the degradation of proteins, the samples should be used within 1 mo.
2. Before electrophoresis, boil the samples for 4 min, place on ice for 5 min, and centrifuge at 12,000*g* for 3 min.
3. Separate the total proteins by SDS-PAGE with a 4% stacking gel and a 10% separating gel for 20 min at 56 V and 4.5 h at 110 V, respectively (*see* **Note 4**).

For electrophoresis and protein transfer, the Bio-Rad miniprotein system is recommended.

4. Wash the gel with transfer buffer and wash the nitrocellulose membrane with ddH_2O for 3 min each.
5. The total protein of the sample is electrophoretically transferred onto nitrocellulose membrane for 2 h, 200 mA, at 4°C.
6. Wash the membrane with TBS twice, 5 min each.
7. Block the membrane overnight at 4°C in TBST buffer containing 5% low-fat or skimmed milk.
8. To detect both p90rsk and ERK1/2, blots are cut into two parts containing the proteins above and below the 68-kDa molecular-weight marker (prestained molecular marker is recommended) and incubated separately for 1 h in TBST with 1:300 polyclonal rabbit anti-mouse p90rsk antibodies for the upper part of the membrane and with 1:500 mouse anti-p-ERK1/2 antibody for the lower part of the membrane.
9. After three washes of 10 min each in TBST, the upper and lower parts of the membrane are incubated for 1 h at 37°C with horseradish peroxidase (HRP)-conjugated goat anti-rabbit IgG and HRP-conjugated rabbit anti-mouse IgG diluted 1:1000 in TBST, respectively.
10. Detection of proteins: Incubate membrane with 10 mL LumiGLO with gentle agitation for 1 min at room temperature. Drain membrane of excess developing solution (do not let it dry). Wrap it in plastic wrap and expose to X-ray film. An initial 10-s exposure should indicate the proper exposure time (*see* **Note 5**).
11. For reprobing of total ERK2, the lower part of the membrane is washed in stripping buffer (100 m*M* μ-mercaptoethanol, 20% SDS, 62.5 m*M* Tris-HCl, pH 6.7) at 50°C for 30 min with occasional agitation to strip off bound antibody after ECL detection,.
12. The membrane is reprobed with polyclonal rabbit anti-ERK2 antibody diluted 1:300, incubated with HRP-labeled goat anti-rabbit IgG diluted 1:1000, and finally processed as described above.

3.3. Confocal Microscopy of ERK2 and p90rsk

The presence of a signaling molecule, such as a kinase, in a special location does not demonstrate activity of the kinase. In fact, the active form of a kinase may represent only a small fraction of the total kinase in a cell, and active kinase could be present in only part of the area occupied by the total kinase. This has been recently demonstrated for MAPK in mitotic cells ***(10)*** and in mouse ***(11)*** and pig ***(5)*** oocytes. Using confocal microscopy, investigators can map the distribution of the total MAPK and compare it with the distribution of the active kinase using two different types of antibody: one type that recognizes all forms of MAPK in cells and a second type that recognizes the phosphorylated form. In this way, the investigators are able to determine where the kinase functions in the cell. In the methods introduced here, confocal microscopy for both total p90rsk or ERK2 and active MAPK is described.

3.3.1. Confocal Microscopy of Total ERK2 and p90rsk

1. Remove the zona pellucida of the eggs by a short exposure (less than 30 s) to acidified Tyrode's solution (pH 2.5).
2. Fix the eggs in a drop of fixation solution 1 (or fixation solution 2) for 30 min at room temperature (RT). Both fixation solutions work well in detection of ERK2 and p90rsk.
3. Fixed eggs are permeabilized in incubation buffer (0.5% Triton X-100 in 20 m*M* HEPES, pH 7.4, 3 m*M* $MgCl_2$, 50 m*M* NaCl, 300 m*M* sucrose, 0.02% NaN_3) for 30 min, which is a step to increase the permeability of plasma membrane (*see* **Note 6**).
4. Wash the eggs in washing buffer for three times 5 min each and then incubate the eggs with 1: 100 diluted rabbit anti-ERK2 or anti-p90rsk antibody (Santa Cruz Biotechnology, Santa Cruz, CA) for 1 h at RT (*see* **Note 7**).
5. Stain the nuclear DNA with 10 μg/mL propidium iodide (PI) for 10 min at RT.
6. Mount the samples between a coverslip and a glass slide supported by four columns of a mixture of vaseline and paraffin (9:1). Then, fill the space between the coverslip and the slide with DABCO. The slides are sealed with nail polish and preserved in a dark box at –20°C.
7. Observe the samples using confocal laser scanning microscopy within 1 wk.

3.3.2. Confocal Microscopy of Active MAPK

1. Fix the eggs at the desired stage for 30 min in 2.0% paraformaldehyde in ICB and permeabilize the cell for 30 min in 2% paraformaldehyde plus 1% Tween-20 in ICB.
2. The permeabilization is followed by three 15-min washes in ICB with 1% BSA prior to introducing primary antibodies.
3. Anti-active ERK1/2 is diluted 1:10 in ICB with 1% BSA and applied to eggs overnight at 4°C.
4. Following the overnight incubation, eggs are washed four times in ICB containing 1% BSA for 30 min each and then placed in secondary antibody (FITC- or tetramethylrhodamine isothiocyanate [TRITC]-conjugated rabbit anti-mouse IgG), also diluted in 1% BSA in ICB, for 1 h at RT.
5. Wash the eggs for four times: twice in ICB containing 1% BSA and the remaining two washes in ICB for 30 min.
6. To visualize DNA, DAPI (0.5 μg/mL, 15 min incubation) is included in the last ICB wash.

3.4. In Vitro MAPK Assay

3.4.1. Method 1 (**8**)

1. Ten oocytes are lysed by repeated cycles of freezing and thawing in H1 kinase buffer and stored at –80°C until use.
2. The reaction is started by the addition of 1 mg/mL myelin basic protein (MBP), 0.7 m*M* ATP, and 50 μCi of [γ-^{32}P]ATP to 5 μL of the crude oocyte lysate, and run for 30 min at 30°C.

3. The reaction is stopped by adding twice the concentrated sample buffer and boiling for 3 min.
4. The samples are analyzed by electrophoresis in 15% SDS-PAGE.
5. Phosphorylation levels of MBP is detected by autoradiography.

3.4.2. Method 2 (12)

1. Ten denuded oocytes are lysed in 10 µL lysate buffer and frozen at –80°C until use.
2. Assay is started by the addition of 5 µL of 2.5 µ*M* cAMP-dependent protein kinase inhibitor, 5 µL of 0.1 m*M* [γ-^{32}P]ATP, and 5 µL of 2 mg/mL MBP, and performed at 30°C for 40 min.
3. The reaction is stopped with 0.4 mL of 20% (w/v) trichloroacetic acid (TCA) solution and 0.1 mL of 1% (w/v) BSA as a carrier protein.
4. Centrifuge at 12,000*g* for 5 min; the precipitates are washed twice with 0.4 mL of 20% TCA and dissolved in 0.2 mL of 1 *M* NaOH.
5. Radioactivity is counted using a liquid scintillation counter. The value of blank tubes containing all materials except for oocytes should be subtracted when kinase activity is determined.

3.5. MAPK Inhibition and Activation in Eggs by Chemicals

To evaluate the roles of MAPK signaling pathway at some specific developmental stages, it is necessary to inhibit or activate the pathway in cells. Some chemicals may serve as effective activators or inhibitors of this kinase cascade.

3.5.1. MAPK Inhibition in Eggs

1. U0126: As a potent and specific inhibitor of MEK1 and MEK2, U0126 has little effect on other kinases such as Cdk2, ERK, JNK, MEKK, and Raf. Because it is an effective reagent to disturb the MOS/MEK/MAPK pathway, U0126 has been widely used in recent research on meiotic progression of oocytes. Dilute the stock solution to 10–15 µ*M* with the medium used for in vitro culture just prior to use and incubate the eggs in this medium containing U0126. The medium containing U0126 should be renewed every 12 h, because the U0126 is unstable in water and it may also resolve in the paraffin oil covering the medium during culture. This chemical can inhibit the activation of MAPK in oocytes at the GV stage and induce the inactivation of MAPK in eggs at the MII stage. The biologically inactive U0126 analog U0124 can be used in experiments as the negative control.
2. PD98059: this acts by inhibiting the activation of MAPK and subsequent phosphorylation of MAPK substrate. It is effective in inhibiting the activation of MAPK in *Xenopus* oocytes at a concentration of 10 µ*M*. (*See* **Note 8**.)

3.5.2. MAPK Activation in Eggs

Okadaic acid (OA) is a selective inhibitor of protein phosphatase (PP) 1 and 2A, which may be involved in the downregulation of MAP kinase activity. Therefore, OA may lead to the increase of MAPK activity in cells, including mammalian eggs.

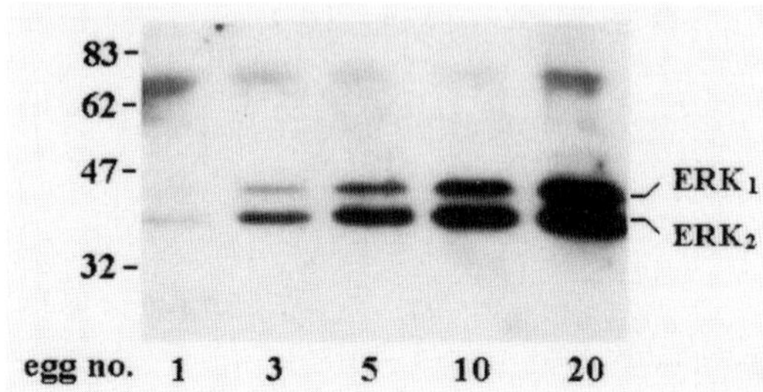

Fig. 1. MAPK expression in different numbers of mouse eggs as revealed by Western blot. The existence of MAPK can be detected even in a single mouse egg at the MII stage.

Dilute the stock solution to 2 μ*M* in the medium used for in vitro culture just prior to use and incubate the eggs in this medium containing OA. In pig oocytes at the GV stage, this chemical may lead to the activation of MAPK as quickly as 5 min after treatment.

4. Notes

1. Heat PBS to 60°C and then add a drop of 1 *M* NaOH in PBS to facilitate dissolving of paraformaldehyde.
2. This method for Western blot analysis of MAPK is highly sensitive. Even the MAPK in a single mouse MII egg can be detected. However, in order to guarantee the success of experiments, the investigators should use 20 or 30 cells in each lane. **Figure 1** shows the MAPK expression in different numbers of mouse eggs as revealed by Western blot.
3. This sample volume is suitable for SDS-PAGE using 10-well gel that is thicker than 0.5 mm. For the PAGE system using a 15-well gel of 0.5 mm thickness, the investigators can adjust the sample volume to 10 μL by adding 5.5 μL double-strength sample buffer and 4.5 μL ddH_2O.
4. This protocol of electrophoresis is optimized for detection of both MAPK and p90rsk on a single membrane. If the investigators only want to detect MAPK, the protocol can be changed to 20 min in stacking gel at 56 V and 2 h in separating gel at 120 V, respectively, which is a time-saving protocol. The detection of total MAPK, active MAPK, and p90rsk on the same membrane is shown in **Fig. 2**.
5. LumiGLO substrate can be further diluted if signal response is too fast. Because of the kinetics of the detection reaction, the signal is most intense immediately following LumiGLO incubation and declines over the following 2 h.
6. The protocol of confocal microscopy described here is optimized for MAPK detection in mouse and rat eggs, which are relatively small in size and less fatty when compared to pig oocytes. In studying the subcellular localization of MAPK in pig eggs, the cells should be permeabilized in 1% Triton X-100 in PBS at 37°C

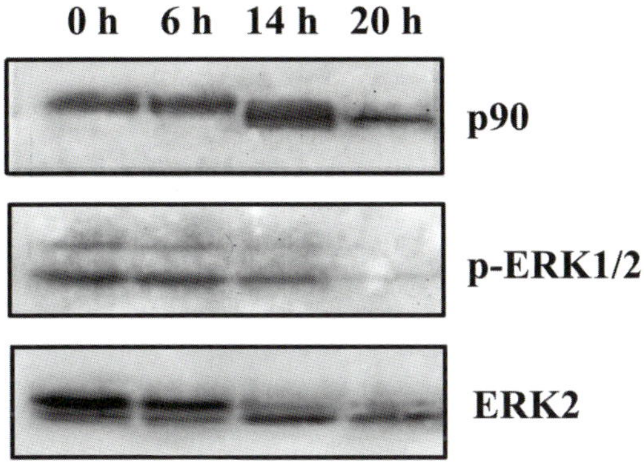

Fig. 2. The detection of total MAPK, active MAPK, and p90rsk during electrical activation of pig eggs on the same membrane. Fifty eggs are loaded in each lane.

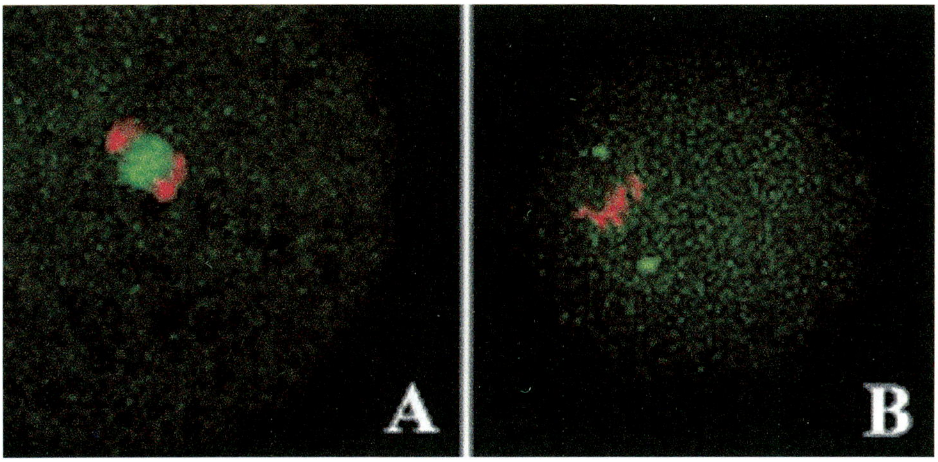

Fig. 3. Results of confocal microscopy of MAPK in pig and mouse eggs: **(A)** localization of total ERK2 in pig oocytes at the stage of anaphase I; **(B)** localization of active ERK2 in mouse oocytes at the stage of metaphase I. Green: ERK2 **(A)** or active ERK2 **(B)**; red: DNA.

overnight. No satisfactory results can be obtained if the cells are not sufficiently permeabilized because the antibodies cannot bind to their targets easily when fat-rich droplets are in the cytoplasm. The other steps for confocal microscopy of pig eggs are the same as those of mouse. The detection of total ERK2 in pig oocytes and active ERK2 in mouse oocytes is shown in **Fig. 3**.

7. The incubation time of antibodies can be extended to overnight at 4°C. Because the cells at a specific cell cycle stage may be harvested during the night and the

subsequent manipulation of confocal microscopy is time-consuming, the investigators can temporarily stop the experiment at this step and continue the following day.

8. In the research on mammalian eggs, U0126, instead of PD98059, is recommended by the authors, because, according to our experience, U0126 is more potent than PD98059 in inhibiting MAP kinase signaling pathway in mammalian cells.

References

1. Chen, D. Y., ed. (2000) *Biology of Fertilization.* Science Press, Beijing.
2. Sun, Q. Y., Breitbart, H., and Schatten, H. (1999) Role of the MAPK cascade in mammalian germ cells. *Reprod. Fertil. Dev.* **11,** 443–450.
3. Kalab, P., Kubiak, J. Z., Verlhac, M. H., et al. (1996) Activation of $p90^{rsk}$ during meiotic maturation and first mitosis in mouse oocytes and eggs: MAP kinase-independent and -dependent activation. *Development* **122,** 1957–1964.
4. Gordo, A. C., He, C. L., Smith, S., et al. (2001) Mitogen-activated protein kinase plays a significant role in metaphase II arrest, spindle morphology, and maintenance of maturation promoting factor activity in bovine oocytes. *Mol. Reprod. Dev.* **59,** 106–114.
5. Inoue, M., Naito, K., Nakayama, T., et al. (1998) Mitogen-activated protein kinase translocates into the germinal vesicle and induces germinal vesicle breakdown in porcine oocytes. *Biol. Reprod.* **58,** 130–136.
6. Verlhac, M. H., Pennart, H. D., Maro, B., et al. (1993) MAP kinase becomes stably activated at metaphase and in associated with microtubule-organizing centers during meiotic maturation of mouse oocytes. *Dev. Biol.* **158,** 330–340.
7. Lee, J., Miyano, T., and Moor, R. M. (2000) Localization of phosphorylated MAP kinase during the transition from meiosis I to meiosis II in pig oocytes. *Zygote* **8,** 119–125.
8. Verlhac, M. H., Pennart, H. D., Maro, B., et al. (1993) MAP kinase becomes stably associated at metaphase and is associated with microtubule-organizing centers during meiotic maturation of mouse oocytes. *Dev. Biol.* **158,** 33–340.
9. Verlhac, M. H., Kubiak, J. Z., Clarke, H. J., et al. (1994) Microtubule and chromatin behavior follow MAP kinase activity but not MPF activity during meiosis in mouse oocytes. *Development* **120,** 1017–1025.
10. Shapiro, P. S., Vaisberg, E., Hunt, A. J., et al. (1998) Activation of the MEK/ERK pathway during somatic cell mitosis: direct interaction of active ERK with kinetochores and regulation of the mitotic 3F3/2 phosphoantigen. *J. Cell. Biol.* **142,** 1533–1545.
11. Hatch, K. R. and Capco, D. G. (2001) Colocalization of CaM KII and MAP kinase on architectural elements of the mouse egg: potentiation of MAP kinase activity by CaM KII. *Mol. Reprod. Dev.* **58,** 69–77.
12. Tatemoto, H. and Muto, N. (2001) Mitogen-activated protein kinase regulates normal transition from metaphase to interphase following parthenogenetic activation in porcine oocytes. *Zygote* **9,** 15–23.

Index